Plant Biotic Interactions

Ajit Varma • Swati Tripathi • Ram Prasad
Editors

Plant Biotic Interactions

State of the Art

Springer

Editors
Ajit Varma
Amity Institute of Microbial Technology
Amity University
Noida, Uttar Pradesh, India

Swati Tripathi
Amity Institute of Microbial Technology
Amity University
Noida, Uttar Pradesh, India

Ram Prasad
Department of Botany
Mahatma Gandhi Central University
Motihari, Bihar, India

ISBN 978-3-030-26656-1 ISBN 978-3-030-26657-8 (eBook)
https://doi.org/10.1007/978-3-030-26657-8

This Springer imprint is published by the registered company Springer Nature Switzerland AG.
The registered company address is: Gewerbestrasse 11, 6330 Cham, Switzerland

Contents

Chapter 1
Endophytic Microorganisms as Biological Control Agents for Plant Pathogens: A Panacea for Sustainable Agriculture

Charles Oluwaseun Adetunji, Deepak Kumar, Meenakshi Raina, Olawale Arogundade, and Neera Bhalla Sarin

Abstract The utilization of endophytic microorganisms as natural biological control agents could serve as a permanent replacement to the synthetic chemicals used in the management of plant pathogens. The constant usage of pesticides has led not only to pest and disease resistance but also bioaccumulation and biomagnification of the chemicals, subsequently leading to environmental pollution and health hazards. Mitigating the effects of the long-term usage of chemical pesticides remains one of the greatest challenges in achieving sustainable agriculture. Thus, to achieve sustainable goals, which include safe environment, food safety, healthy lives, and sustainable agriculture, there is an increased need for alternatives in pest management. Utilizing beneficial microorganisms has been considered an alternative. In view of these, this chapter intends to report the recent trends in the application of endophytic microorganisms in the management of plant pathogens and also the use of nanotechnology in the production of biologically active compounds from the endophytic microorganisms.

C. O. Adetunji (✉)
Applied Microbiology, Biotechnology and Nanotechnology Laboratory, Department of Microbiology, Edo University, Iyamho, Edo State, Nigeria

School of Life Sciences, Jawaharlal Nehru University, New Delhi, India

D. Kumar (✉) · M. Raina
Department of Botany, Central University of Jammu, Samba, Jammu and Kashmir, India
e-mail: deepakkumar@cujammu.ac.in

O. Arogundade
National Horticultural Research Institute, Ibadan, Oyo State, Nigeria

N. B. Sarin
School of Life Sciences, Jawaharlal Nehru University, New Delhi, India

A. Varma et al. (eds.), *Plant Biotic Interactions*,
https://doi.org/10.1007/978-3-030-26657-8_1

1.1 Introduction

World population has been speculated to reach 9 billion by the year 2050. In order to match this astronomical figure, the FAO has warned that food production will have to increase at least by 60% in proportion (FAO 2009). Food has been described a daily necessity for the survival of mankind throughout the globe (DESA 2015; FAO 2012). In view of this, there is a clarion call on all scientists to tackle the following problems: (1) the need for agriculture to undergo a significant transformation in order to feed the ever-increasing population; (2) the issue of climatic changes, especially in the developing world; (3) health challenges; (4) pests and diseases. All these problems have been identified as a major factor militating against an increase in productivity toward the achievement of sustainable agriculture (Cocq et al. 2017). Pests and diseases have been highlighted as one of the major problems encountered in international and local food systems, preventing the achievement of sustainable agriculture (Srivastava et al. 2016).

Many farmers utilize chemical pesticides for the management of pests (Adetunji et al. 2017). However, the application of chemical pesticides has a lot of demerits, including their high cost to local farmers, inaccessibility to rural farmers, the problem of pesticide residue, high rates of contamination in the environment, and disruption of the ecosystem (Priyanka et al. 2018). In order to face these challenges, there is a need to adopt a simple, clean, healthy, and eco-friendly, technology (Liu et al. 2017). The application of these chemical pesticides has led to many deleterious environmental effects (Adetunji et al. 2018a). Also, various anthropogenic activities have been highlighted to cause emission of greenhouse gases, particularly carbon dioxide, methane, and nitrous oxide, thereby increasing global mean temperature (Miraglia et al. 2009). The effect of biotic factors coupled with the issue of climate change has enhanced the susceptibility of many crop species to the adverse effect of pathogenic microorganisms, which are a major factor responsible for low crop production year in year out (Kõiv et al. 2015). The application of beneficial microorganisms with disease resistance attributes as well as those that could help the plant to withstand drought, extreme temperature, and adverse environmental conditions is important (Lata et al. 2015). The beneficial microorganisms must be able to give high crop productivity in terms of rapid growth, enhanced development, and maximum yield, which is urgently required to feed the ever increasing world population (FAO 2012; Tian et al. 2015).

Moreover, it has been shown that variation in environmental CO_2 and temperature, coupled with the type of soil and plant physiological health status, all have an influence on the microbiome variation of endophytic microorganism communities (Ferrando and Scavino 2015). When all these factors are pooled together, they form a factor that influences the observable changes in climate (Ren et al. 2015a). Therefore, there is need to search for microorganisms that could secrete natural metabolite that could become a permanent replacement for chemically synthesized pesticides for the management of pests and diseases (Deepika et al. 2016). This will help to proffer a substantive solution towards the food system target for 2050 (DESA 2015).

Endophytes are beneficial microorganisms that exhibit a healthy relationship between plant cells and secrete valuable bioactive compounds, enzymes, and secondary metabolites with a wide range of applications (Shahzad et al. 2018). It has been found that endophytic microorganisms serve as a good candidate for the management of plant disease and pests (Marasco et al. 2012) and environmental stress (Suman et al. 2016). Endophytes have been documented as excellent plant growth enhancers because they induce plant growth hormones. The phytohormones produced by these endophytes play an indispensable role in plant growth and drought resistance. Moreover, their ability to adapt and tolerate various environmental stresses as well as their eco-friendly attributes validates the exploitation of their secondary metabolites for the management of pests and diseases (Brader et al. 2017). The environmental friendly nature of the endophytic microorganism premise on the symbiotic relationship they exhibit between their plant host and their direct environment which increases their unlimited ability to produce bioactive metabolites. These metabolites have been linked to natural plant defense mechanisms due to their ability to inhibit and prevent pathogens (Casella et al. 2013).

This chapter focuses on the recent trends in the application of endophytic microorganism as a new food paradigm, which is required to meet these global challenges and to achieve global food security, food production, eco-friendliness, and desired human health in order to foster healthy solutions for a sustainable future.

1.2 Endophytic Microorganisms and Their Usefulness

Endophytic microorganisms have been described as organisms that have potential to regulate metabolic and environmental interactions towards the production of useful secondary metabolites (Strobel et al. 2004). These organisms include bacteria, fungus, yeast, and actinomycetes (Leitão and Enguita 2016; Farrer and Suding 2016). These organisms live in inimitable biological niches of plant tissue and constitute a pool of natural secondary metabolites with useful properties for utilization in agriculture, industry, environmental remediation, and their utilization in medical services (Schutz 2001). The myriad of factors found in any plant colonized by endophytes forms the basis for the selection of plant as well as endophytic microorganism responsible for the production of unique metabolites for various biological activities. These biological activities include biocontrol of pests and diseases (Eljounaidi et al. 2016), bioprospecting of bioactive compounds (Sharma et al. 2016), bioremediation of agricultural polluted soils (Manohari and Yogalakshmi 2016), enzymatic production (Kumar and Belur 2016), plant enhancement attributes (Afzal et al. 2016), seed germination enhancer (Shearin et al. 2017), and stress relievers of host plants (Tiwari et al. 2017a). It has been reported that as many as 300,000 plant species have the capacity to provide a biological niche to many endophytic microorganisms (Aly et al. 2010). However, little out of these endophytes have been exploited from a diversity of plants and ecosystems (Ryan et al. 2008). The biological control of the attributes of endophytic bacteria is shown in Fig. 1.1.

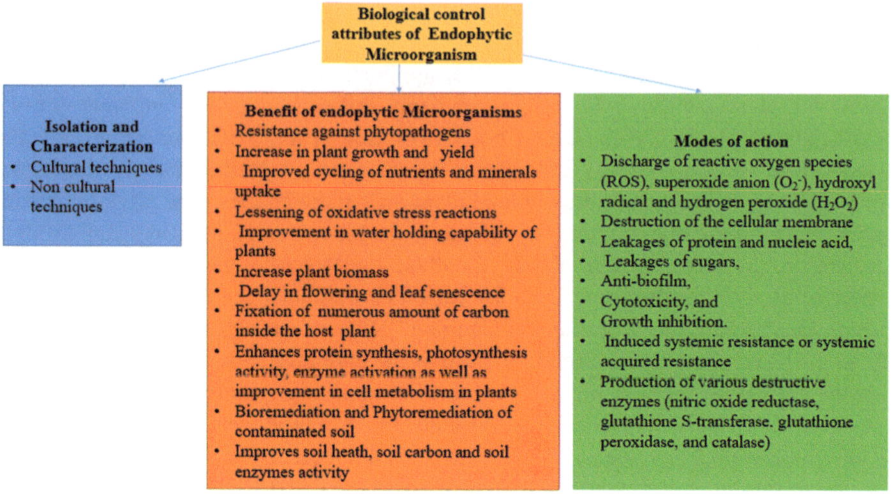

Fig. 1.1 Schematic model showing the process involved in the biological control of plant pathogens by endophytic bacteria

1.3 Isolation and Characterization Techniques of Endophytes Available from Different Ecological Niches

The isolation and characterization of endophytic microorganisms involves the utilization of culture-dependent and culture-independent techniques. The former involve morphological analysis, microscopic analysis, physiological analysis (growth characteristics, simple tests; biochemical characterization (assimilation and fermentation patterns (API, BIOLOG etc.), protein profiling (sodium dodecyl sulfate-polyacrylamide gel electrophoresis of cellular proteins) Murmu (2016), used API and BIOLOG kit for the biochemical characterization of some endophytic microorganisms from finger millet (*Eleusine coracana* (L.) Gaertn.) genotype JWM1.

Narayan et al. (2012), studied the morphological characteristics of *Heterobasidion araucariae* CNU081069 with the help of light microscopy. Verma et al. (2013), isolated an endophytic actinomycetes *Saccharomonospora* sp., from the root surface of *Azadirachta indica A. Juss.* Moreover, they utilized sodium dodecyl sulfate-polyacrylamide gel electrophoresis to further characterize the proteins obtained from the endophytic actinomycetes. Nongkhlaw et al. (2017) utilized transmission electron microscopy (TEM) to observe the colonization of some endophytic bacteria associated with ethnomedicinal plants. The TEM gave a better microscopic visualization for proper understanding about the isolated endophytic bacteria.

The culture-independent techniques include the following: polymerase chain reaction; repetitive sequence-based polymerase chain reaction; reverse transcription polymerase chain reaction; length heterogeneity polymerase chain reaction; reverse transcription quantitative polymerase chain reaction; quantitative polymerase chain reaction; separation by length polymorphism, i.e., restriction fragment length polymorphism; terminal restriction fragment length polymorphism; amplified ribosomal DNA restriction analysis; denaturing gradient gel electrophoresis/temperature gradient gel electrophoresis; ribosomal RNA (rRNA) intergenic spacer analysis/automated ribosomal intergenic spacer analysis; single-strand conformation polymorphism; suppression subtractive hybridization; random amplified polymorphic; amplified fragment length polymorphism; pulsed-field gel electrophoresis; DNA–DNA hybridization using labeled probes; microscopic counting; flow cytometry in situ hybridization; fluorescence in situ hybridization, *tyramide* signal amplification, DNA reassociation kinetics, DNA association. Ding et al. (2013) characterized and identified the diversity and distribution of endophytic bacteria present in five plant species. The fragments of the bacterial 16S rDNA genes present in the endophytic bacteria were characterized by terminal restriction fragment length polymorphism.

Mada and Bruce (2017) isolated, characterized, and identified 33 endophytic bacteria present in monocotyledonous and dicotyledonous plants. The endophytic bacteria present were characterized using cultural, biochemical, and molecular techniques by carrying out polymerase chain reaction with the partial 16S-rRNA gene from the isolated endophytic bacteria. Tellenbach et al. (2010) applied single-copy qPCR assay and a multicopy locus-based quantitative polymerase chain reaction so as to determine the amount of biomass present in endophytic fungus after the colonization of *Picea abies* seedlings. Rabha et al. (2016) isolated and characterized 30 endophytic fungi from tea plant using cultural and molecular identification using Internal Transcribed Spacer r-DNA Sequence Analysis. They also employed Random Amplified Polymorphic DNA molecular markers so as to determine the level of genetic variation among all the isolates. Martini et al. (2009) applied ITS1 and ITS2. *Aureobasidium pullulans* and *Epicoccum nigrum* available in the grapevine varieties (Merlot and Prosecco) were identified by the characterization of variable regions of ITS1 and ITS2 using polymerase chain reaction. These two endophytes from grapevine were later identified and confirmed by restriction fragment length polymorphism (RFLP) when compared to reference strains. Shen (2013) utilized culture-dependent and culture-independent methods (denaturing gradient gel electrophoresis and terminal restriction fragment length polymorphism analysis) for the characterization of endophytic bacteria present in the branches of some tree grown over a period of three seasons in a hydrocarbon-contaminated site. They discovered that there was great variation in the level of the endophytic colonization among the same species on plants in different seasons. Finyom (2012) applied denaturing gradient gel electrophoresis and terminal restriction fragment length polymorphism techniques for the isolation and characterization of endophytic bacteria present in various farm sites of sorghum crop in South Africa for the characterization of endophytic bacteria in these communities. The molecular analysis carried out revealed that the endophytic bacteria showed the same level of colonization among sorghum plants with the same community but showed a high level of variation when compared to different sampling sites. They concluded that the isolated endophytic bacteria could be used for the improvement and increase in the yield of sorghum.

Tejesvi et al. (2007) explored the application of RAPD for rapid detection of genetic diversity in endophytic fungus available in four medicinal plants in order to select the best strain with antifungal activity. They discovered that amongst all the strains isolated, *Terminalia arjuna* demonstrated the best antifungal activity against all six tested isolates. Ardanov et al. (2012) tested the efficacy of *Methylobacterium* spp. Endophytic bacteria could be used in the management of different phytopathogens affecting potatoes (*Solanum tuberosum* L.) and pine (*Pinus sylvestris* L.). They utilized terminal restriction fragment length polymorphism to study the effect of their interaction on the dominant endophytic microflora available in this community where changes in their relative abundance were determined by terminal fragments. They discovered that the application of *Methylobacterium* spp. to induce resistance and antagonism traits on the tested phytopathogens depended on the cultivar, pathogen, and the concentration of the endophytic bacteria and the particular type of endophyte community of the host.

1.4 Benefits and Mechanism of Plant–Microbe Interaction Among Endophytic Microorganisms

There are a lot of mutual benefits whenever there is plant–microbe interaction, especially with endophytic features which include resistance against phytopathogens (Mishra et al. 2015), increase in plant growth and yield (Shatrupa et al. 2018), improved cycling of nutrients and mineral uptake (Malinowski and Belesky 2004), lessening of oxidative stress reactions (White and Torres 2010), improvement in water holding capability of plants (Bailey et al. 2006), and increase in biomass accomplished by a delay in flowering and leaf senescence. This might improve the fixation of a large amount of carbon inside the host plant (Owen and Hundley 2004), and also result in forest regeneration and phytoremediation of contaminated soils (Ma et al. 2016). The plant-growth-promoting attributes observed when endophytes colonize plant are a result of various mechanisms, which include production of a siderophore (Costa and Loper 1994), indole acetic acid production (Lee et al. 2004), phosphate solubilization activity (Wakelin et al. 2004), provision of essential vitamins to plants (Compant et al. 2005a, b). The release of volatile substances including acetoin and 2–3 butanediol has also been described as another mechanism responsible for the rapid development of plants (Ryu et al. 2003).

Apart from these benefits, the level of mineral ions available with the plants colonized by endophytes has also been found to influence and contribute significantly to the development of plants. Fe (III) has been utilized by some endophytic bacteria in the formation of complexes whenever siderophores are released. The endophytic bacteria possess a Ton-B-dependent membrane that can adsorb the ferric-siderophore complexes. The release of siderophores is normally used as a mode of action for inhibition of phytopathogens (Hurek and Hurek 2011). Also, the presence of K^+ has been discovered to enhance the protein synthesis, photosynthesis activity, enzyme activation, and improvement in cell metabolism (Shabala and Cuin 2008).

1.5 Role of Genes Regulating the Activities of Endophytic Microorganisms

A better understanding of recent trends in the genomic study will give a better understanding of the role played by endophytic microorganisms during their mutualistic interaction with their host plant (Santoyo et al. 2016). This will give an insight into and a better knowledge of the array of genes regulating various attributes that have been reported for endophytic microorganisms necessary to increase sustainable agricultural practices in regards to plant growth enhancement capability, management, and prevention of pests and diseases and cleaning up processes of contaminated agricultural lands (Bruto et al. 2014).

Osbourn and Field (2009) observed that gene cluster is related to histone alterations as a result of chromatin remodeling complexes in most microorganisms. The closeness and arrangement of clusters play a significant role in the regulation of genes controlling the metabolic pathway in microorganisms (Nei 2003) Furthermore, whenever there is interference in the activities of the gene required for the synthesis of these active metabolites, the production of toxic metabolites is triggered and this might lead to the gathering of lethal pathway intermediates. This forms the basis for the biological control activities against phytopathogens and pests and enhances cluster selection (Sproul et al. 2005).

In this regard, different researchers have identified the significance of genes regulating the biological life cycle of endophytes, which attribute them as a unique group of microorganisms with useful potential, especially in the area of agriculture. Some of the unique functions reported include biocontrol of pathogen and pest (Martinez-Garcia et al. 2015), nitrogen fixation (Fouts et al. 2008); plant growth promotion (Firrincieli et al. 2015), phytohormone production (Tiwari et al. 2017b); induction of systemic resistance in plants (Nadarajah (2017), and abiotic stress tolerance (Abd El-Daim et al. 2018). Enhancement in the rate of chemical signaling between endophytes and their host (Chagas et al. 2018) 1-aminocyclopropane-1-carboxylate deaminase and antifungal, (Kwak et al. 2012), indole 3-acetic acid synthesis (Weilharter et al. 2011).

Zhao et al. (2010) isolated strain YY-11 of *Chaetomium globosum* from rape seedlings as well as some endophytic bacteria *Enterobacter* sp. and *Bacillus* subtilis, which were utilized in the expression of *Pinellia ternate* agglutinin (PtA) gene. The result obtained from their study shows that deduced from those recombinant endophytes expressing PtA gene enhanced the biological control activities of sap-sucking pests affecting many crop seedlings. Similarly, Zhang et al. (2011) expressed (PtA) gene in recombinant endophytic bacteria *Enterobacter cloacae*, and insecticidal activities were observed on white backed plant hopper.

1.6 Mode of Action of Endophytic Microorganisms Against Biotic and Abiotic Stress

Endophytic microorganisms utilize a lot of mechanisms in order to overcome any type of biotic and abiotic stress. Whenever the endophytic microorganisms sense the presence of biotic and abiotic stress, they discharge reactive oxygen species (ROS), like superoxide anion (O_2^-), hydroxyl radical, and hydrogen peroxide (H_2O_2) (Apel and Hirt 2004; Gill et al. 2016; Nath et al. 2016, 2017). The presence of ROS induces a form of pressure which leads to the destruction of the cellular membrane of the host, which consequently leads to leakages of protein, nucleic acid, and sugars and results in their anti-biofilm, cytotoxicity, and growth inhibition on these pests and pathogens (White and Torres 2010; Halliwell and Gutteridge 1999). Furthermore, it has been discovered that endophytic microorganisms colonizing their plant hosts have the capability to quench the reactive oxygen species whenever they encounter biotic and abiotic stress. The capability of endophytic microorganisms to carry out their biological control activities has been linked to their capability to move into their specific host, population dynamics, and level of colonizing the host, and the potential of endophytes to induce systemic resistance or systemic acquired resistance (Benhamou et al. 1998; Van Loon et al. 1998). Endophytic microorganism has many arrays of enzymes (nitric oxide reductase, glutathione S-transferase, glutathione peroxidase, and catalase) capable of alleviating any triggered reactive oxygen species as well as for the management of abiotic and biotic stress (Hardoim et al. 2015).

Matsouri et al. (2010) observed that *Trichoderma harzianum* prevents biotic and abiotic stress in germinating seeds and seedlings. Also, Baltruschat et al. (2008) discovered that there was an increase in the level of salt tolerance of induced barley plants after inoculating with *Piriformospora indica*, an endophytic microorganism. Similarly, Rodriguez et al. (2008) depicted that *Leymus mollis* not inoculated with endophytic microorganism showed a high level of severe wilting and desiccation after seven days of inducing with brine solution, and died after 14 days. But these symptoms were absent in the *Leymus mollis*, whether or not inoculated with *Fusarium culmorum*, an endophytic fungus. Bae et al. (2009) discovered that an isolate DIS 219b of *Trichoderma hamatum* has the capability to enhance the development of *Theobroma cacao* and prevent the resistance of the plant when exposed to a severe drought when inoculated with the endophytic fungus. Their activities could be linked to the ability of the endophytic fungus to increase the level of some amino acids like proline, which serve as osmoprotectants when exposed to abiotic stress due to drought. Ray et al. (2016) discovered that the endophytic bacteria *Alcaligenes faecalis* was able to prevent an oxidative stress when inoculated with *Sclerotium rolfsii*, a pathogenic fungus.

1.7 Exploiting the Virus–Fungus–Plant Interaction

Plant viruses are interesting plant pathogens, because the host plant properties play an important role in insect vector transmission of viruses from one host to another. The effect of host metabolites and effects of plant quality on insect vectors directly or indirectly affect transmitting the viruses from host to host (Lehtonen et al. 2006). There is a dearth of information on the specificity of any virus able to infect both fungal and plant hosts in nature. However, reports show that some plant viruses like RdRPs can replicate in trans of genomic or DI RNAs in the yeast *Saccharomyces cerevisiae* experimentally. Also, recently, Tobacco mosaic virus was also shown to be able to replicate in a filamentous ascomycetous fungus. This shows that some plant viruses can infect some fungi on experimental basis (Nerva et al. 2017).

Fungi are very frequently infected with viruses, and endophytic fungi are no exception. The thermo-tolerance of the plant–fungus–virus holobiont that allows a panic grass to grow in geothermal soils in Yellowstone National Park requires the curvularia thermal tolerance virus. Several endophytic fungi have been reported to contain viruses and these fungi are from many plants.

Viruses that infect fungi are referred to as mycoviruses. These viruses have the potential to control fungal diseases of crops when associated with hypovirulence. Hypovirulence is the most important quality of mycoviruses, which can be utilized to manage fungal diseases, and this means the reduction of virulence of plant pathogenic fungi. Mycoviruses infect fungal species, but are present in latent stage in them. They rarely cause visible disease symptom (Abbas 2016), although there are a few mycoviruses which are involved in causing considerable changes in fungal hosts. The most obvious changes in their fungal hosts are irregular growth, abnormal pigmentation, and mutated sexual reproduction (Jiang et al. 2013). This quality has attracted a lot of attention recently because several fungal diseases of crops can be managed.

The use of chemical pesticides is the main method for controlling diseases of economically important crops, especially when resistant cultivars are not available. To reduce the dependence on chemical pesticides, efficient and environmentally friendly alternative methods to control diseases are desirable. Mycovirus-mediated hypovirulence, a phenomenon in which the virulence of fungal pathogens is reduced or even completely lost as a result of virus infection, can be exploited (Yu et al. 2010). Hypovirulence plays a role in counterbalancing plant diseases in nature, and has been used successfully in Europe in the control of chestnut blight (caused by the fungus *Cryphonectria parasitica*) (Yu et al. 2010). Several hypovirulent strains have been discovered in other fungi due to the successful utilization of hypoviruses for biological control of the chestnut blight fungus.

Plant viruses cause major losses to agricultural crops around the world. Chemical agents are not effective to control virus diseases, hence, the strategies for virus management are mostly aimed at destroying the source of infection to prevent it from reaching the crop and or controlling of vectors to prevent the spread of the disease (Arogundade et al. 2012). However, endophytic fungi could be an alternative that can be utilized in the management of diseases caused by viruses. Lehtonen and his team in 2006 observed that endophyte-infected meadow ryegrass plants harbored less viral infections both in natural and common garden conditions than uninfected plants. Also, they noted that the reproduction of bird cherry oat aphids was decreased on endophyte-infected plants compared to uninfected plants. This, if properly developed, could be the key to safe and sustainable food threatened by the menace caused by difficult-to-control plant pathogenic organisms like viruses.

1.8 Naturally Synthesized Compounds from Endophytic Microorganisms

Endophytic microorganisms are capable of producing novel and active secondary metabolites which are of low molecular weight. The production of these secondary metabolites is regulated by some genes which are always produced in a cluster. The cluster has been discovered to play a significant role in the production of natural compounds that could induce resistance to pests and diseases (Shimura et al. 2007). Nature has endowed these endophytic microorganisms and their host plant with different structurally diverse active compounds with various applications. Recently, the application of metabolomics has revealed that bioprospecting of these active metabolites from endophytic microorganisms could be used in the management of pests and diseases (Hyde and Soytong 2008). These metabolites contain some important enzymes of biotechnological significance due to their diversity as well as complex structure, which is difficult for phytopathogens to digest or metabolize. This feature plays a major role in the biological control of pests and diseases (Kusari and Spiteller 2011). Many researchers have reported various metabolites from endophytic microorganisms. Liang et al. (2015) identified a compound called Podophyllotoxin from *Alternaria* sp. isolated from *Sinopodophyllum emodi* while Kusari et al. (2012) isolated Azadirachtin from *Eupenicillium parvum*, an endophytic fungus which colonized *Azadirachta indica* plant. Moreover, Mohana Kumara et al. (2012) discovered a natural compound called Rohitukine from an endophytic fungus called *Fusarium proliferatum* from *Dysoxylum binectariferum*.

1.9 Nontarget Effect of Endophytic Microorganisms

Endophytic microorganisms have been primed for their great potential for the management of pests and diseases for sustainable agriculture, but there is a need to critically review their nontarget effects (Adetunji et al. 2018b). There is now great awareness that the high rate of pesticide utilization by farmers has caused a lot of demerits on various Agricultural practices. They have also caused a lot of changes in the level of soil microbial communities (Seghers et al. 2004). Therefore, the application of plant endophytic microorganisms will be another alternative for the production of natural, safe, and eco-friendly biocontrol agents (Dai et al. 2016).

This could be linked to their capability to improve rapid growth in plants; maintenance of good health status, improvement of plant defense, and high tolerance to abiotic and biotic stress (Tétard-Jones and Edwards 2016). Moreover, a good biological control agent should not cause any disruption of microbial processes, which are important for the normal operation of soil ecosystem functioning (Winding et al. 2004). They should only perform the activity they are meant to carry out without any following nontarget effects, which include antagonism effects on other beneficial microorganisms, decrease or adverse effect on soil quality after their application, reduction of soil enzymes, soil carbon, and soil respiration (Adetunji et al. 2018a).

Scherwinski et al. (2008) discovered that the endophytic bacteria *Serratia plymuthica* 3Re4–18, *Pseudomonas trivialis* 3Re2–7, and rhizobacterium *Pseudomonas fluorescens* L13–6-12 exhibited very strong biological control activities on *Rhizoctonia solani*, without any observable nontarget effect on other microorganisms during the field trial.

1.10 The Beneficial Role of Endophytic Microorganisms in Bioremediation and Phytoremediation

Intensive agricultural practices have been highlighted as one of the major factors responsible for the high record of heavy metal pollution in an agricultural system. The long-term effects of these heavy metals include disruption and contamination of the ecosystem and effects on human health (Kidd et al. 2012). This is due to the complexity in their structure and their nonbiodegradability. There is a need to search for eco-friendly microorganisms with high phytoremediation capacity because the utilization of physiochemical remediation techniques is not economically feasible and eco-friendly (Dharni et al. 2014). Also, there is a need to discover an isolate that could promote rapid development of plants, which has a potential to metabolize toxic metal phytotoxicity into useful metabolites for plant development and at the same time to perform the role of phytoremediation in heavy-metal-polluted soils (Ma et al. 2015). Therefore, the utilization of endophytic microorganisms will go a long way in the sanitization of heavy-metal-contaminated soils and allow plant development under heavy metal stress (Babu et al. 2015). The endophytic microorganisms have been documented to enhance plant growth under metallic stress by their ability to transform mineral nutrients, secretion of biosurfactants, siderophores, organic acids, biodegradative enzymes, and phytohormones as well as to have the potential to induce systemic resistance against pests and diseases in their host plants (Ma et al. 2016) Also, it has been discovered that the great microbial diversity of endophytic microorganisms enables them to perform bioremediation and biodegradation of harmful and complex substances from the environment turning them into simple and useful compounds (Russell et al. 2011; Prasad 2017, 2018).

Endophytes have a powerful ability to break down complex compounds. Bioremediation is a method of removal of pollutants and wastes from the environment by the use of microorganisms (Prasad 2017, 2018). It relies on the biological processes of microbes to break down these wastes. This is made possible due to the great microbial diversity (Mastretta et al. 2009; Russell et al. 2011). Also, endophytic microorganisms have been documented for their potential to carry out phytostimulation of important elements that plants need for their rapid growth and development (Malinowski et al. 2000). This could be linked to their capability to produce various phytohormones like gibberellic acids, auxins, and cytokinins (Xin et al. 2009; Hamayun et al. 2009).

1.11 Role of Nanobiotechnology in the Utilization of Endophytic Microorganisms in the Management of Plant Diseases

The application of nanotechnology has been discovered as a newly emerging field that could help achieving sustainable agricultural productivity, especially in the management of pests and diseases (Parisi et al. 2015; Prasad et al. 2014, 2017). The nontoxic, environment and human health friendly, and nonpathogenic nature of the active metabolites from endophytic microorganisms guarantee their utilization in the biomanufacturing of nanopesticides for effective management of pests and diseases (Bhattacharyya et al. 2016; Ismail et al. 2017; Iavicoli et al. 2017; Adetunji and Sarin 2017; Gupta et al. 2018). They have a target delivery of their active ingredient through remote activation and monitoring of the synthesized nanoformulation that will enable the farmers to reduce the application of pesticides (Abd-Elsalam and Prasad 2018). The application of nanoparticle-mediated gene transfer into endophytic microorganisms will bring about a drastic agricultural output because of high resistance to pests and diseases (de Oliveira et al. 2014). Furthermore, the development of biosensors from these endophytic microorganisms will assist farmers, especially in developing countries, to monitor the emergence of pests and disease outbreaks on their farms (Adetunji et al. 2018b).

1.12 Conclusion

This chapter provides a detailed insight into the roles of endophytic microorganisms for the management of pests and diseases. The study has shown the recent trends in the utilization of endophytic microorganisms as a new source of natural, safe, and eco-friendly pesticide for disease management as well as to enhance plant growth and reduce stress. More emphasis has been laid also on their utilization for bioremediation of polluted environments without losing their biological control and plant-growth-promoting attributes. Moreover, various isolation and characterization techniques have been highlighted. This will have a far better effect by providing an opportunity to isolate new endophytic microorganisms that have not been exploited for their great attributes. On the whole, the application of innovative biotechnological techniques will open new opportunities to gain better knowledge of plant–microbe interaction in search for a new active compound that increases plant growth and enhances biological activity and is also involved in the biodegradation of environmental contaminants, thereby maintaining a clean environment. In view of all this, the utilization of endophytic microorganisms has tremendous potential that will assist in the achievement of sustainable agriculture, clean environment, and health security for humans. Therefore there is a need to formulate a policy that will prevent increase in global warming, that will also support the utilization of natural endophytic microorganisms that can prevent fluctuation in the level of CO_2, temperature and increase the plant-growth-promoting feature of agricultural crops (Ren et al. 2015b).

Acknowledgments The authors are grateful to the Department of Biotechnology (DBT), New Delhi, India, The World Academy of Science (TWAS) for TWAS-DBT post-doctoral given Dr. Adetunji. FR number: 3240293141.

References

Abbas A (2016) A review paper on Mycoviruses. J Plant Pathol Microbiol 7:390. https://doi.org/10.4172/2157-7471.1000390

Abd El-Daim IA, Bejai S, Fridborg I, Meijer J, Pineda A (2018) Identifying potential molecular factors involved in *Bacillus amyloliquefaciens* 5113 mediated abiotic stress tolerance in wheat. Plant Biol 20(2):271–279

Abd-Elsalam KA, Prasad R (2018) Nanobiotechnology applications in plant protection. Springer, Cham. https://www.springer.com/us/book/9783319911601

Adetunji CO, Sarin NB (2017) Impacts of biogenic nanoparticle on the biological control of plant pathogens. Adv Biotechnol Microbiol 7(3):555711. https://doi.org/10.19080/AIBM.2017.07.555711

Adetunji CO, Oloke J, Kumar A, Swaranjit S, Akpor B (2017) Synergetic effect of rhamnolipid from *Pseudomonas aeruginosa* C1501 and phytotoxic metabolite from *Lasiodiplodia pseudotheobromae* C1136 on *Amaranthus hybridus* L. and *Echinochloa crus-galli* weeds. Environ Sci Pollut Res 24:13700–13709

Adetunji CO, Oloke JK, Osemwegie OO (2018a) Environmental fate and effects of granular pest formulation from strains of *Pseudomonas aeruginosa* C1501 and *Lasiodiplodia pseudotheobromae* C1136 on soil activity and weeds. Chemosphere 195(2018):98–107. https://doi.org/10.1016/j.chemosphere.2017.12.056

Adetunji CO, Phazang P, Sarin NB (2018b) Biosensors: a fast-growing technology for pathogen detection in agriculture and food sector. InTechOpen:1–17. https://doi.org/10.5772/intechopen.74668. In Biosensing technologies for the detection of pathogens—a prospective way for rapid analysis, book edited by Toonika Rinken, Kairi Kivirand, ISBN 978-953-51-3916-4, Print ISBN 978-953-51-3915-7

Afzal I, Iqrar I, Shinwari ZK, Yasmin A (2016) Plant growth-promoting potential of endophytic bacteria isolated from roots of wild Dodonaea viscosa L. J Plant Growth Regul:1–10

Aly AH, Debbab A, Kjer J, Proksch P (2010) Fungal endophytes from higher plants: a prolific source of phytochemicals and other bioactive natural products. Fungal Divers 41:1–16

Apel K, Hirt H (2004) Reactive oxygen species: metabolism, oxidative stress, and signal transduction. Annu Rev Plant Biol 55:373–399

Ardanov P, Sessitsch A, Häggman H, Kozyrovska N, Pirttila AM (2012) Methylobacterium-induced endophyte community changes correspond with protection of plants against pathogen attack. PLoS One 7(10):e46802. https://doi.org/10.1371/journal.pone.0046802

Arogundade O, Balogun OS, Kareem KT (2012) Occurrence and distribution of pepper veinal mottle virus and cucumber mosaic virus in pepper in Ibadan, Nigeria. Virol J 9:79

Babu AG, Shea PJ, Sudhakar D, Jung IB, Oh BT (2015) Potential use of *Pseudomonas koreensis* AGB-1 in association with *Miscanthus sinensis* to remediate heavy metal(loid)-contaminated mining site soil. J Environ Manag 151:160–166

Bae H, Sicher RC, Moon SK, Kim SH, Strem MD, Melnick RL, Bailey BA (2009) The beneficial endophyte *Trichoderma hamatum* isolate DIS 219b promotes growth and delays the onset of the drought response in *Theobroma cacao*. Exp Bot 60(11):3279–2395

Bailey BA, Bae H, Strem MD, Roberts DP, Thomas SE, Crozier J, Samuels GJ, Choi IY, Holmes KA (2006) Fungal and plant gene expression during the colonization of cacao seedlings by endophytic isolates of four Trichoderma species. Planta 224:1449–1464

Baltruschat H, Fodor J, Harrach BD, Niemczyk E, Barna B, Gullner G, Janeczko A, Kogel KH, Schäfer P, Schwarczinger I, Zuccaro A, Skoczowski A (2008) Salt tolerance of barley induced by the root endophyte *Piriformospora indica* is associated with a strong increase in antioxidants. New Phytol 180:501–510

Benhamou N, Kloepper JW, Tuzun S (1998) Induction of resistance against Fusarium wilt of tomato by combination of chitosan with an endophytic bacterial strain: ultrastructure and cytochemistry of the host response. Planta 204(2):153–168

Bhattacharyya A, Duraisamy P, Govindarajan M, Buhroo AA, Prasad R (2016) Nano-biofungicides: emerging trend in insect pest control. In: Prasad R (ed) Advances and applications through fungal nanobiotechnology. Springer International Publishing, Switzerland, pp 307–319

Brader G, Compant S, Vescio K, Mitter B, Trognitz F, Ma LJ, Sessitsch A (2017) Ecology and genomic insights into plant-pathogenic and plant-nonpathogenic endophytes. Annu Rev Phytopathol 55:61–83. https://doi.org/10.1146/annurev-phyto-080516-035641

Bruto M, Prigent-Combaret C, Muller D, Moenne-Loccoz Y (2014) Analysis of genes contributing to plant-beneficial functions in plant growth-promoting rhizobacteria and related proteobacteria. Sci Rep 4:6261. https://doi.org/10.1038/srep06261

Casella TM, Eparvier V, Mandavid H, Bendelac A, Odonne G, Dayan L, Duplais C, Espindola LS, Stien D (2013) Antimicrobial and cytotoxic secondary metabolites from tropical leaf endophytes: isolation of antibacterial agent pyrrocidine C from *Lewia infectoria* SNB-GTC2402. Phytochemistry 96:370–377

Chagas FO, Pessotti RC, Caraballo-Rodríguez AM, Mônica T (2018) Chemical signaling involved in plant–microbe interactions. Chem Soc Rev 47(5):1652–1704

Cocq KL, Gurr SJ, Hirsch PR, Mauchline TH (2017) Exploitation of endophytes for sustainable agricultural intensification. Mol Plant Pathol 18(3):469–473

Compant S, Duffy B, Nowak J, Clément C, Barka EA (2005a) Use of plant growth-promoting bacteria for biocontrol of plant diseases: principles, mechanisms of action, and future prospects. Appl Environ Microbiol 71:4951–4959

Compant S, Reiter B, Sessitsch A, Nowak J, Clément C, Barka EA (2005b) Endophytic colonization of *Vitis vinifera* L. by a plant growth-promoting bacterium, Burkholderia sp. strain PsJN. Appl Environ Microbiol 71:1685–1693

Costa JM, Loper JE (1994) Characterization of siderophore production by the biological-control agent *Enterobacter cloacae*. Mol Plant-Microbe Interact 7:440–448

Dai ZC, Fu W, Wan LY, Cai HH, Wang N, Qi SS, Du D-L (2016) Different growth promoting effects of endophytic bacteria on invasive and native clonal plants. Front Plant Sci 7:706. https://doi.org/10.3389/fpls.2016.00706

De Oliveira JL, Campos EVR, Bakshi M, Abhilash PC, Fraceto LF (2014) Application of nanotechnology for the encapsulation of botanical insecticides for sustainable agriculture: prospects and promises. Biotechnol Adv 32(2014):1550–1561

Deepika VB, Murali TS, Satyamoorth K (2016) Modulation of genetic clusters for synthesis of bioactive molecules in fungal endophytes: a review. Microbiol Res 182:125–140. https://doi.org/10.1016/j.micres.2015.10.009

DESA (2015) World population prospects: the 2015 revision, key findings and advance tables. Working paper no ESA/P/WP 241. United Nations, Department of Economic and Social Affairs (DESA), Population Division, New York

Dharni S, Srivastava AK, Samad A, Patra DD (2014) Impact of plant growth promoting *Pseudomonas monteilii* PsF84 and *Pseudomonas plecoglossicida* PsF610 on metal uptake and production of secondary metabolite (monoterpenes) by rose-scented geranium (*Pelargonium graveolens cv. bourbon*) grown on tannery sludge amended soil. Chemosphere 117:433–439

Ding T, Palmer MW, Melcher U (2013) Community terminal restriction fragment length polymorphisms reveal insights into the diversity and dynamics of leaf endophytic bacteria. BMC Microbiol 13:1. https://doi.org/10.1186/1471-2180-13-1

Eljounaidi K, Kyu-Lee S, Bae H (2016) Bacterial endophytes as potential biocontrol agents of vascular wilt diseases—review and future prospects. Biol Control 103:62–68

FAO (2009) Global agriculture towards (2050). How to feed the World 2050. High-level Expert Forum. http://www.fao.org/fileadmin/templates/wsfs/docs/Issues_papers/HLEF2050_Global_Agriculture.pdf

FAO (2012) IFAD (2012) The state of food insecurity in the world 2012. Economic growth is necessary but not sufficient to accelerate reduction of hunger and malnutrition. FAO, Rome, pp 1–61

Farrer EC, Suding KN (2016) Teasing apart plant community responses to N enrichment: the roles of resource limitation, competition and soil microbes. Ecol Lett 19:1287–1296. https://doi.org/10.1111/ele.12665

Ferrando L, Scavino AF (2015) Strong shift in the diazotrophic endophytic bacterial community inhabiting rice (*Oryza sativa*) plants after flooding. FEMS Microbiol Ecol 91:fiv104. https://doi.org/10.1093/femsec/fiv104

Finyom CWB (2012) Characterisation of the endophytic bacterial communities associated with South African sorghum plants: looking for potential plant growth-promoting endophytes. MSc thesis, pp 1–137

Firrincieli A, Otillar R, Salamov A, Schmutz J, Khan Z, Redman RS, Fleck ND, Lindquist E, Grigoriev IV, Doty SL (2015) Genome sequence of the plant growth promoting endophytic yeast *Rhodotorula graminis* WP1. Front Microbiol 6:978. https://doi.org/10.3389/fmicb.2015.00978

Fouts DE, Tyler HL, DeBoy RT, Daugherty S, Ren Q, Badger JH, Durkin AS, Huot H, Shrivastava S, Kothari S, Dodson RJ, Mohamoud Y, Khouri H, Roesch LF, Krogfelt KA, Struve C, Triplett EW, Methé BA (2008) Complete genome sequence of the N2-fixing broad host range endophyte *Klebsiella pneumoniae* 342 and virulence predictions verified in mice. PLoS Genet 4:e1000141. https://doi.org/10.1371/journal.pgen.1000141

Gill SS, Gill R, Trivedi DK, Anjum NA, Sharma KK, Ansari MW, Ansari AA, Johri AK, Prasad R, Pereira E, Varma A, Tuteja N (2016) *Piriformospora indica*: potential and significance in plant stress tolerance. Front Microbiol 7:332. https://doi.org/10.3389/fmicb.2016.00332

Gupta N, Upadhyaya CP, Singh A, Abd-Elsalam KA, Prasad R (2018) Applications of silver nanoparticles in plant protection. In: Abd-Elsalam K, Prasad R (eds) Nanobiotechnology applications in plant protection. Springer International Publishing AG, Switzerland

Halliwell B, Gutteridge JM (1999) Oxidative stress and antioxidant protection: some special cases. In: Halliwell B, Gutteridge JMC (eds) Free radicals in biology and medicine. Oxford University Press, Oxford, pp 485–543

Hamayun M, Afzal Khan S, Ahmad N (2009) *Cladosporium sphaerospermum* as a new plant growth-promoting endophyte from the roots of *Glycine max* (L.) Merr. World J Microbiol Biotechnol 25(4):627–632

Hardoim PR, Van Overbeek LS, Berg G, Pirttilä AM, Compant S, Campisano A, Döring M, Sessitsch A (2015) The hidden world within plants: ecological and evolutionary considerations for defining functioning of microbial endophytes. Microbiol Mol Biol Rev 79(3):293–320

Hurek BR, Hurek T (2011) Living inside plants: bacterial endophytes. Curr Opin Plant Biol 4 (4):435–443

Hyde KD, Soytong K (2008) The fungal endophyte dilemma. Fungal Divers 33:163–173

Iavicoli I, Leso V, Beezhold DH, Shvedova AA (2017) Nanotechnology in agriculture: opportunities, toxicological implications, and occupational risks. Toxicol Appl Pharmacol 329 (2017):96–111

Ismail M, Prasad R, Ibrahim AIM, Ahmed ISA (2017) Modern prospects of nanotechnology in plant pathology. In: Prasad R, Kumar M, Kumar V (eds), Nanotechnology. Springer Nature. Singapore, pp. 305–317

Jiang D, Fu Y, Ghabrial SA (2013) Mycoviruses: Chapter eight—viruses of the plant pathogenic fungus *Sclerotinia sclerotiorum*. Adv Virus Res 86:215–248

Kidd KA, Muir DCG, Evans MS, Wang XW, Whittle M, Swanson HK, Johnston T, Guildford S (2012) Biomagnification of mercury through lake trout (*Salvelinus namaycush*) food webs of lakes with different physical, chemical and biological characteristics. Sci Total Environ 438:135–143

Kõiv V, Roosaare M, Vedler E, Kivistik PA, Toppi K, Schryer DW, Remm M, Tenson T, Mäe A (2015) Microbial population dynamics in response to *Pectobacterium atrosepticum* infection in potato tubers. Sci Rep 5:11606. https://doi.org/10.1038/srep11606

Kumar K, Belur PD (2016) New extracellular thermostable oxalate oxidase produced from endophytic *Ochrobactrum intermedium* CL6: purification and biochemical characterization. Prep Biochem Biotechnol 46:734–739

Kusari S, Spiteller M (2011) Are we ready for industrial production of bioactive plant secondary metabolites utilizing endophytes? Nat Prod Rep 28:1203–1207

Kusari S, Verma VC, Lamshoeft M, Spiteller M (2012) An endophytic fungus from Azadirachta indica a. Juss. That produces azadirachtin. World J Microbiol Biotechnol 28(3):1287–1294. https://doi.org/10.1007/s11274-011-0876-2

Kwak MJ, Song JY, Kim SY, Jeong H, Kang SG, Kim BK, Kwon SK, Lee CH, Yu DS, Park SH, Kim JF (2012) Complete genome sequence of the endophytic bacterium *Burkholderia* sp. strain KJ006. J Bacteriol 194:4432–4433

Lata C, Muthamilarasan M, Prasad M (2015) Drought stress responses and signal transduction in plants. In: Pandey GK (ed) Elucidation of abiotic stress signaling in plants. Springer, New York, NY, pp 195–225. https://doi.org/10.1007/978-1-4939-2540-7_7

Lee S, Flores-Encarnacion M, Contreras-Zentella M, Garcia-Flores L, Escamilla JE, Kennedy C (2004) Indole-3-acetic acid biosynthesis is deficient in *Gluconacetobacter diazotrophicus* strains with mutations in cytochrome C biogenesis genes. J Bacteriol 186:5384–5391

Lehtonen PT, Helander M, Siddiqui SA, Lehto K, Saikkonen K (2006) Endophytic fungus decreases plant virus infections in meadow ryegrass (Loliumpratense). Biol Lett 2(4):620–623. https://doi.org/10.1098/rsbl.2006.0499

Leitão AL, Enguita FJ (2016) Gibberellins in *Penicillium* strains: challenges for endophyte-plant host interactions under salinity stress. Microbiol Res 183:8–18. https://doi.org/10.1016/j.micres.2015.11.004

Liang Z, Zhang J, Zhang X, Li J, Zhang X, Zhao C (2015) Endophytic fungus from *Sinopodophyllum emodi* (wall.) Ying that produces Podophyllotoxin. J Chromatogr Sci 54(2):175–178. https://doi.org/10.1093/chromsci/bmv124

Liu H, Carvalhais LC, Crawford M, Singh E, Dennis PG, Pieterse CMJ, Schenk PM (2017) Inner plant values: diversity, colonization and benefits from endophytic bacteria. Front Microbiol 8:2552. https://doi.org/10.3389/fmicb.2017.02552

Ma Y, Oliveira RS, Nai FJ, Rajkumar M, Luo YM, Rocha I, Freitas H (2015) The hyperaccumulator *Sedum plumbizincicola* harbors metal-resistant endophytic bacteria that improve its phytoextraction capacity in multi-metal contaminated soil. J Environ Manag 156:62–69

Ma Y, Rajkumar M, Zhang C, Freitas H (2016) Beneficial role of bacterial endophytes in heavy metal phytoremediation. J Environ Manag 174(2016):14–25

Mada FA, Bruce B (2017) Isolation, characterization and identification of putative bacterial endophytes from some plants in hot springs, South Dakota. Int J Curr Microbiol Appl Sci 6 (6):756–767. https://doi.org/10.20546/ijcmas.606.089

Malinowski DP, Belesky DP (2004) Adaptations of endophyte-infected cool-season grasses to environmental stresses: mechanisms of drought and mineral stress tolerance. Crop Sci 40 (4):923–940

Malinowski DP, Alloush GA, Belesky DP (2000) Leaf endophyte *Neotyphodium coenophialum* modifies mineral uptake in tall fescue. Plant Soil 227:115–126

Manohari R, Yogalakshmi KN (2016) Optimization of copper (II) removal by response surface methodology using root nodule endophytic bacteria isolated from *Vigna unguiculata*. Water Air Soil Pollut 227:285

Marasco R, Rolli E, Ettoumi B, Vigani G, Mapelli F, Borin S, Abou-Hadid AF, El-Behairy UA, Sorlini C, Cherif A, Zocchi G, Daffonchio D (2012) A drought resistance-promoting microbiome is selected by root system under desert farming. PLoS One 7:48479. https://doi.org/10.1371/journal.pone.0048479

Martinez-Garcia PM, Ruano-Rosa D, Schiliro E, Prieto P, Ramos C, Rodríguez-Palenzuela P, Mercado-Blanco J (2015) Complete genome sequence of *Pseudomonas fluorescens* strain PICF7, an indigenous root endophyte from olive (*Olea europaea* L.) and effective biocontrol agent against *Verticillium dahlia*. Stand Genomic Sci 10:10. https://doi.org/10.1186/1944-3277-10-10

Martini M, Musetti R, Grisan S, Polizzotto R, Borselli S, Pavan F, Osler R (2009) DNA- dependent detection of the grapevine fungal Endophytes *Aureobasidium pullulans* and *Epicoccum nigrum*. Plant Dis 93(10):993–998. https://doi.org/10.1094/PDIS-93-10-0993

Mastretta C, Taghavi S, van der Lelie D (2009) Endophytic bacteria from seeds of *Nicotiana tabacum* can reduce cadmium phytotoxicity. Int J Phytoremediation 11(3):251–267

Matsouri F, Björkman T, Harman GE (2010) Seed treatment with *Trichoderma harzianum* alleviates biotic, abiotic, and physiological stresses in germinating seeds and seedlings. Phytopathology 100(11):1213–1221

Miraglia M, Marvin HJP, Kleter GA, Battilani P, Brera C, Coni E, Cubadda F, Croci L, De Santis B, Dekkers S, Filippi L, Hutjes RW, Noordam MY, Pisante M, Piva G, Prandini A, Toti L, van den Born GJ, Vespermann A (2009) Climate change and food safety: an emerging issue with special focus on Europe. Food Chem Toxicol 47:1009–1021. https://doi.org/10.1016/j.fct.2009.02.005

Mishra S, Singh A, Keswani C, Saxena A, Sharma BK, Singh HB (2015) Harnessing plant-microbe interactions for enhanced protection against phytopathogens. In: Arora NK (ed) Plant microbes symbiosis: applied facets. Springer, India, pp 111–125

Mohana Kumara P, Zuehlke S, Priti V, Ramesha BT, Shweta S, Ravikanth G, Vasudeva R, Santhoshkumar TR, Spiteller M, Uma Shaanker R (2012) Fusarium proliferatum, an endophytic fungus from Dysoxylum binectariferum Hook.F, produces rohitukine, a chromane alkaloid possessing anti-cancer activity. Antonie Van Leeuwenhoek 101(2):323–329. https://doi.org/10.1007/s10482-011-9638-2

Murmu N (2016) Characterization of culturable endophytes of bfinger millet (*Eleusine coracana* (L.) Gaertn.) genotype JWM1, MSc thesis, pp 25–32

Nadarajah KK (2017) Induction of systemic resistance for disease suppression. In: Abdullah S, Chai-Ling H, Wagstaff C (eds) Crop improvement. Springer, Cham. https://doi.org/10.1007/978-3-319-65079-1_15

Narayan CP, Jian XD, Kyu SS, Seung HY (2012) Molecular and morphological characterization of Endophytic *Heterobasidion araucariae* from roots of *Capsicum annuum* L. in Korea. Mycobiology 40(2):85–90. https://doi.org/10.5941/MYCO.2012.40.2.85

Nath M, Bhatt D, Prasad R, Gill SS, Anjum NA, Tuteja N (2016) Reactive oxygen species generation-scavenging and signaling during plant-arbuscular mycorrhizal and Piriformospora indica interaction under stress condition. Front Plant Sci 7:1574. https://doi.org/10.3389/fpls.2016.01574

Nath M, Bhatt D, Prasad R, Tuteja N (2017) Reactive oxygen species (ROS) metabolism and signaling in plant-mycorrhizal association under biotic and abiotic stress conditions. In: Varma A, Prasad R, Tuteja N (eds) Mycorrhiza. Springer International Publishing AG, Switzerland, pp 223–232

Nei M (2003) Genome evolution: let's stick together. Heredity 90:411–412

Nerva L, Varese GC, Falk BW, Turina M (2017) Mycoviruses of an endophytic fungus can replicate in plant cells: evolutionary implications. Sci Rep 7:1908. https://doi.org/10.1038/s41598-017-02017-3

Nongkhlaw F, Mary W, Joshi SR (2017) Microscopic study on colonization and antimicrobial property of endophytic bacteria associated with ethnomedicinal plants of Meghalaya. J Microsc Ultrastructure 5:132–113

Osbourn AE, Field B (2009) Operons. Cell Mol Life Sci 66:3755–3775

Owen NL, Hundley N (2004) Endophytes-the chemical synthesizers inside plants. Sci Prog 87 (pt 2):79–99

Parisi C, Vigani M, Rodríguez-Cerezo E (2015) Agricultural nanotechnologies: what are the current possibilities? Nano Today 10:124–127

Prasad R (2017) Mycoremediation and environmental sustainability, vol 1. Springer, Switzerland. isbn:978-3-319-68957-9. https://link.springer.com/book/10.1007/978-3-319-68957-9

Prasad R (2018) Mycoremediation and environmental sustainability, vol 2. Springer, Switzerland. isbn:978-3-319-77386-5. https://www.springer.com/us/book/9783319773858

Prasad R, Kumar V, Prasad KS (2014) Nanotechnology in sustainable agriculture: present concerns and future aspects. Afr J Biotechnol 13(6):705–713

Prasad R, Bhattacharyya A, Nguyen QD (2017) Nanotechnology in sustainable agriculture: recent developments, challenges, and perspectives. Front Microbiol 8:1014. https://doi.org/10.3389/fmicb.2017.01014

Priyanka B, Monika B, Umang A, Rekha S, Tanisha S, Leela W (2018) Endophytes: an environmental friendly Bacteria for plant growth promotion. Int J Curr Microbiol Appl Sci 7 (2):1899–1911. https://doi.org/10.20546/ijcmas.2018.702.229

Rabha AJ, Naglot A, Sharma GD, Gogoi HK, Gupta VK, Shreemali DD, Veer V (2016) Morphological and molecular diversity of endophytic *Colletotrichum gloeosporioides* from tea plant, *Camellia sinensis* (L.) O. Kuntze of Assam, India. J Genet Eng Biotechnol 14:181–187

Ray S, Singh V, Singh S, Sarma BK, Singh HB (2016) Biochemical and histochemical analyses revealing endophytic *Alcaligenes faecalis* mediated suppression of oxidative stress in *Abelmoschus esculentus* challenged with *Sclerotium rolfsii*. Plant Physiol Biochem 109 (1):430–441

Ren G, Zhang H, Lin X, Zhu J, Jia Z (2015a) Response of leaf endophytic bacterial community to elevated CO_2 at different growth stages of rice plant. Front Microbiol 6:855. https://doi.org/10.3389/fmicb.2015.00855

Ren G, Zhu C, Alam MS, Tokida T, Sakai H, Nakamura H, Jia Z (2015b) Response of soil, leaf endosphere and phyllosphere bacterial communities to elevated CO_2 and soil temperature in a rice paddy. Plant Soil 392:27–44

Rodriguez RJ, Henson J, Van Volkenburgh E, Hoy M, Wright L, Beckwith F, Kim YO, Redman RS (2008) Stress tolerance in plants via habitat-adapted symbiosis. ISME 2:404–416

Russell JR, Huang J, Anand P (2011) Biodegradation of polyester polyurethane by endophytic fungi. Appl Environ Microbiol 77(17):6076–6084

Ryan RP, Germaine K, Franks A, Ryan DJ, Dowling DN (2008) Bacterial endophytes: recent developments and applications. FEMS Microbiol Lett 278:1–9

Ryu CM, Farag MA, Hu CH, Reddy MS, Wei HX, Paré PW, Kloepper JW (2003) Bacterial volatiles promote growth in Arabidopsis. Proc Natl Acad Sci U S A 100(8):4927–4932

Santoyo G, Moreno-Hagelsieb G, Orozco-Mosqueda MC, Glick BR (2016) Plant growth-promoting bacterial endophytes. Microbiol Res 183:92–99

Scherwinski K, Grosch R, Berg G (2008) Effect of bacterial antagonists on lettuce: active biocontrol of *Rhizoctonia solani* and negligible, short-term effects on non-target microorganisms. FEMS Microbiol Ecol 64(1):106–116. https://doi.org/10.1111/j.1574-6941.2007.00421.x

Schutz B (2001) Bioactive fungal metabolites-impact and exploitation. In: International symposium proceedings. University of Wales, British Mycological Society, Swansea, UK, p 20

Seghers D, Wittebolle L, Top EM, Verstraete W, Siciliano SD (2004) Impact of agricultural practices on the *Zea mays* L. endophytic community. Appl Environ Microbiol 70(3):1475–1482. https://doi.org/10.1128/AEM.70.3.1475-1482.2004

Shabala S, Cuin TA (2008) Potassium transport and plant salt tolerance. Physiol Plant 133 (4):651–669

Shahzad R, Khan AL, Bilal S, Asaf S, Lee I-J (2018) What is there in seeds? Vertically transmitted endophytic resources for sustainable improvement in plant growth. Front Plant Sci 9:24. https://doi.org/10.3389/fpls.2018.00024

Sharma D, Pramanik A, Agrawal PK (2016) Evaluation of bioactive secondary metabolites from endophytic fungus *Pestalotiopsis neglecta* BAB-5510 isolated from leaves of *Cupressus torulosa* D.Don. 3. Biotech 6:210

Shatrupa R, Jyoti S, Rahul SR, Harikesh BS, Surendra S (2018) Endophytic bacteria: an essential requirement of Phyto nutrition. Nutr Food Sci Int J 5(2):555657. https://doi.org/10.19080/NFSIJ.2018.05.555657

Shearin ZRC, Filipek M, Desai R, Bickford WA, Kowalski KP, Clay K (2017) Fungal endophytes from seeds of invasive, non-native *Phragmites australis* and their potential role in germination and seedling growth. Plant Soil 2017:1–12. https://doi.org/10.1007/s11104-017-3241-x

Shen SY (2013) Bacterial endophytes: exploration of methods and analysis of community variation. MSc thesis, pp 1–183

Shimura K, Okada A, Okada K, Jikumaru Y, Ko K, Toyomasu T (2007) Identification of a biosynthetic gene cluster in rice for momilactones. J Biol Chem 282:34013–34018

Sproul D, Gilbert N, Bickmore WA (2005) The role of chromatin structure in regulating the expression of clustered genes. Nat Rev Genet 6:775–781

Srivastava S, Bist V, Srivastava S, Singh PC, Trivedi PK, Asif MH, Chauhan PS, Nautiyal CS (2016) Unraveling aspects of *Bacillus amyloliquefaciens* mediated enhanced production of rice under biotic stress of *Rhizoctonia solani*. Front Plant Sci 7:587. https://doi.org/10.3389/fpls.2016.00587

Strobel G, Daisy B, Castillo U, Harper J (2004) Natural products from endophytic microorganisms. J Nat Prod 2004(67):257–268

Suman A, Yadav AN, Verma P (2016) Endophytic microbes in crops: diversity and beneficial impact for sustainable agriculture. In: Singh DP, Singh HB, Prabha R (eds) Microbial inoculants in sustainable agricultural productivity, vol 1, research perspectives. Springer, New Delhi, pp 117–143

Tejesvi MV, Kini KR, Prakash HS, Subbiah V, Shetty HS (2007) Genetic diversity and antifungal activity of species of *Pestalotiopsis* isolated as endophytes from medicinal plants. Fungal Divers 24:37–54

Tellenbach C, Grunig CR, Sieber TN (2010) Suitability of quantitative real-time PCR to estimate the biomass of fungal root endophytes. Appl Environ Microbiol 76(17):5764–5772. https://doi.org/10.1128/AEM.00907-10

Tétard-Jones C, Edwards R (2016) Potential roles for microbial endophytes in herbicide tolerance in plants. Pest Manag Sci 72(2):203–209. https://doi.org/10.1002/ps.4147

Tian BY, Cao Y, Zhang KQ (2015) Metagenomic insights into communities, functions of endophytes, and their associates with infection by root-knot nematode, Meloidogyne incognita, in tomato roots. Sci Rep 5:17087. https://doi.org/10.1038/srep17087

Tiwari S, Lata C, Prasad M, Prasad V, Chauhan PS (2017a) A functional genomic perspective on drought signalling and its cross-talk with various phytohormone-mediated signalling pathways in plants. Curr Genomics. https://doi.org/10.2174/1389202918666170605083319

Tiwari S, Prasad V, Chauhan PS, Lata C (2017b) *Bacillus amyloliquefaciens* confers tolerance to various abiotic stresses and modulates plant response to Phytohormones through osmoprotection and gene expression regulation in rice. Front Plant Sci 8:1510. https://doi.org/10.3389/fpls.2017.01510

Van Loon LC, Bakker PA, Pieterse CM (1998) Systemic resistance induced by rhizosphere bacteria. Annu Rev Phytopathol 36(1):453–483

Verma VC, Anand S, Ulrichs C (2013) Biogenic gold nanotriangles from *Saccharomonospora* sp., an endophytic actinomycetes of *Azadirachta indica* A. Juss. Int Nano Lett 3:21. https://doi.org/10.1186/2228-5326-3-21

Wakelin S, Warren R, Harvey P, Ryder M (2004) Phosphate solubilization by Penicillium spp. closely associated with wheat roots. Biol Fertil Soils 40:36–43

Weilharter A, Mitter B, Shin MV, Chain PS, Nowak J, Sessitsch A (2011) Complete genome sequence of the plant growth-promoting endophyte *Burkholderia phytofirmans* strain PsJN. J Bacteriol 193:3383–3384

White JF, Torres MS (2010) Is plant endophyte-mediated defensive mutualism the result of oxidative stress protection? Physiol Plant 138(4):440–446

Winding A, Binnerup SJ, Pritchard H (2004) Non-target effects of bacterial biological control agents suppressing root pathogenic fungi. FEMS Microbiol Ecol 47(2):129–141. https://doi.org/10.1016/S0168-6496(03)00261-7

Xin G, Zhang G, Kang JW, Staley JT, Doty SL (2009) A diazotrophic, indole-3-acetic acid-producing endophyte from wild cottonwood. Biol Fertil Soils 45(6):669–674

Yu X, Li B, Fu Y, Jiang D, Ghabrial SA, Li G, Peng Y, Xie J, Cheng J, Huang J, Yi X (2010) A geminivirus-related DNA mycovirus that confers hypovirulence to a plant pathogenic fungus. Proc Natl Acad Sci U S A 107(18):8387–8392. https://doi.org/10.1073/pnas.0913535107

Zhang X, Li J, Qi G, Wen K, Lu J, Zhao V (2011) Insecticidal effect of recombinant endophytic bacterium containing *Pinellia ternata* agglutinin against white backed plant hopper, *Sogatella furcifera*. Crop Prot 30(11):1478–1484

Zhao X, Qi G, Zhang X, Lan N, Ma X (2010) Controlling sap-sucking insect pests with recombinant endophytes expressing plant lectin. Nature Preceding's. 21, article 21

Chapter 2
Plant–*Phytophthora* Interaction Proteomics

M. Anandaraj and P. Umadevi

Abstract Study on plant–pathogen interactions is to identify the plant factors that favor or resist pathogen development and to target those factors responsible for abolishing pathogen development. In this focus, the strategies such as genomics, transcriptomics, proteomics, and metabolomics are adopted. Among these strategies, the differential proteomics is gaining attention. The differential protein profiling of sugar pine to white pine blister rust using a gel-based approach by Ekramoddoullah and Hunt in 1993 was the first study of proteomics in plant–pathogen interactions, Following this, until now hundreds of proteomic experiments have been successfully carried out in the area of molecular plant–microbe interaction. *Phytophthora* is a major pathogen affecting many economically important crops. Information on plant proteins upon *Phytophthora* interaction will pave the way for improving plant resistance. This chapter highlights the progress made so far in this area to bring out the current scenario of proteomics in plant–*Phytophthora* interactions and its successful application in plant breeding.

2.1 Introduction

Plants have innate immunity, which is recognized and activated by the invading organism. Recognition of non-self molecules is the first step towards an active immune response and is mediated by pattern-recognition receptors (PRRs) in the host cells. These PRRs are able to recognize microbe-associated molecular patterns (MAMPs), which are also often termed as pathogen-associated molecular patterns (PAMPs) (Boller and Felix 2009). The recognition of MAMPs/PAMPs by plant PRRs leads to so-called PAMP-triggered immunity/pattern triggered immunity (PTI)/non-host resistance, which provides a first line of defense against most nonadapted pathogens (Jones and Dangl 2006). When PTI is suppressed by

M. Anandaraj
ICAR-Indian Institute of Spices Research, Kozhikode, Kerala, India

P. Umadevi (✉)
Division of Crop Improvement and Biotechnology, ICAR-Indian Institute of Spices Research, Kozhikode, Kerala, India

A. Varma et al. (eds.), *Plant Biotic Interactions*,
https://doi.org/10.1007/978-3-030-26657-8_2

pathogen effectors delivered into the host cell, plants can overcome pathogen suppression of PTI and bring back the pathogen resistance through effector-triggered immunity (ETI). This branch of innate immune signaling is activated when individual pathogen effectors are recognized by disease resistance (R) proteins (Nomura et al. 2011). In addition to PTI and ETI, plant immune responses are regulated by several plant hormones, including salicylic acid (SA), jasmonic acid (JA), and ethylene. Based on plant response, there are two major patterns, viz., compatible and incompatible interactions. In compatible interactions, plants allow the pathogens to complete their life cycle while in incompatible interactions, plants exert efficient defense responses to forestall pathogen growth. The defense responses induced upon pathogen attack are local and systemic defense responses.

In the past few years there have been remarkable efforts in the application of proteomics for studying plant–pathogen interactions (reviewed in Mehta et al. 2008; Quirino et al. 2010; Lodha et al. 2013). *Phytophthora* the major soil-borne oomycete causes disease/yield loss in crops such as potato (Widmark et al. 2007), soybean (Kaufmann and Gerdeman 1958), tomato (Jupe et al. 2013), capsicum (Leonian 1922), and black pepper (Anandaraj 1997). The hemibiotrophic nature of the pathogen makes the plant–*Phytophthora* interaction a complex network. The proteomics brings out the dynamic changes of the proteome to understand the entire signal perception and transduction upon *Phytophthora* interaction in plants. This chapter summarizes the progress made so far in this area to bring out the current scenario of proteomics in plant–*Phytophthora* interactions. The proteomics studies in plant–*Phytophthora* is given in Table 2.1. The identification of peptides from mass spectrometry data is the core in gel-based or gel-free proteomics (Fig. 2.1). This has been done successfully by using specific search criteria on the whole

Table 2.1 List of published proteomics studies on plant–*Phytophthora* interaction

Plant	Nature of plant genotype	Pathogen	Method	References
Fagus sylvatica	Susceptible	*Phytophthora citricola*	2DE	Valcu et al. (2009)
Soybean	Reference genotype infected with virulent and avirulent isolate	*Phytophthora sojae*	2DE	Jing et al. (2015)
Potato	Resistant and Susceptible	*Phytophthora infestans*	Label free	Ali et al. (2014)
	Resistant and Susceptible	*Phytophthora infestans*	Label free	Chawade et al. (2016)
	Resistant, moderately resistant, and susceptible	*Phytophthora infestans*	Label free	Larsen et al. (2016)
Tomato	Susceptible	*Phytophthora capsici*	Label free	Howden et al. (2017)
Black pepper	Moderately tolerant	*Phytophthora capsici*	Label free	Anandaraj and Umadevi (2016)
	Susceptible	*Phytophthora capsici*	Label free	Mahadevan et al. (2016)
	Moderately tolerant	*Phytophthora capsici*	Label free	Umadevi et al. (2018)

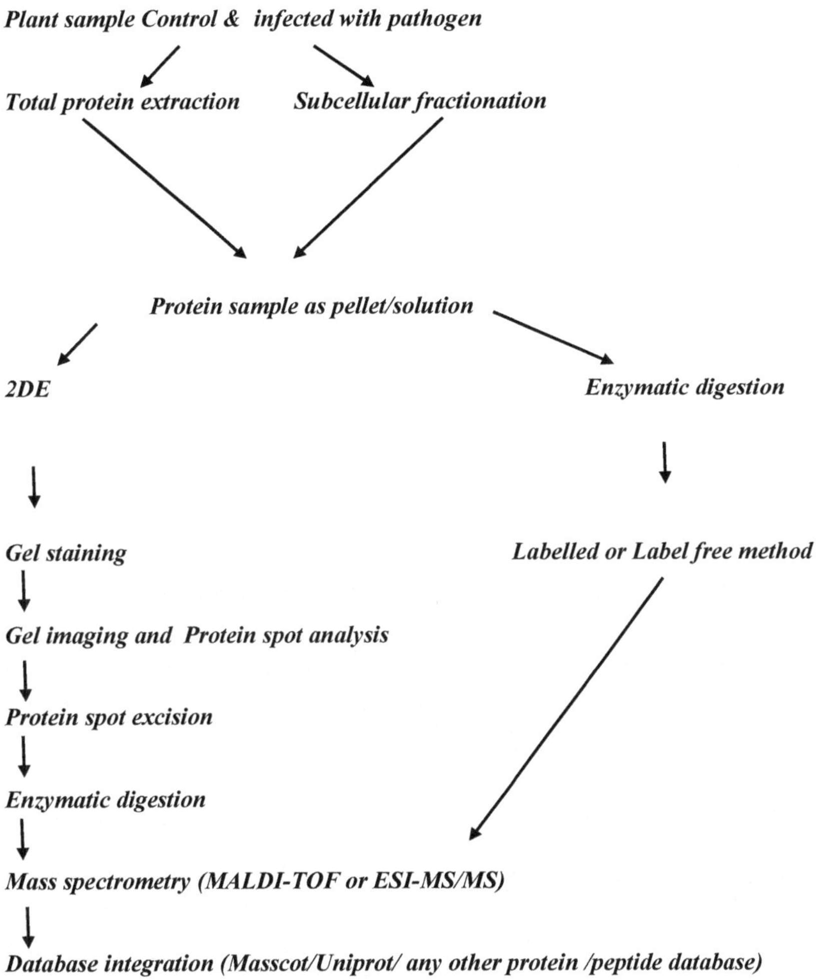

Fig. 2.1 Common steps involved in proteomics strategy

genome and EST datasets (Jing et al. 2015; Ali et al. 2014; Chawade et al. 2016; Larsen et al. 2016; Howden et al. 2017) and the homology-driven proteomics to characterize proteomes of organisms with unsequenced genomes (Valcu et al. 2009; Anandaraj and Umadevi 2016; Umadevi et al. 2018; Mahadevan et al. 2016) in plant–*Phytophthora* interactions.

2.2 Gel-Based Proteomics

Two-dimensional gel electrophoresis (2-DE) followed by MALDI-TOF-TOF study demonstrated the interactions between soybean reference cultivar Williams 82 with the avirulent isolate P6497 (incompatible) and the virulent isolate P7076 (compatible) (Jing et al. 2015). The network for soybean infected by virulent and avirulent pathogens was also proposed (Wang 2015). The infected soybean displayed resistance to P6497 and susceptibility to P7076. A total of 95 protein spots with 1.5-fold or less than 0.66-fold expression difference (P\0.05) in at least one infected time point when compared to the control showed the involvement of upregulated proteins in reactive oxygen species (ROS) burst, salicylic acid (SA) signal pathway, and biosynthesis of isoflavones as defense factors in soybean for *P. sojae.*

Two-dimensional gel electrophores is followed by homology-driven mass spectrometric analysis of proteins from *Fagus sylvatica* infected with the root pathogen *Phytophthora citricola* showed the at local and systemic response (Valcu et al. 2009) of a compatible interaction. The 68 protein spots identified represented 51 proteins involved in protein synthesis and processing, energy, signal transduction, stress and defense, and primary and secondary metabolism. The local response (root) was found to be weaker than the systemic (leaf) response based on the presence of upregulated proteins. This suggested that in *Fagus*, the defense reactions of the seedlings are suppressed by the pathogen. The systemic response to infections was by downregulation of carbon fixation, upregulation of auxin and ethylene synthesis, and phenylpropanoid and flavonoid biosynthesis pathway. The degree of systemic defense response in *Fagus* was relatively low at early stages of infection.

2.3 Label-Free Proteomics

Label-free proteomics using 1D gel-separated apoplastic proteins from leaves of three potato clones Desiree (Susceptible), Sapro Mira, and SW93-1015 (Resistant) infected with *P. infestans* at 6, 24, and 72 hpi along with control demonstrated the compatible and incompatible response of potato to *P. infestans* (Ali et al. 2014). The hypersensitive reaction (HR), symptom development, and putative effectors were elucidated in this study. The proteins with differential abundance proteins in at least one point of pathogen infection was found to be enriched with 517 secretory proteins of subtilase, peroxidase, protease inhibitors, GDSL-lipase, and pectin esterase families. Comparative analysis showed that there was a decrease in abundance of a large number of proteins in compatible interaction than in incompatible. Around 59, 131, and 66 proteins were in increased abundance at 6, 24, and 72 hpi, respectively, in incompatible reactions. These proteins were regarded as HR and early-resistance-related proteins in potato against *P. infestans*. Based on a minimum fourfold increased abundance in resistant genotypes, 49 candidate proteins for HR were derived. The HR candidate proteins were found to increase at 6 and 24 hpi with 23 and 43 proteins, respectively, while it was reduced to 9 at 72 hpi. The glucanases and glucosidase which were in abundance in susceptible clone were attributed as symptom-development-related proteins. The downregulated proteins with a minimum of eightfold abundance in compatible interaction were regarded as putative effector targets. Forty single proteins from large gene families were identified in which 9 were specific for 6 hpi time point and 7 specific for 24 hpi, 29 proteins were common. The datasets by Ali et al. (2014) were applied in translational research by Chawade et al. (2016) to develop the selective reaction monitoring (SRM) assay system.

In cases with complex traits involving protein modification and changes in abundance, the quantitative protein estimation plays an important role. This proof of concept has been demonstrated successfully by them. In their study, the targeted quantitative proteomics of selected peptides was applied to SW93-1015 (resistant) and Desiré (susceptible) crossing populations. The presence of four important peptides in close proximity indicated that chromosome 6 is the putative hot spot for tuber resistance. Two peptides of Kunitz trypsin inhibitor, cystatin, nonspecific lipid transfer protein, serine carboxypeptidase III with a low level and peptides of beta-d glucan exohydroase, and peroxidase with a high level were identified as markers for tuber resistance. This highly applicable marker panel at the field level is an important line of proteomics work, in which the markers for the complex and important phenotype that lacks good DNA-based markers is successfully demonstrated.

Larsen et al. (2016) has developed the most extensive label-free proteomics datasets on potato interactions with mixed strain culture of *P. infestans*. The developed raw data and MaxQuant results of the time course study over 258 h on leaf proteomics from commercial cultivars of differential resistance to pathogen, viz., Kuras (moderate), SarpoMira (highly resistant), and Bintje (very susceptible), is available in ProteomeXchange Consortium-PRIDE partner repository with the dataset identifier PXD0002767.

Total nuclear proteins were studied in tomato upon *P. capsici* infection (Howden et al. 2017). A number of proteins were found to have dynamics in abundance 24 h post infection relative to noninfected samples (507 with increased and 173 with decreased abundance). Members of the AT-Hook-Like (AHL) protein family are demonstrated to contribute to immunity (Kim et al. 2007). This study on nuclear proteome had successfully demonstrated the impact of AHL Types I and II proteins in PTI responses thereby enhancing immunity to *Phytophthora capsici*. This suggests that modification of AHL protein abundance could be used to enhance disease resistance.

Panniyur 1 variety of black pepper (*P. nigrum*), which is highly susceptible to *Phytophthora*, was analyzed using detached leaf inoculation method. The label-free quantification showed 22 proteins with upregulation (two fold) and 134 proteins with downregulation (0.8 fold). The downregulation of the proteins involved in 14 major pathways including carbon fixation, photosynthesis, amino acid metabolism, glutathione metabolism, and phenylpropanoid biosynthesis and the upregulation of aspartyl protease, pectin esterase, polyphenoloxidase, and mlp-like proteins were the identified as susceptibility factors in black pepper (Mahadevan et al. 2016). The

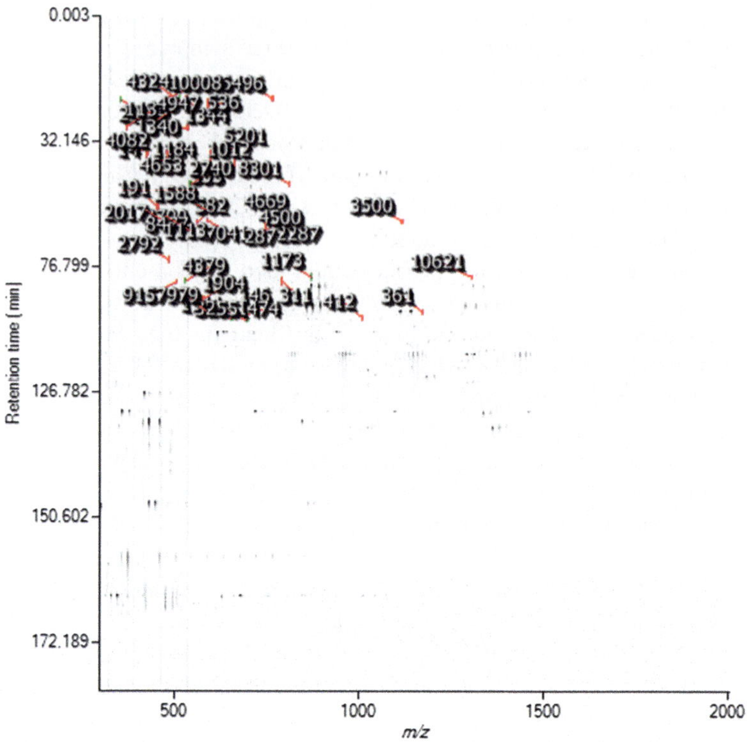

Fig. 2.2 Ion intensity map showing black pepper peptides under quantification as spots

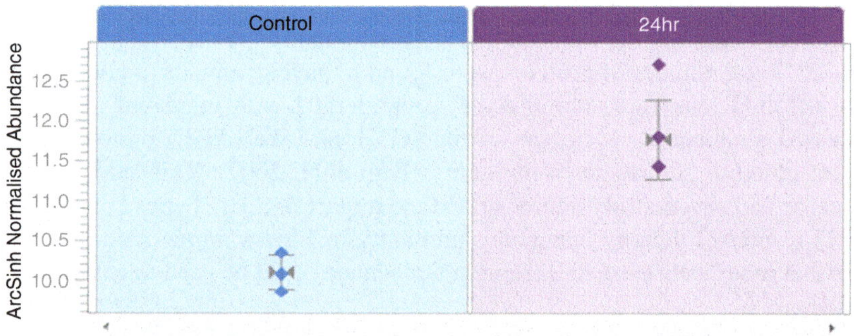

Fig. 2.3 Representative graph with Average Normalized Abundance showing relative quantification of black pepper peptide from three biological replicates at 24 h post infection

upregulation of certain R genes is regarded as S genes in susceptible variety of black pepper (Subhakara) for *P. capsici* (Umadevi and Anandaraj 2017).

Label-free proteomics strategy (Figs. 2.2 and 2.3) on tolerant (IISR Shakthi) and susceptible (Subhakara) genotypes showed 299 proteins, out of them 84 proteins were found to have above 4-fold to 973-fold increase in expression upregulated at

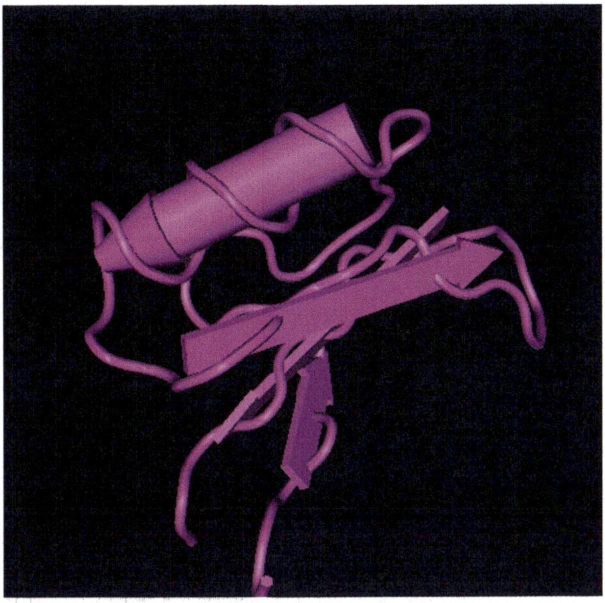

Fig. 2.4 Structure of defensin-type AMP from black pepper

24 hpi. The tolerant genotype suppressed the pathogen by pattern-triggered immunity (PTI), which was attributed by pattern recognition receptor against *Phytophthora* in black pepper, viz., receptor-like kinases (RLKs), RPP13 (R gene), and Salicylic acid (SA)-mediated SAR. PR proteins (with antifungal activity) production were found to be more tolerant in black pepper. The susceptible genotype recorded a weak PTI. Effector-triggered immunity (ETI) was found working with the presence of PRM1 (R gene) with jasmonic-acid-mediated suppression of SA, production of PR5 (Thaumatin), and protein (Anandaraj and Umadevi 2016).

The integration of label-free proteomics data with the *Piper* transcriptome DB on IISR Shakthi Illumina GLX 2X platform, annotated with Blast2Go, identified the R gene families in black pepper for *P. capsici* (Umadevi et al. 2015). The phylogeny of important R genes was derived and the differential regulation of R genes was also attributed to host-specific resistance (Umadevi and Anandaraj 2017). The involvement of AMPs in black pepper resistance to *Phytophthora* has been demonstrated using label-free proteomics. The differential abundance of AMP signature peptides from resistant variety (IISR Shakthi) signifies the presence of AMPs in the defense network. The 24 AMP signatures belonged to the cationic, anionic, cysteine-rich, and cysteine-free group and also had conserved domains in the small peptide sequences (Umadevi et al. 2018). The important plant AMP group defensin with antifungal activity identified in black pepper is shown in the Fig. 2.4.

2.4 Conclusion and Future Prospects

Proteomics is an efficient strategy to bring global or specific data on differential proteome during plant–pathogen interactions. Apart from understanding the pathogenicity, the application of peptide data as SRM assay in potato sends a strong message that the proteomics data is a useful targets towards translational research, which can be applied at the field level.

Acknowledgments The authors acknowledge with thanks the financial support from the ICAR-Outreach Programme on *Phytophthora, Fusarium*, and *Ralstonia* diseases of horticultural and field crops (PhytoFuRa).

References

Ali A, Alexandersson E, Sandin M, Resjö S, Lenman M, Hedley P, Lavander F, Andreasson E (2014) Quantitative proteomics and transcriptomics of potato in response to *Phytophthora infestans* in compatible and incompatible interactions. BMC Genomics 15:497

Anandaraj M (1997) Ecology of *Phytophthora capsici* causal organism of foot rot of black pepper (*P. nigrum* L.), PhD thesis, University of Calicut

Anandaraj M, Umadevi P (2016) The post genomic era: novel approaches for studying plant diseases and their management. Indian Phytopathol 69(4s):260–265

Boller T, Felix G (2009) A renaissance of elicitors: perception of microbe-associated molecular patterns and danger signals by pattern-recognition receptors. Annu Rev Plant Biol 60:379–406

Chawade A, Alexandersson E, Bengtsson T, Erik Andreasson E, Levander F (2016) Targeted proteomics approach for precision plant breeding. J Proteome Res 15:638–646

Howden AJM, Stam R, Heredia VM, Motion GB, Have S, Hodge K, Marques Monteiro Amaro TM, Edgar Huitema E (2017) Quantitative analysis of the tomato nuclear proteome during *Phytophthora capsici* infection unveils regulators of immunity. New Phytol 215:309–322

Jing M, Ma H, Li H, Guo B, Zhang X, Ye W, Wang H, Wang Q, Wang Y (2015) Differential regulation of defense-related proteins in soybean during compatible and incompatible interactions between *Phytophthora sojae* and soybean by comparative proteomic analysis. Plant Cell Rep 34:1263–1280

Jones JDG, Dangl JL (2006) The plant immune system. Nature 444:323–329

Jupe J, Stam R, Howden A, Morris J, Zhang R, Hedley PE, Huitama E (2013) *Phytophthora capsici*-tomato interaction features dramatic shifts in gene expression associated with a hemi-biotrophic lifestyle. Genome Biol 14:R63

Kaufmann MJ, Gerdemann JW (1958) Root and stem rot of soybean caused by *Phytophthora sojae* n.sp. Phytopathology 48:201–208

Kim SY, Kim YC, Seong ES, Lee YH, Park JM, Choi D (2007) The chili pepper CaATL1: an AT-hook motif-containing transcription factor implicated in defence responses against pathogens. Mol Plant Pathol 8:761–771

Larsen MKG, Jørgensen MM, Bennike TB, Stensballe A (2016) Time-course investigation of *Phytophthora infestans* infection of potato leaf from three cultivars by quantitative proteomics. Data Brief 6:238–248

Leonian LH (1922) Stem and fruit blight of peppers caused by *Phytophthora capsici* sp. nov. Plant Dis 12:401–408

Lodha TD, Hembram P, Tep N, Basak J (2013) Proteomics: a successful approach to understand the molecular mechanism of plant-pathogen interaction. Am J Plant Sci 4:1212–1226

Mahadevan C, Krishnan A, Saraswathy GG, Surendran A, Jaleel A, Sakuntala M (2016) Transcriptome- assisted label-free quantitative proteomics analysis reveals novel insights into *Piper nigrum-Phytophthora capsici* Phytopathosystem. Front Plant Sci 7:785

Mehta A, Brasileiro ACM, Souza DSL, Romano E, Campos MA, Grossi-de-Sa MF, Silva MS, Franco OL, Fragoso RR, Rosangela Bevitori R, Rocha TL (2008) Plant–pathogen interactions: what is proteomics telling us? FEBS J 275:3731–3746

Nomura K, Mecey C, Lee YN, Imboden LA, Chang JH, He SY (2011) Effector-triggered immunity blocks pathogen degradation of an immunity-associated vesicle traffic regulator in Arabidopsis. PNAS 108(26):10774–10779

Quirino BF, Candido ES, Campos PF, Franco OL, Krüger RH (2010) Proteomic approaches to study plant-pathogen interactions. Phytochemistry 71:351–362

Umadevi P, Anandaraj M (2017) Genotype specific host resistance for *Phytophthora* in black pepper (*Piper nigrum* L.). Physiol Mol Plant Pathol 100:237–241

Umadevi P, Anandaraj M, George JK (2015) Towards understanding the black pepper-*Phytophthora* pathosystem using integrated transcriptome and proteome dataset. In: International symposium on *Phytophthora* at IIHR, Bangalore, 9–12 September

Umadevi P, Soumya M, George JK, Anandaraj M (2018) Proteomics assisted profiling of antimicrobial peptide signatures from black pepper (*Piper nigrum* L.). Physiol Mol Biol Plants 24 (3):379–387

Valcu C, Junqueira M, Andrej Shevchenko A, Katja Schlink K (2009) Comparative proteomic analysis of responses to pathogen infection and wounding in *Fagus sylvatica*. J Proteome Res 8 (8):4077–4091

Wang Y (2015) Differential regulation of defense-related proteins in soybean during compatible and incompatible interactions between *Phytophthora sojae* and soybean by comparative proteomic analysis. Plant Cell Rep 34:1263–1280

Widmark AK, Andersson B, Cassel Lundhagen A, Sandström M, Yuen J (2007) *Phytophthora infestans* in a single field in Southwest Sweden early in spring: symptoms, spatial distribution and genotypic variation. Plant Pathol 56(4):573–579

Chapter 3
Impact of Climate Change on Soil Microbial Community

Srikanth Mekala and Srilatha Polepongu

Abstract As climate changes endlessly, it becomes more important to understand possible reactions from soils to the climate system. It is a known fact that microorganisms, which are associated with plant, may stimulate plant growth and enhance resistance to disease and abiotic stresses. The effects of climate change factors, such as elevated CO_2, drought, and temperature on beneficial plant–microorganism interactions are increasingly being explored. Organisms live in concert with thousands of other species, such as some beneficial and pathogenic species which have little to no effect on complex communities. Since natural communities are composed of organisms with very different life history traits and dispersal ability, it is unlikely that all of the microbial community will respond to climatic change factors in a similar way. Among the different factors related to climate change, elevated CO_2 had a positive influence on the abundance of arbuscular and ectomycorrhizal fungi, whereas the effects on plant-growth-promoting bacteria and endophytic fungi were more variable. The rise in temperature effects on beneficial plant-associated microorganisms were more variable, positive, neutral, and negative, which were equally common and varied considerably with the temperature range. Likewise, plant-growth-promoting microorganisms (i.e., bacteria and fungi) positively affected plants subjected to drought stress. In this chapter, we explore how climatic change affects soil microbes and plant-associated microorganisms.

3.1 Introduction

Microbial communities are groups of microorganisms that share a common living space. The microbial populations that form the community can interact in different ways, for example, as predators and prey or as symbionts. The community includes both positive (like symbiosis) and negative (like antibiosis) interactions.

S. Mekala (✉)
Department of Vegetable Science, CCSHAU, Hisar, Haryana, India

S. Polepongu
Department of Plant Pathology, PJSTAU, Aswaraopeta, Telangana, India

A. Varma et al. (eds.), *Plant Biotic Interactions*,
https://doi.org/10.1007/978-3-030-26657-8_3

Microorganisms are the backbone of all ecosystems, even then in some zones, they are unable to process photosynthesis due to absence of light. In such zones, chemosynthetic microorganisms provide carbon and energy to the other organisms. Soil microbial communities play an important role in agro-ecosystem functioning and are on the field scale essential for plant nutrition and health. Moreover, they are involved in turnover processes of organic matter, breakdown of xenobiotics and formation of soil aggregates. These soil microbial communities are responsible for the cycling of carbon (C) and nutrients in the ecosystem and their activities are regulated by biotic and abiotic factors such as the quantity and quality of litter inputs, temperature, and moisture. Atmospheric and climatic changes will affect both abiotic and biotic drivers in ecosystems and the response of ecosystems to these changes. Soil microbial communities may also regulate feedbacks from ecosystem to the atmosphere. The response of the microbial communities that regulate ecosystem processes is less predictable. These microbes play an important role in biogeochemical cycles. The nitrogen cycle, the phosphorus cycle, the sulfur cycle and the carbon cycle all depend on microorganisms. The nitrogen gas, which makes up 78% of the earth's atmosphere, is unavailable to most organisms, until it is converted to a biologically available form by the microbial process of nitrogen fixation.

Climate change factors, such as atmospheric CO_2 concentrations, temperature, and altered precipitation regimes, can potentially have both direct and indirect impacts on soil microbial communities. The response of soil microbial communities to changes in atmospheric CO_2 concentrations can be positive or negative. Increasing temperatures can increase microbial activity, processing, and turnover, causing the microbial community to shift in favor of representatives adapted to higher temperatures and faster growth rates (Bradford et al. 2008). How multiple climate change factors interact with each other to influence microbial community responses is poorly understood. For example, elevated atmospheric CO_2 and precipitation changes might increase soil moisture in an ecosystem, but this increase may be counteracted by temperature (Dermody et al. 2007). Similarly, temperature may increase microbial activity in an ecosystem, but this increase may be eliminated if changes in precipitation lead to a drier soil condition or reduced litter quantity, quality, and turnover. Similarly, changes in soil moisture and ecosystem temperature do not always lead to predictable or significant changes in bacterial and fungal abundance (Chen et al. 2007; Kandeler et al. 2006).

3.2 Impact of Climate Change on Soil Microbial Communities

Climate change is a change in the statistical distribution of weather patterns that lasts for an extended period of time (i.e., decades to millions of years). Impact of climate change on soil microbial communities results in increasing atmospheric CO_2 concentration, temperature, and drought. Climate change has both positive and negative effects on soil microbial communities. Majority of studies showed that elevated CO_2 had a positive influence on the abundance of arbuscular and ectomycorrhizal fungi whereas the effects on plant-growth-promoting bacteria and endophytic fungi were more variable. In most cases, plant-associated microorganisms had a beneficial effect on plants under elevated CO_2.

Climate change is caused by various factors such as biotic processes, variations in solar radiation received by earth, plate tectonics, and volcanic eruptions. Certain human activities are also responsible for ongoing climate change, often referred to as global temperature. It will show direct and indirect impacts on terrestrial ecosystems, both above and below the ground. On the surface of the soil, the effects of global change will be largely direct: elevated atmospheric carbon dioxide as well as changes in temperature, precipitation, and nitrogen availability, which will all alter the growth of plant species. Below the surface of the soil, disturbances often act quickly like microbial metabolic activities can be changed by many ecosystem-scale factors such as nitrogen deposition, elevated carbon dioxide (Dhillion et al. 1996; Ajwa et al. 1999; Mayr et al. 1999). The response to climate change is more complex. Plant responses affect the type and amount of carbon entering the soil system as well as the physical architecture of the plant root zone.

Through their diverse metabolic activities, soil microbial communities are the major drivers of soil nutrient cycling and their importance in mediating climate change and ecosystem functioning should not be underestimated (Balser et al. 2001). Climatic change involves increasing atmospheric CO_2 concentration, temperature, precipitation, and drought. It can alter the relative abundance and function of soil communities because soil community members differ in their physiology, temperature sensitivity, and growth rates (Lennon et al. 2012; Briones et al. 2014; Delgado-Baquerizo et al. 2014).

3.2.1 *Impact of Elevated CO_2*

Altered environmental conditions due to climate change are likely to induce changes in plant physiology and root exudation. Particularly elevated CO_2 leads to increase in C allocation at the root zone and potentially alters root exudation compositions. Alterations might include changes in the availability of chemo attractants or signal compounds as well as a different C/N ratio or nutrient availability (Kandeler et al. 2006; Haase et al. 2007). Photosynthetically derived carbon (C) into the rhizosphere through root exudation, making root exudates, is a key factor in increasing microbial abundance and activity in the rhizosphere compared to bulk soil (Lynch 1990; Kapoor and Mukerji 2006). Low-molecular-weight C compounds present in root exudates, including sugars, organic acids (OAs), and amino acids are readily assimilated by microorganisms and are proposed to play a primary role in regulating microbial community dynamics in the rhizosphere (Bais et al. 2006).

Root exudates play a key role in plant–microorganism interactions by influencing the structure and function of soil microbial communities. Model exudate solutions, based on organic acids (OAs) (quinic, lactic, maleic acids) and sugars (glucose, sucrose, fructose), previously identified in the rhizosphere of *Pinus radiata*, were applied to soil miniature. OAs caused significantly greater increases than sugars in the detectable richness of the soil bacterial community. The greater response of bacteria to OAs may be due to the higher amounts of added carbon, solubilization of soil organic matter, or shifts in soil pH. The impact of climate change on root exudates like model solutions such as OAs, which plays a significant role in shaping soil bacterial communities therefore, this may have a significant impact on plant growth (Shi et al. 2011).

Climate change may significantly impact the diversity and activities of such plant-associated microbial communities (Drigo et al. 2008). Consequently, microorganisms known for their beneficial effects on plant growth or health might also be impaired, in terms of exhibiting their desirable properties and their colonization capacity under certain conditions. The majority of studies performed so far have indicated that elevated CO_2 conditions will lead to increased colonization of PGPF (plant-growth-promoting fungi). It is also important to note that elevated CO_2 concentrations may induce AMF (arbuscular mycorrhizal fungi) community composition changes (Klironomos et al. 2005). AMF are known to enhance plant nutrient uptake (mostly phosphate) or confer other benefits in exchange for rhizosphere carbohydrate compounds (Koide 1991; Newsham et al. 1995).

The composition of microbial communities correlates with plant physiology and is likely to be driven by root exudation or metabolite patterns. This indicates that the colonization of plant-associated microorganisms depends on the availability of certain compounds produced by the plant acting as the primary nutrient source, as chemo attractants or signal molecules. Consequently, at elevated CO_2 concentrations in particular, but also under conditions of increased temperature or drought, different genotypes of PGPF or PGPB show potential for different functional activities in the plant environment (Waldon et al. 1989; Marilley et al. 1999; Drigo et al. 2009). However, because of altered community structures, beneficial microorganisms might have to compete with different microbial communities and therefore might show an altered colonization behavior. In addition to the potential effects of climate change on microbial colonization characteristics, microbial activities may be affected (Kandeler et al. 2006; Haase et al. 2008). Microbial communities showing different activities or producing altered signals in the long term either may result in the establishment of altered communities or in the elicitation of different plant responses.

Considering the fact that plant-beneficial microorganisms such as mycorrhizal fungi and nitrogen-fixing bacteria provide up to 80% N and up to 75% P and that an extremely high number of plant species are completely dependent on microbial symbionts for growth and survival, it is evident that alterations in the plant-beneficial microbial communities may ultimately influence plant diversity and functioning of soil microbiota. Furthermore, the abundance or the activity of microorganisms with biocontrol activities may be altered and thereby plant pathogen populations may be affected (Compant et al. 2010). Under elevated CO_2 conditions, nutrients such as N might be limiting, leading to the need for enhanced fertilizer input in agriculture, in such cases plant growth-promoting microorganisms supporting nutrient acquisition such as mycorrhizae or N-fixing bacteria may gain increasing importance.

Overall, these results suggest that elevated CO_2 may interact with the plant–fungal symbiosis and may lead to increased endophyte infection frequency, although with lowered toxin production (Brosi et al. 2009). The impact of CO_2 on the host plant and its endosymbionts may additionally change the plant carbohydrate content.

3.2.2 Impact of Temperature

Temperature changes are often coupled with changes in soil moisture, which may explain some inconsistent results from experiments exploring how microbial communities respond to climatic change. For example, rates of microbial activity at warmer temperatures can be limited by diffusion and microbial contact with available substrates (Zak et al. 1993). While bacterial communities may respond rapidly to moisture pulses, the slower-growing fungal community may lag in their response (Bell et al. 2008; Cregger et al. 2012, 2014). Global changes such as temperature are directly altering microbial soil respiration rates because soil microorganisms, and the processes they mediate are temperature sensitive. The role of elevated temperature in microbial metabolism has received considerable attention (Bradford 2013; Frey et al. 2013; Hagerty et al. 2014; Karhu et al. 2014). Q_{10} is often used in climate change models to account for microbial temperature sensitivity; however, using this relationship masks many of the interactions that influence the temperature sensitivity of microbial processes, such as decomposition. Therefore, using only Q_{10} to account for temperature sensitivity in models may lead to poor predictions. Further, while decomposition of soil organic matter, soil respiration, and growth of microbial biomass generally increase with temperature (Bradford et al. 2008). The transient effects of temperature on soil communities have been hypothesized. The soil carbon substrates are depleted by increased microbial activity and because of trade-offs as microbial communities either acclimate, shift in composition, or constrain their biomass to respond to altered conditions and substrate availability (Allison and Martiny 2008; Bradford 2013). Experimental temperature can initially alter the composition of microbial communities, and shift the abundance of Gram-positive and Gram-negative bacteria (Zogg et al. 1997).

While plant species migrations in response to climate change are well described (Grabherr et al. 1994; Walther et al. 2002; Parmesan and Yohe 2003) most studies fail to address the ability of associated soil microorganisms to shift their range to maintain the positive or negative relationship between the plant and the soil community (van der Putten 2012). Shifts in microbial activity can in turn lead to changes in decomposition, nitrogen mineralization and organic carbon storage. Nitrogen deposition is likely to decrease mycorrhizal fungal biomass while increasing bacterial and saprotrophic fungal biomass (Treseder 2004; Rinnan et al. 2007) and has the potential to increase carbon cycling by increasing the activity of microbial enzymes related to carbon cycling. Soil biota may be poor dispersers, therefore they may respond to climate change at a different rate than plants (van der Putten 2012).

Increased temperature causes the soil water to decrease in some areas (Le Houerou 1996), leading to enhanced drought in several regions of the world, whereas others are known to support plant growth and to increase plant tolerance to biotic and abiotic stresses (Bent 2006). Many of these plant-growth-promoting microorganisms colonize the rhizosphere, the portion of soil attached to the root surface and influenced by root exudates and by microorganisms (Bent 2006; Lugtenberg and Kamilova 2009; Prasad et al. 2015). Some microorganisms can also enter the root system of their hosts and enhance their beneficial effects with an endophytic lifestyle (Stone et al. 2000).

In the majority of cases, the responses of AMF to increased temperature had a positive impact on their colonization and hyphal length. In some cases however, no effects or negative effects of enhanced temperature on AMF were reported (Compant et al. 2010). AMF may respond to higher temperatures with enhanced growth and plant colonization for majority of strains (Furlan and Fortin 1973; Graham et al. 1982; Fitter et al. 2000). This was demonstrated, for instance, in strains of *Glomus intraradices*, *Glomus mossae*, and many others (Baon et al. 1994; Monz et al. 1994). Temperature may also significantly alter the structure of the AMF hyphal network and induce a switch from more vesicles responsible for storage in cooler soils to more extensive mycorrhizal hyphal networks, indicating growth in warmer soils (Hawkes et al. 2008). This is linked to faster carbon allocation to the rhizosphere and an increased respiration of the extra mycorrhizal mycelium at a high soil temperature (Heinemeyer et al. 2006).

3.2.3 Impact of Drought

Drought stress might be an additional consequence of climate change. High temperature leads to drought or soil moisture stress. Soil moisture related to the microbial community is more highly unpredictable and complex than temperature, and less investigated. Drought amplifies the differential temperature sensitivity of fungal and bacterial groups (Briones et al. 2014). Even with small changes in soil moisture availability (30% reduction in water holding capacity), soil fungal communities may shift from one dominant member to another while bacterial communities remain constant. Interactions among microbes and background temperature and moisture regimes in any given location influence microbial composition and function with changing climate.

Despite logical mechanisms by which microbial communities may be altered by changes in soil moisture. Generally, drought reduces AMF colonization, but in some cases this kind of response might be dependent on strains as reported by Davies et al. (2002), who demonstrated that drought enhanced arbuscular formation and hyphae development of the *Glomus* sp. strain ZAC-19, whereas colonization by a *Glomus fasciculatum* strain was reduced.

Drought is frequently responsible for reduced plant growth in roots and aerial plant parts, which makes the plant susceptible to other pathogens attack. This may lead to changes in the allocation of photosynthates in the rhizosphere as well as in ectomycorrhizal mycelium (ECM) formation. However, plants may change the type of mycorrhiza colonizing their hosts during longer exposure to drought as it was shown by Querejeta et al. (2009). The beneficial association between some strains of AMF and plants may thus reduce the severe effects of water limitation to plants. Moreover, some AMF may resist drought stress better than others.

Different mechanisms may thus be responsible for ECM-induced stress tolerance. Improved performance of mycorrhizal seedlings under drought stress conditions may also be linked to better P and K nutrition as well as to a more extensive root system with mycelial strands as demonstrated with *Picea sitchensis* and *P. involutus* (Lehto 1992). The consequences of drought stress tolerance induced by ECM may in addition affect belowground litter quality and quantity as well as accumulation of organic matter in soils.

There are some microsymbionts, such as the strains of the taxa *Atkinsonella, Balansia, Balansiopsis, Echinodothis, Epichloë, Myriogenospora* (White 1994), *Parepichloë* (White and Reddy 1998), *Neotyphodium* (Glenn et al. 1996), and certain endophytic fungi that can colonize entirely within plant tissues and may grow within roots, stems, and/or leaves, emerging to sporulate at plant or host-tissue senescence (Sherwood and Carroll 1974; Carroll 1988; Bacon and De Battista 1991).

Water may act in concert with nitrogen addition to increase decomposition of plant tissues (Henry et al. 2005). Increased moisture, or alleviation of water stress, can also alter the lignification of plant cell walls (Henry et al. 2005), increase grassland productivity, and impact on soil carbon. Soil moisture coupled with elevated carbon dioxide also decreases abundance of ammonium oxidizing bacteria, potentially altering the soil nitrogen cycle (Horz et al. 2004). Temperature with elevated carbon dioxide may act additively to increase soil respiration (Niinisto et al. 2004; Van Veen et al. 1991; Hungate et al. 1997). Although there have been few reports on the interactions between elevated temperature and moisture, there is evidence to suggest that together they may lead to shifts in the structure of methane-oxidizing bacterial communities (Horz et al. 2005).

3.3 Conclusion and Future Prospects

Climate change will affect soil microbial communities' structure and functions both directly and indirectly. Directly through changing the physical structure of the soil and carbon allocation and indirectly by changing land use. Soil microbes are essential components in the agricultural ecosystem responses to climate change through which the process of cycling of nutrients and soil carbon allocation occurs. Temperature interacts with changes in water and nitrogen availability. All climate change factors such as temperature, increasing CO_2 concentration, precipitation, and drought impact will be both positive and negative. However, most of the cases it has a positive effect on the microbial community. Although, in order to clearly understand the exact mechanism of impact on climate change on soil microbial community, some other factors which show impact on soil microbial communities and models need to be studied.

References

Ajwa H, Dell CJ, Rice CW (1999) Changes in enzyme activities and microbial biomass of tallgrass prairie soil as related to burning and nitrogen fertilization. Soil Biol Biochem 31:769–777

Allison SD, Martiny JBH (2008) Resistance, resilience, and redundancy in microbial communities. Proc Natl Acad Sci USA 105:11512–11519

Bacon CW, De Battista J (1991) Endophytic fungi of grasses. In: Arora DK, Rai B, Mukerji KG, Knudsen GR (eds) Handbook of applied mycology Vol. 1. Soil and plants. Dekker, New York, pp 231–256

Bais HP, Weir TL, Perry LG, Gilroy S, Vivanco JM (2006) The role of root exudates in rhizosphere interactions with plants and other organisms. Annu Rev Plant Biol 57:233–266

Balser TC, Kinzig A, Firestone MK (2001) Linking soil microbial communities and ecosystem functioning. In: Kinzig A, Pacala S, Tilman D (eds) The functional consequences of biodiversity: empirical progress and theoretical extensions. Princeton University Press, Princeton, NJ, pp 265–294

Baon JB, Smith SE, Alston AM (1994) Phosphorus uptake and growth of barley as affected by soil temperature and mycorrhizal infection. J Plant Nutr 17:479–492

Bell C, McIntyre N, Cox S, Tissue D, Zak J (2008) Soil microbial responses to temporal variations of moisture and temperature in a Chihuahuan Desert grassland. Microb Ecol 56:153–167

Bent E (2006) Induced systemic resistance mediated by plant growth-promoting rhizobacteria (PGPR) and fungi (PGPF). In: Tuzun S, Bent E (eds) Multigenic and induced systemic resistance in plants. Springer, Berlin, pp 225–258

Bradford MA (2013) Thermal adaptation of decomposer communities in temperature soils. Front Microbiol 4:333. https://doi.org/10.3389/fmicb.2013.00333

Bradford MA, Davies CA, Frey SD, Maddox TR, Melillo JM, Mohan JE, Reynolds JF, Treseder KK, Wallenstein MD (2008) Thermal adaptation of soil microbial respiration to elevated temperature. Ecol Lett 11:1316–1327

Briones MJI, McNamara NP, Poskitt J, Crow SE, Ostle NJ (2014) Interactive biotic and abiotic regulators of soil carbon cycling: evidence from controlled climate experiments on peatland and boreal soils. Glob Chang Biol 20:2971–2982

Brosi GB, Nelson JA, McCulley RL, Classen AT, Norby R (2009) PS 45–40: Global change factors interact with fungal endophyte symbiosis to determine tall fescue litter chemistry. The 94th ESA annual meeting, PS 45–40

Carroll G (1988) Fungal endophytes in stems and leaves—from latent pathogen to mutualistic symbiont. Ecology 69:2–9

Chen X, Tu C, Burton MG, Watson DM, Burkey KO, Hu S (2007) Plant nitrogen acquisition and interactions under elevated carbon dioxide: impact of endophytes and mycorrhizae. Glob Chang Biol 13:1238–1249

Compant S, Cl'ement C, Sessitsch A (2010) Colonization of plant growth-promoting bacteria in the rhizo- and endosphere of plants: importance, mechanisms involved and future prospects. Soil Biol Biochem 42:669–678

Cregger MA, Schadt CW, Mc Dowell NG, Pockman WT, Classen AT (2012) Response of the soil microbial community to changes in precipitation in a semiarid ecosystem. Appl Environ Microbiol 78:8587–8594

Cregger MA, Sanders NJ, Dunn RR, Classen AT (2014) Microbial communities respond to experimental temperature, but site matters. Peer J 2

Davies FT Jr, Olalde-Portugal V, Aguilera-Gómez L, Alvarado MJ, Ferrera-Cerrato RC, Boutton TW (2002) Alleviation of drought stress of Chile ancho pepper (Capsicum annuum L. cv. San Luis) with arbuscular mycorrhiza indigenous to Mexico. Sci Hortic 92:347–359

Delgado-Baquerizo M, Maestre FT, Escolar C, Gallardo A, Ochoa V, Gozalo B, Prado Comesana A (2014) Direct and indirect impacts of climate change on microbial and biocrust communities alter the resistance of the N cycle in a semiarid grassland. J Ecol 102:1592–1605

Dermody O, Weltzin JF, Engel EC, Allen P, Norby RJ (2007) How do elevated CO_2, warming, and reduced precipitation interact to affect soil moisture and LAI in an old field ecosystem? Plant Soil 301:255–266. https://doi.org/10.1007/s11104-007-9443-x
Dhillion S, Roy J, Abrams M (1996) Assessing the impact of elevated CO_2 in a Mediterranean model ecosystem. Plant Soil 187:333–342
Drigo B, Kowalchuk GA, van Veen JA (2008) Climate change goes underground: effects of elevated atmospheric CO_2 on microbial community structure and activities in the rhizosphere. Biol Fertil Soils 44:667–679
Drigo B, van Veen JA, Kowalchuk GA (2009) Specific rhizosphere bacterial and fungal groups respond to elevated atmospheric CO_2. ISME J 3:1204–1217
Fitter AH, Heinemeyer A, Staddon PL (2000) The impact of elevated CO_2 and global climate change on arbuscular mycorrhizas: a mycocentric approach. New Phytol 147:179–187
Frey SD, Lee J, Melillo JM, Six J (2013) The temperature response of soil microbial efficiency and its feedback to climate. Nat Clim Chang 3:395–398
Furlan V, Fortin J-A (1973) Formation of endomycorrhizae by Endogone calospora on *Allium cepa* under three temperature regimes. Nat Can 100:467–477
Glenn AE, Bacon CW, Price R, Hanlin RT (1996) Molecular phylogeny of Acremonium and its taxonomic implications. Mycologia 88:369–383
Grabherr G, Gottfried M, Pauli H (1994) Climate effects on mountain plants. Nature 369:448–448
Graham JH, Leonard RT, Menge JA (1982) Interaction of light and soil temperature with phosphorus inhibition of vesicular–arbuscular mycorrhiza formation. New Phytol 91:683–690
Haase S, Neumann G, Kania A, Kuzyakov Y, Römheld V, Kandeler E (2007) Elevation of atmospheric CO_2 and Nnutritional status modify nodulation, nodule-carbon supply, and root exudation of *Phaseolus vulgaris* L. Soil Biol Biochem 39:2208–2221
Haase S, Philippot L, Neumann G, Marhan S, Kandeler E (2008) Local response of bacterial densities and enzyme activities to elevated atmospheric CO_2 and different N supply in the rhizosphere of *Phaseolus vulgaris* L. Soil Biol Biochem 40:1225–1234
Hagerty SB, van Groenigen KJ, Allison SD, Hungate BA, Schwartz E, Koch GW, Kolka RK, Dijkstra P (2014) Accelerated microbial turnover but constant growth efficiency with temperature in soil. Nat Clim Chang 4:903–906
Hawkes CV, Hartley IP, Ineson P, Fitter AH (2008) Soil temperature affects allocation within arbuscular mycorrhizal networks and carbon transport from plant to fungus. Glob Chang Biol 14:1181–1190
Heinemeyer A, Ineson P, Ostle N, Fitter AH (2006) Respiration of the external mycelium in the arbuscular mycorrhizal symbiosis shows strong dependence on recent photosynthates and acclimation to temperature. New Phytol 171:159–170
Henry HA, Clelad EE, Field CB, Vitousek PM (2005) Interactive effects of CO_2, N deposition, and climate change on plant litter quality in a California annual grassland. Oecologia 142:465–473
Horz HP, Barbrook A, Field CB, Bohannan BJM (2004) Ammonia-oxidizing bacteria respond to multifactorial global change. Proc Natl Acad Sci USA 101:15136–15141
Horz HP, Rich V, Avrahami S, Bohannan BJM (2005) Methane-oxidizing bacteria in a California upland grassland soil: diversity and response to simulated global change. Appl Environ Microbiol 71:2642–2652
Hungate BA, Lund CP, Pearson HL, Chapin FS (1997) Elevated CO_2 and nutrient addition alter soil N cycling and N trace gas fluxes with early season wet-up in a California annual grassland. Biogeochemistry 37:89–109
Kandeler E, Mosier AR, Morgan JA, Milchunas DG, King JY, Rudolph S, Tscherko D (2006) Response of soil microbial biomass and enzyme activities to the transient elevation of carbon dioxide in a semi-arid grassland. Soil Biol Biochem 38:2448–2460
Kapoor R, Mukerji KG (2006) Rhizosphere microbial community dynamics. In: Mukerji KG, Manoharachary C, Singh J (eds) Microbial activity in the rhizosphere. Springer, Berlin, pp 55–66

Karhu K et al (2014) Temperature sensitivity of soil respiration rates enhanced by microbial community response. Nature 513:81–84

Klironomos JN, Allen MF, Rillig MC, Piotrowski J, MakvandiNejad S, Wolfe BE, Powell JR (2005) Abrupt rise in atmospheric CO_2 overestimates community response in a model plant–soil system. Nature 433:621–624

Koide R (1991) Nutrient supply, nutrient demand and plant response to mycorrhizal infection. New Phytol 117:365–386

Le Houérou HN (1996) Climate change, drought and desertification. J Arid Environ 34:133–185

Lehto T (1992) Mycorrhizas and drought resistance of Picea sitchensis (bong) Carr. I. in conditions of nutrient deficiency. New Phytol 122:669–673

Lennon JT, Aanderud ZT, Lehmkuhl BK, Schoolmaster DR Jr (2012) Mapping the niche space of soil microorganisms using taxonomy and traits. Ecology 93:1867–1879

Lugtenberg B, Kamilova F (2009) Plant-growth-promorting Rhizobacteria. Annu Rev Microbiol 63 (1):541–556

Lynch JM (1990) Introduction: some consequences of microbial rhizosphere competence for plant and soil. In: Lynch JM (ed) The rhizosphere. Wiley, West Sussex, pp 1–10

Marilley L, Hartwig UA, Aragno M (1999) Influence of an elevated atmospheric CO_2 content on soil and rhizosphere bacterial communities beneath Lolium perenne and Trifolium repens under field conditions. Microb Ecol 38:39–49

Mayr C, Miller M, Insam H (1999) Elevated CO_2 alters community-level physiological profiles and enzyme activities in alpine grassland. J Microbiol Methods 36:35–43

Monz CA, Kunt HW, Reeves FB, Elliot ET (1994) The response ofmycorrhizal colonization to elevated CO_2 and climate change in *Pascopyrum smithii* and *Bouteloua gracilis*. Plant Soil 165:75–80

Newsham KK, Fitter AH, Watkinson AR (1995) Multifunctionality and biodiversity in arbuscular mycorrhizas. Trends Ecol Evol 10:407–411

Niinisto SM, Silvola J, Kellomaki S (2004) Soil CO_2 efflux in a boreal pine forest under atmospheric CO_2 enrichment and air temperature. Glob Chang Biol 10:1–14

Parmesan C, Yohe G (2003) A globally coherent fingerprint of climate change impacts across natural systems. Nature 421:37–42

Prasad R, Kumar M, Varma A (2015) Role of PGPR in soil fertility and plant health. In: Egamberdieva D, Shrivastava S, Varma A (eds) Plant growth-promoting Rhizobacteria (PGPR) and medicinal plants. Springer International Publishing, Switzerland, pp 247–260

Querejeta JI, Egerton-Warburton LM, Allen MF (2009) Topographic position modulates the mycorrhizal response of oak trees to interannual rainfall variability. Ecology 90:649–662

Rinnan R, Michelsen A, Baath E, Jonasson S (2007) Fifteen years of climate change manipulations alter soil microbial communities in a subarctic heath ecosystem. Glob Chang Biol 13:28–39

Sherwood M, Carroll G (1974) Fungal succession on needles and young twigs of old-growth Douglas fir. Mycologia 66:499–506

Shi S, Condron L, Larsen S, Richardson AE, Jones E, Jiao J, O'Callaghan M, Stewart A (2011) In situ sampling of low molecular weight organic anions from rhizosphere of *Pinus radiata* grown in a rhizotron system. Environ Exp Bot 70:131–142

Stone JK, Bacon CW, White JF (2000) An overview of endophytic microbes: endophytism definded. In: Bacon CW, White JF (eds) Microbial endophytes. Dekker, New York, pp 3–29

Treseder KK (2004) A meta-analysis of mycorrhizal responses to nitrogen, phosphorus, and atmospheric CO2 in field studies. New Phytol 164:347–355

van der Putten WH (2012) Climate change, aboveground-belowground interactions and species range shifts. Annu Rev Ecol Evol Syst 43:365–383

van Veen K, Liljeroth E, Lekkerkerk J (1991) Carbon fluxes in plant-soil systems at elevated atmospheric carbon dioxide levels. Ecol Appl 1:175–181

Waldon HB, Jenkins MB, Virginia RA, Harding EE (1989) Characteristics of woodland rhizobial population from surface- and deep-soil environments of the Sonoran Desert. Appl Environ Microbiol 55:3058–3064

Walther GR, Post E, Convey P, Menzel A, Parmesan C, Beebee TJC, Fromentin JM, Hoegh-Guldberg O, Bairlein F (2002) Ecological responses to recent climate change. Nature 416:389–395

White JF Jr (1994) Taxonomic relationships among the members of the Balansiae (Clavicipitales). In: Bacon CW, White JF Jr (eds) Biotechnology of endophytic Fungi of Grasse. CRC Press, Boca Raton, FL, pp 3–20

White JF Jr, Reddy PV (1998) Examination of structure and molecular phylogenetic relationships of some graminicolous symbionts in genera Epichloë and Parepichloë. Mycologia 90:226–234

Zak DR, Pregitzer KS, Curtis PS, Teeri JA, Fogel R, Randlett DL (1993) Elevated atmospheric CO_2 and feedback between carbon and nitrogen cycles. Plant Soil 151:105–117

Zogg GP, Zak DR, Ringelberg DB, MacDonald NW, Pregitzer KS, White DC (1997) Compositional and functional shifts in microbial communities due to soil temperature. Soil Sci Soc Am J 61:475–481

Chapter 4
Industrial Effluents: Impact on Agricultural Soils and Microbial Diversity

Sujata Mani, Pankaj Chowdhary, and Vishvas Hare

Abstract One of the most important alternative water resources in regions where scarcity of freshwater is common is the application of industrial effluents. The application of different, treated industrial wastewaters/effluents and sludge on agricultural fields offers an alternative to disposal by utilizing it in the production of crops. Industrial effluents could provide sufficient water and essential nutrients required for plants since they are very rich in organic matter, minerals, metals, etc. These effluents added to the soil in sufficient quantities would improve the soil's physical condition and render it a more favorable environment to manage water and its nutrient content. Irrigation with such a kind of water might affect the diversity and function of the soil microbial community and alter the structure of soil. However, unlike manufactured fertilizers in which nutrient properties are managed to suit the crop requirements, the nutrients in the effluents are totally uncontrolled. Thus, before application to agricultural lands, the effluents should be treated at agronomic rates for satisfying the requirements of nutrients to be in excessive or in deficient amounts. The fate and transport of potentially harmful constituents in the environment are also of great concern. If the constituents from effluents are not immobilized in the soil surface, they might escape the root zone and leach groundwater. Thus, this chapter reviews the possible physical and chemical changes on agricultural soil as well as on crops as a result of wastewater application for irrigation. This chapter also improves our understanding on how irrigation with wastewater changes the activity of soil's microbial process.

S. Mani
Department of Biochemistry, Gramin Science (Vocational) College, Vishnupuri, Nanded, Maharashtra, India

P. Chowdhary (✉) · V. Hare
Department of Microbiology, Babasaheb Bhimrao Ambedkar University (A Central University), Lucknow, Uttar Pradesh, India

A. Varma et al. (eds.), *Plant Biotic Interactions*,
https://doi.org/10.1007/978-3-030-26657-8_4

4.1 Introduction

The most important system of terrestrial ecosystem is soil the pollutants in which have direct impact on ion minerals, organic matters, and soil microbial community. The utilization of industrial effluents is a worldwide practice in irrigating agricultural lands. It is the most common practice in developing countries where the cost of wastewater treatment is unaffordable (Feigin et al. 1991). An economic alternative for disposal of industrial effluents than surface water are agricultural fields or sites where they can contribute to nutrient cycling. The irrigation of agricultural lands from wastewater provides water, nitrogen, phosphorus, and organic materials to the soil (Chowdhary and Bharagava 2019; Chowdhary et al. 2018c). All these factors provide beneficial effects on soil microbiota, but pollutants such as salts, heavy metals, and surfactants from these effluents may accumulate in the agricultural soils and can cause potential hazard to soil quality as well as its productivity for a long term (Chowdhary et al. 2018a, b).

Several human and animal health problems can occur due to the consumption of plants growing in metal-contaminated soils. Many researchers have stated that these heavy metals get accumulated in the soil due to continuous irrigation with industrial effluents (Hare et al. 2017; Chowdhary et al. 2017b, 2018b; Bharagava and Chandra 2010). The use of wastewater, on the other hand, could solve problems related to disposals, minimize human health hazards, and can also prevent the pollution of groundwater. These effluents from agricultural and urban sources have great potential for reuse as a source of organic matter, nutrients, soil-conditioning agents, and water.

The impact of the application of industrial effluents varies with effluents' characteristics, type of soil, and the vegetation of the soil that is irrigated. Irrigation with different industrial wastewaters may change soil's physical, chemical, and biological properties. These properties of soil play a very important role in the transformation of nutrients present in the wastewater applied for irrigation (Chowdhary and Bharagava 2019; Magesan et al. 1999). Since wastewaters can improve the physical properties as well as the nutrient content of the soil, the application of these wastewaters to cropland and forested lands is an attractive approach for their disposal. The wastewaters used for irrigation purposes provide water, nitrogen, phosphorus, and organic matter to the soil (Chowdhary and Bharagava 2019). But there is a great concern on the accumulation of some possibly toxic elements such as cadmium, copper, iron, manganese, lead, and zinc, from both domestic as well as industrial sources. Thus, checking the concentration of such elements in the wastewater is crucial before application. The main public concern regarding crops and forage plants is their potential uptake of trace elements which are present in the wastewater applied for irrigation purposes (Chowdhary et al. 2018a, b; Chowdhary and Bharagava 2019). Apart from this, another major concern regarding toxicity is the ingestion of these elements by grazing animals from soil under wastewater irrigation. Therefore, the accumulation of these toxic elements by the plants grown and their uptake by grazing animals must be cautiously evaluated.

This chapter systematically reviews the effects of different industrial wastewaters, such as textile, distillery, pulp and paper, tannery, etc., on physical, chemical, and biological properties of soil when applied for irrigation on agricultural fields. This chapter also puts forward suggestions on future prospects for the application of industrial effluents on irrigational fields.

4.2 Sources of Effluents to Agricultural Fields and Their Properties

In the recent decades, the utilization of industrial effluents in agricultural fields has become a common practice in waste disposal. It may be beneficial to agricultural soils as well as in crop productivity since it contains several plant nutrients and organic materials. Conversely, depending on its sources, wastewater can often contain a considerable amount of organic toxicants and several other toxic metals. Industries such as distillery, tannery, textile, pulp and paper, paint, pharmaceutical, etc., use a substantial amount of water in their processes and so release as much water as well (Chowdhary et al. 2018a, b) (Fig. 4.1).

Distillery effluents are dark brown in color with a strong odor, low pH, and extremely high nutrient contents. These effluents have high BOD, COD, and nutrients

Fig. 4.1 Different sources of wastewater utilized for irrigation purposes

level, including nitrogen, phosphorus, potassium, etc. The toxic pollutants released from distillery effluents affect the food chain and thus disturb the ecological balance of the soil (Chowdhary et al. 2017a; Pal and Vimala 2012; Singh and Patel 2012).

Effluents from tanneries produce a huge amount of wastewater which is generally discharged to fields for irrigation. This type of wastewater is yellowish brown in color with high TDS, TSS, TS, BOD, COD and contains a large amount of salts, heavy metals, and various other toxic pollutants. When these pollutants of tannery effluents accumulate in the irrigated soils, reduction in soil fertility occurs through effluents' adverse effect on heterogeneous microbial communities inhabiting soils, causing less crop productivity (Alvarez-Bernal et al. 2006). These heavy metals may also cause hazards to humans and plants via accumulating in the food chain (Takahashi et al. 2012; Dutton and Fisher 2011).

One of the largest industries which release huge volumes of effluents during processing is the textile industry. Approximately 15% of the total textile dyes are predicted to be directly lost with the effluent (Mani and Bharagava 2016; Zollinger 1987). The wastewater emerging from textile industries generally has high TDS, TSS, TS, BOD, and COD. The effluents are extensively rich in micro- and macronutrients with high salinity and excessive concentrations of carbonate ions, sodium, and alkalinity (pH 10–11.5), but reversely are low in calcium concentrations (Chowdhary et al. 2018a, b, c; Garg and Kaushik 2008). Textile effluents consist of a broad spectra of chemicals such as sodium sulfite, sodium nitrate, sodium carbonate, sodium hydroxide, sodium oxychloride, NaCl, H_2SO_4, sodium peroxide, sodium silicate, H_2O_2, acetic acid, bleaching powder, tannic acid, malt, detergents or enzymes, mordants, dyes, gum, etc. These chemicals-containing effluents utilized for irrigation purposes disturb the normal weathering and leaching activities of soil. They also affect soil's water movement activity by altering the chemical and biological nature of soil.

Among the most tainted industries of the environment is pulp and paper, which requires large amounts of freshwater and lignocellulosic material for its processing. Large quantities of effluents are generated after the paper production process (Kumar and Chopra 2011, 2013a, b, 2015). The widely used chemicals in pulp and paper industries are sodium hydroxide, sodium carbonate, sodium sulfate, Mg (HSO_4), and $CaClO_2$. The pulp and paper industry effluents are dark in color with a foul smell, high BOD, COD, pH, and organic content (Kumar and Chopra 2014a, b, c; Kumar et al. 2010; El-Bestawy et al. 2008; Deilek and Bese 2001).

4.3 Effects of Effluents on Soil Physicochemical Properties

Industrial effluents discharged without proper treatment from different industries may have a reflective impact on the physicochemical as well as biological properties of soils, especially related to its fertility. Several reports are available on the occurrence of changes in properties of soils due to the discharge of effluents from industries such as distillery industry, dairy industry, cotton mill, etc. (David 2010; Nizamuddin et al. 2008; Nagaraju et al. 2007; Narasimba et al. 1999). Conventionally, industrial wastes are disposed in the surroundings or nearby fields without adequate and proper recycling or treatment. These large quantities of unutilized and rejected chemicals, such as calcium, carbonates, $MgSO_4$, Fe_2O_3, $FeCl_3$, fly ashes, sludge, saw dust, bottles, plastic, radioactive wastes, etc., are dumped on the soil surface by almost all the industries. Larger the production rate of industries, larger is their waste generation rate.

All wastes dumped on soil surfaces by industries tend to change the physical, chemical, and biological nature of soil and thus contribute a large amount of toxic substances or pollutants to underground freshwater. The deposition of these chemicals on agricultural lands disrupts the normal activities of leaching and weathering of agricultural soil and also affects the soil's water movement activity. The microorganisms responsible for maintaining the ecological equilibrium of soil tend to be affected the most. The long-term application of these effluents would badly affect the physical, chemical, and biological properties of soil (Antil et al. 2007).

4.3.1 *Effects of Wastewater on Soil Physical Properties*

The irrigation of agricultural fields by industrial effluents has improved the physical as well as chemical assets of the soils by increasing soil microorganisms' activity. Continuous irrigation with these effluents is also expected to increase soil nutrient levels. Though very rich in organic content and nutrient supplements, soil's haphazard usage for an extended period of time risks deteriorating its physicochemical properties (Tripathi et al. 2011). The wastewater containing pollutants may slowly get introduced and accumulated in the soil and pose a risk to its quality as well as productivity rate (Friedel et al. 2000). In spite of the strong possibility of agronomic and economic benefits of effluent irrigation, the long-term application effects of effluents have been less investigated. Industrial effluents pose different physical effects on the soil ecosystem, which are discussed next.

4.3.1.1 Organic Matter

The organic materials in soil enhance its structural properties by binding the soil particles into aggregates or knobs and creating noncapillary pores through which air, water, and allow other essentials. Soil organic matter is gradually lost as the land is repeatedly utilized for cropping, leading to deterioration of soil's physical properties. Agricultural lands with continuous cultivation are highly deficient in organic materials because the rate at which organic matter returns from crop residues is lower as compared to the decomposition rate of organic matter in soil. The area lacking organic matter results in less stable soil aggregates which could easily fall apart in the presence of rain water or penetrating water. Despite cultivation, the larger soil pores are lost, soil air decreases with restricted water movement, and the soil becomes more closely packed with increased bulk density.

4.3.1.2 Water Retention Properties

Usually, the application of industrial effluents to agricultural fields increases soil's capacity to retain water. But it has been reported that the organic content of the effluents may affect water retention capability either by directly affecting the organic particles of soil or by indirectly influencing the other physical properties of soil, such as porosity, bulk density, or pore size distribution. Nevertheless, several other researchers have reported an increase in water retention capacity of soil at field capacity and at wilting point by effluent application on soil (Metzger and Yaron 1987; Chang et al. 1983).

4.3.1.3 Structure and Aggregation

Industrial effluents are full of organic materials which can improve the quality and structure of soil by reducing bulk density and promoting soil aggregation. Long-term application of effluents on agricultural lands may damage the natural balance on soil, causing ecological deterioration. The direct effect of effluents on soil's physical properties is structural damage, functional disturbances, and soil hardening (Wang and Lin 2003). One of the most important parameters for quantifying soil physical properties is its bulk density, which has a great impact on soil aeration, water holding capacity, absorption capacity, erosion resistance ability, infiltration, and solute migration (Huang 2000). The changes in soil density expose the porosity of soil (Neves et al. 2003). A study on the long-term application of wastewater for irrigation showed a significant change in the structure of soil. Bulk density and soil porosity had a close association with effluent irrigation as soil porosity decreases with increase in bulk density over time.

Soil hardening resulted in soil resistance to penetration, which is an important index in the measurement of crop roots elongation resistance which is related to the aggregation characteristics and spatial arrangement of soil (Franzluebbers et al. 2000; Barber 1994; Mullins et al. 1994; Ahmed et al. 1987). Further, irrigation from effluents with high salinity could made soil secondary salinization and also enhanced total alkalinity and sodium alkalinity in soil causing its hardening and decreasing its permeability (Li et al. 2006, 2003; Martens and Frankenberger 1992; Pagliai et al. 1981).

4.3.1.4 Water Transmission Properties

The organic materials present in wastewater can delay infiltration and aeration temporarily by plugging the soil surface. However, the net effect of organic matter on soil aggregation is that it improves soil structure, which enhanced water transmission, water infiltration, and also reduces soil susceptibility to erosion. The levels of sodium, calcium, and magnesium, under certain conditions, could adversely affect soil structure and worsen its tillage, infiltration, and friability characteristics. Sodium can cause dispersion of soil aggregates, leading to permeability and infiltration, when present in high concentration as compared to calcium and magnesium. The soil structure is not only associated with concentration of Ca and Mg but also to salinity of effluent, at which level different concentration of Na may affect soil structure. The effect of sodium on infiltration and permeability of water is predicted by the technique based on the sodium adsorption ratio (SAR), a ratio of concentration of sodium in water to the square root of the sum of the concentrations of calcium plus magnesium (mM/L) (Table 4.1).

High-SAR water with low salinity disperses the soil, making it harder to work on, less friable, and permeable to water (Table 4.2). Plants vary in their tolerance power to soil salinity as it affects the growth of plants by affecting water availability to the plants (Maas 1990; Bouwer and Idelovitch 1987).

4.3.2 *Effects of Wastewater on Soil Chemical Properties*

Biologically stabilized industrial effluents contain approximately an average of 50% organic materials. When added to agricultural fields, the wastewater undergoes decomposition to carbon dioxide, low-molecular-weight soluble organic acids, inorganic constituents, and residual organic matter (Table 4.3).

Though most of the organic materials in wastewater get converted into water and carbon dioxide, some become a stable layer of the soil humus, which tends to increase the cationic exchange capacity net negative charge of the soil (Hernandez et al. 1990; Boyd et al. 1980; Thompson et al. 1989; National Research Council

Table 4.1 Effect of salt concentration on sodium adsorption ratio

	Magnesium		Sodium		Calcium	
Sodium adsorption ratio	(mg/L)	(mmol$_c$/L)	(mg/L)	(mmol$_c$/L)	(mg/L)	(mmol$_c$/L)
2.39	50	4.11	100	4.35	50	2.50
3.39	100	8.22	200	8.70	100	4.99
4.79	200	16.45	400	17.40	200	9.98
10.70	1000	82.26	2000	87.00	1000	49.90

Table 4.2 Guidelines for the interpretation of water quality for irrigation

	EC(dS/m)		
SAR	No problem	Slight to moderate problem	Severe problem
0–3	>0.9	0.9–0.2	<0.2
3–6	>1.3	1.3–0.25	<0.25
6–12	>2.0	2.0–0.35	<0.35
12–20	>3.1	3.1–0.9	<0.9
20+	>5.6	5.6–1.8	<1.8

Table 4.3 Chemical characteristics of freshwater and wastewater used for irrigation (Silva et al. 2016)

Property	Freshwater	Wastewater	Threshold levels for irrigation
pH	7.13 ± 0.61	7.55 ± 0.53	6.5–8.4
COD	ne	169 ± 21	–
N-total	ne	89	–
N–NH_4	ne	82 ± 4	–
N–NO_3	ne	0.31 ± 0.38	<30
P-total	ne	4.3 ± 0.06	<02
EC (dS m^{-1})	0.78 ± 0.17	1.66 ± 0.24	0–3
Calcium (mg/L)	33.16 ± 6.33	34.52 ± 6.87	<400
Magnesium (mg/L)	26.58 ± 3.96	44.54 ± 9.07	<61
Sodium (mg/L)	76.25 ± 3.61	157.21 ± 11.21	0–69
Potassium (mg/L)	5.32 ± 1.02	21.66 ± 0.65	<2
SAR $(mmol/L)^{0.5}$	2.39 ± 0.18	4.16 ± 0.30	<9
Sulfates (mg/L)	Absent	Present	<960
Bicarbonates (mg/L)	116.39 ± 21.65	436.76 ± 97.96	<610
Chlorides (mg/L)	176.28 ± 2.53	286.36 ± 11.21	<350
Boron (mg/L)	0.24 ± 0.01	0.26 ± 0.10	<0.70
Copper (mg/L)	<0.05	<0.05	0.20
Iron (mg/L)	<0.05	0.46 ± 0.06	5
Manganese (mg/L)	<0.03	0.05 ± 0.01	0.20
Zinc (mg/L)	<0.05	<0.05	2

ne not evaluated

1977). Cationic exchange capacity is the measurement of the capacity of soil to retain cations. The higher rate of cationic exchange capacity prevents loss of essential nutrients by leaching (National Research Council 1977; Broadbent 1973). Commonly, the different constituents released along with wastewater may be categorized under four groups such as (1) the more soluble cations, anions, and molecules; (2) trace elements which form sparingly soluble reaction products; (3) potentially harmful inorganic chemicals; and (4) potentially harmful organics.

4.3.2.1 Soluble Cations, Anions, and Molecules

The anions, cations, and other molecules such as sodium, potassium, calcium, magnesium, chloride, sulfate, nitrate, selenite, bicarbonate, and boron (as borate and boric acid), which are soluble in industrial effluents, are of great concern in agricultural operations even in lower concentrations. These molecules and ions present in effluents are further absorbed by plants subsequent to irrigation. Since boric acid is weakly absorbed and uncharged, they are normally leached to the levels safe for most crops where water is applied in excess of evapotranspiration (Keren and Bingham 1985). Boron may become toxic to sensitive plants if its concentration reaches above about 0.7 mg/L in irrigation water (Maas 1990). Therefore, precautions should be taken to assure that the concentrations of boron in soils should not exceed this critical level through effluent irrigation, which could harm sensitive plants. But some tolerant crops such as cotton plants normally withstand this critical level of boron and also as high as 10 mg/L without damage, and thus can be promoted for cultivation in these soils.

In order to preserve the productivity of irrigation soil, the wastewater used for irrigation should be reclaimed in salts and other nutrient matter so that normal irrigation practices could be carried out. For irrigation purposes, the typical concentrations of salts in wastewater should be within the accepted levels. These salts accumulate in the soil unless removed by plants or through leaching from the root zone; otherwise they could prevent the growth of all but most tolerant plants when they reach a certain critical level. Crop plants can only remove less than 10% of salts from soils, even under the most ideal conditions applied through irrigation water (Oster and Rhodes 1985). Therefore, in order to sustain growth, the salts from the soil should be leached from the root zone of crops. Precipitation is usually an efficient method to leach salts to an acceptable level in temperate and humid regions where irrigation is practiced only during parched periods.

4.3.2.2 Trace Elements

Subsequently to the decomposition of organic matter, trace elements including Ca, As, Co, Cu, Ni, Mo, Pb, and others are also released along with wastewater. These trace elements are utilized by plants in very small amounts and frugally form soluble reaction products. Thus, this tends to accumulate in the soil surface and slowly becomes part of the soil matrix (McGrath et al. 1994). With the repeated application of wastewater for irrigation, these trace elements could accumulate to toxic levels, which could harm soil microorganisms and plants (McGrath et al. 1994; Chang et al. 1992). Furthermore, through crops, they could get into the food chain, affecting humans, domestic animals, and wildlife that consume the crops (Logan and Chaney 1983). Higher concentrations of trace elements are accumulated in tissues of crops grown in acidic soils and become more liable to phytotoxicity than crops grown on neutral or calcareous soils. Crops grown in extremely acidic soils with excess copper, zinc, and nickel have shown visual symptoms of phytotoxicity. The problems associated with the acidity of the soil could be resolved through routine management operations which could neutralize the acidity of soil (Pearson and Adams 1967). As long as the agronomic practices are continued with the application of treated wastewater to agricultural fields, the possibility of trace elements to harm the yield and wholesomeness of crops would barely exist.

4.3.2.3 Accumulation of Potentially Harmful Inorganic Chemicals in Soils

Treated industrial effluents hardly consist of any harmful elements at concentrations exceeding the irrigation water limit. Therefore, the treated wastewater should be permitted to be used for crop irrigation purposes. Depending on the contributions of different industries, the concentrations of trace elements varies in emerging wastewaters. Bray et al. (1985) examined silage crop on land irrigated with municipal waste for 3 years at moderate to high rates (15–90 metric tons/ha). They found increased but not excessive levels of zinc and cadmium on silage crop, but observed no elevated levels of the other 12 elements, including chromium, lead, mercury, arsenic, and selenium.

4.3.2.4 Accumulation of Potentially Harmful Organic Chemicals in Soils

The most significant component of soil is the organic matter, which is also regarded as an important index for measuring the fertility of soil (Rattan et al. 2005). The accumulation of organic matter in soil depends on the means of input of organic matter (Shao et al. 2006; Qin 2003). When industrial effluents consisting of organic chemicals get added to the soil, they may decompose, be adsorbed, or volatilize but only those that are nonvolatile and resistant to decomposition will accumulate in the soil. Industrial wastewater irrigation could solve shortage of water in agricultural field as well as increase soil fertility, but this heft of increase showed agro-type and altitudinal difference (Xue 2012). A research on soil organic matter displayed significant increase in sandy soil by 36.5 g/kg from 0.85 g/kg to 1.16 g/kg and loam soil by 97.1 g/kg from 2.73 g/kg to 5.38 g/kg (Lan et al. 2010). The degree of soil organic matter increases with different soil layers. The extremely noteworthy effects on soil organic matter were easily discovered within 20 cm of topsoil, which substantially reduces with the depth of soil layers.

4.4 Effects of Wastewater Irrigation on Soil Microbial Activities

The most sensible soil quality indicators are the changes in the soil's biological characteristics, as soil microbiota is more vigorous and sensitive than the physico-chemical properties of soil. More detailed studies have been performed on the effect of addition of industrial sludge to soils on soil microorganisms but less research information is available on the effluents' effect on agricultural soils, such as denitrification, enzymatic activities, and microbial respiration. Soil microorganisms, including bacteria, fungi, actinomycetes, and algae, play an important role in the decomposition of organic matter as well as in the cycling of plant nutrients such as phosphorus, nitrogen, and sulfur (Mani and Bharagava 2016; Bharagava et al. 2018). Long-term application of industrial wastewater has shown a detrimental effect on microbial activity and its biomass, nitrogen fixation, and vesicular-arbuscular mycorrhizae due the accumulation of metals in soil (McGrath et al. 1994, 1988; Smith 1991; Giller et al. 1989).

The most astonishing pool of life in the biosphere is the microbial diversity that we have just begun to understand and explore, but this diversity has been exploited by humans in processes such as production of antibiotics, vitamins, and fermentation. The supply of effluents to the irrigated lands has contributed in enhancing the nutrient stocks for microbes (Li et al. 2015; Jain et al. 2005).

The structural association of soil particles provides a dimensional heterogeneous habitat to microorganisms categorized by different nutrients, substrates, concentrations of oxygen, water content, and flexible pH values (Sessitsch et al. 2001). Soils microorganisms not only degrade organic matter but also promote moisture retention and fertility of soil (Kuske et al. 2002). Soil microbiota also plays a very important role in the cycling of nutrients, carbon sequestration, soil creation, and structure. Therefore, soil community structure and function is the most important aspect in assessing soil health.

4.4.1 Effects of Wastewater on the Nitrification Potential of Soil

Wastewater irrigation not only diminishes the water shortage problem but also leads to the potential accumulation of pollutants and causes conforming alterations in denitrifying communities and process of denitrification (Fig. 4.2).

Hence, the long-term application of wastewater for irrigation must be avoided as it could have possible public health and environmental side effects since it may contain pathogens, high levels of dissolved organic C, detergents, sodium, and toxic metals (Wallach et al. 2005). Furthermore, these could cause significant shifts in structure and function of microbial communities, which increases the viability of agricultural soils. The most important groups of microorganisms which may get affected by these effluents are chemolithotropic ammonia oxidizing bacteria (AOB).

AOBs are the first responsible rate-limiting step in the process of nitrification in which ammonia (NH_4) is transformed to nitrate (NO_3^-) via nitrite (NO_2^-), which play a very critical role in natural nitrogen cycle (Rusan et al. 2007). As per previous research, AOB has been demonstrated to be affected by various chemical conditions but are not limited to levels of organic matter, ammonia, oxygen, toxic metals, pH, and salts (Meng et al. 2016; Abaidoo et al. 2010; Zhang and Wang 2007; Rusan et al. 2007; Xia and Wang 2001; Yang 2000; Brown and Halweil 1998). The AOB exhibits slow growth rates, low biomass yields, and limited number of distinguishing phenotypic characteristics, thus traditional biological methods for studying AOB becomes unrepresentative and extremely time consuming (Angelakis et al. 1999). The addition of effluents to the irrigational fields provides additional exogenous food supply, which increases microbial populations. On the basis of C/N ratio, the increased microbial populations demand for nitrogen could reduce the plant-available nitrogen to levels that are deficient for crop growth (Alexander 1977). This process, referred to as immobilization, occurs to the C/N ratio of about 20 or

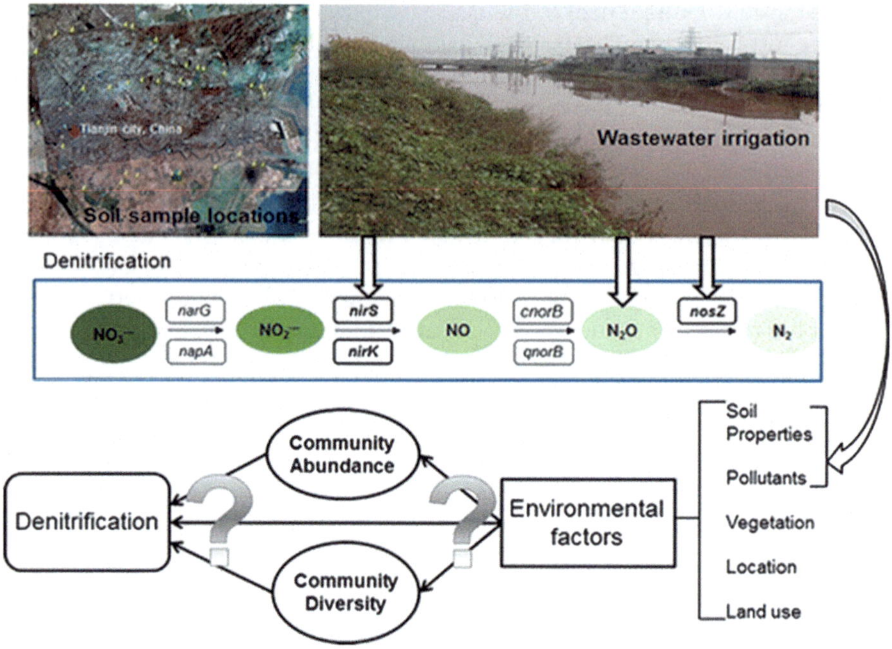

Fig. 4.2 Effect of different factors on nitrification (adopted from Guo et al. 2013)

more in which microbes convert inorganic nitrogen to organic nitrogen, the form unavailable to crops (Alexander 1977). Carbon volatizes to carbon dioxide as decomposition proceeds and the C/N ratio reduces from 20, mineralization of organic nitrogen exceeds immobilization, and an excess availability of inorganic nitrogen to the crops results (Brookes et al. 1986b; Guo et al. 2013).

4.4.2 Effects of Wastewater on Nitrogen Fixation

The adverse effects on the symbiotic relationship between certain strains of rhizobia have been previously reported as being due to the accumulation of metals in soils following long-term application of industrial effluents (McGrath et al. 1994; Giller et al. 1989). An agricultural field which received industrial effluents for over 20 years has shown lesser concentration of nitrogen and decreased yield of white clover (McGrath et al. 1988). The root nodules of clover isolated from high metal-contaminated soil were also found to be smaller, and clover rhizobia were found to be ineffective in nitrogen fixation (Giller et al. 1989). Further, large numbers of effective rhizobium of similar strain was inoculated to the soil for restoring the nitrogen fixation process. It was concluded that clover rhizobia were unable to survive in the free-living state outside the protected root nodules in metal-contaminated soils (McGrath et al. 1994).

Other researchers have shown no conclusion on the inhibition of rhizobium by the application of industrial effluents. Some other studies have shown little or no effect on nitrogen fixation activity. Heckman et al. (1987) studied the effect of industrial sludge on symbiotic nitrogen fixation by soybean. They reported no effect on nitrogen fixation where sludge was not applied but observed a decrease in the nitrogen fixation amount at the site of sludge application. In another study, Kinkle et al. (1987) studied soybean rhizobia in soil which received effluent for 11 years and found no detrimental effect either on soil rhizobial numbers or shift in serogroup distribution. Similar results were also reported by Angle and Chaney (1989). But in contrast to this, reduced nodulation and ineffective symbiosis were reported for alfalfa, white clover, and red clover grown in effluent irrigated soils with pH less than or equal to 5. However, no effect on plant growth, nodulation, and nitrogen fixation were observed when these were grown at pH 6.0 and above.

The long-term applications of wastewater on irrigated soils resulting in metal accumulation have shown reduction in nitrogen fixation by free-living heterotrophic bacteria (Lorenz et al. 1992; Martensson and Witter 1990; Brookes et al. 1986a). Symbiotic relationships between plants and microorganisms are impacted negatively only when soil microbes affect uptake of nutrients by associated plants. Thus, the available information suggests that effluent applications on agricultural soils act to delay or suppress microbial activity (Koomen et al. 1990).

4.5 Conclusion

Industrial effluents, which contain various types of organic compounds and heavy metals, have been shown to have many adverse impacts on aquatic and terrestrial ecosystems. Due to high load of organics and inorganics matters, effluents have been found to have high BOD, COD, TDS, TSS, and TSS values. Similarly, they may also contains few fertilizer elements and exotic compounds and show no adverse effects on agriculture soils, crops, or the ecosystem if used in proper quantities appropriate for agricultural needs. In addition, effluents sludge also contains essential nutrients such as nitrogen, trace elements, organics, etc., which may usually fulfill crops requirements for proper growth and development. But unfortunately, accumulation of potentially toxic elements in sludge-amended soils may prevent the activity of microorganism, like strains of cyanobacteria and rhizobia, and due to this microbial biomass reduces. Adequate industrial effluents at a proper ratio may be used for fertigation and also for soil quality improvement.

Acknowledgments The authors are grateful to Department of Biochemistry, Gramin Science (Voc) College, Nanded, Maharashtra, and Babasaheb Bhimrao Ambedkar University, Lucknow, India, for providing the financial support to Dr. Sujata Mani and Mr. Pankaj Chowdhary for this work.

References

Abaidoo RC, Keraita B, Drechsel P, Dissanayake P, Maxwell AS (2010) Soil and crop contamination through wastewater irrigation and options for risk reduction in developing countries. Springer, London

Ahmed HAM, Gerald ES, Hart RH (1987) Soil bulk density and water infiltration as affected by grazing systems. J Range Manag 40:307–309

Alexander M (1977) Introduction to soil microbiology, 2nd edn. Wiley, New York

Alvarez-Bernal DSM, Olalde-Portugal JT, Contreras-Remos N, Trujillo-Tapia V (2006) Effect of tanneries wastewater on chemical and biological soil characteristics. Appl Soil Ecol 33:267–277

Angelakis AN, Monte MHFMD, Bontoux L, Asano T (1999) The status of wastewater reuse practice in the Mediterranean basin: need for guidelines. Water Res 33:2201–2217

Angle JS, Chaney RL (1989) Cadmium resistance screening in nitrilotriacetate-buffered minimal media. Appl Environ Microbiol 55:2101–2104

Antil RS, Dinesh, Dahiya SS (2007) Utilization of sewer water and its significance in INM. In: Proceedings of ICAR sponsored Winter School on Integrated Nutrient Management, pp 79–83

Barber RG (1994) Persistence of loosened horizons and soybean yield increases in Bolivia. Soil Sci Soc Am J 58:943–950

Bharagava RN, Chandra R (2010) Effect of bacteria treated and untreated post-methanated distillery effluent (PMDE) on seed germination, seedling growth and amylase activity in *Phaseolus mungo* L. J Hazard Mater 180:730–734

Bharagava RN, Mani S, Mulla SI, Saratale GD (2018) Degradation and decolourization potential of an ligninolytic enzyme producing *Aeromonas hydrophila* for crystal violet dye and its phytotoxicity. Ecotoxicol Environ Saf 156:166–175

Bouwer H, Idelovitch E (1987) Quality requirements for irrigation with sewage water. J Irrig Drain Eng 113:516–535

Boyd SA, Sommers LE, Nelson DW (1980) Changes in the humic acid fraction of soil resulting from sludge application. Soil Sci Soc Am J 44:1179–1186

Bray BJ, Dowdy JH, Goodrich RD, Pamp DE (1985) Trace metal accumulations in tissues of goats fed silage produced on sewage sludge amended soil. J Environ Qual 14:114–118

Broadbent FE (1973) Organics in recycling municipal sludges and effluents on land. National Association of State Universities and Land-Grant Colleges, Washington, DC, pp 97–101

Brookes PC, McGrath SP, Heijnen C (1986a) Metal residues in soils previously treated with sewage-sludge and their effects on growth and nitrogen fixation by blue-green algae. Soil Biol Biochem 18:345–353

Brookes PC, Heijnen CE, McGrath SP, Vance ED (1986b) Soil microbiol biomass estimates in soils contaminated with metals. Soil Biol Biochem 18:383–388

Brown LR, Halweil B (1998) China's water shortage could shake world food security. World Watch 11:11–21

Chang AC, Page AL, Warneke JE (1983) Soil conditioning effects of municipal sludge compost. J Environ Eng 109:574–583

Chang AC, Granato TC, Page AL (1992) A methodology for establishing phytotoxicity criteria for chromium, copper, nickel, and zinc in agricultural land application of municipal sewage sludge. J Environ Qual 21:521–536

Chowdhary P, Bharagava RN (2019) Toxicity, beneficial aspects and treatment of alcohol industry wastewater. In: Bharagava R, Chowdhary P (eds) Emerging and eco-friendly approaches for waste management. Springer, Singapore

Chowdhary P, More N, Raj A, Bharagava RN (2017a) Characterization and identification of bacterial pathogens from treated tannery wastewater. Microbiol Res Int 5:30–36

Chowdhary P, Yadav A, Kaithwas G, Bharagava R N (2017b). Distillery wastewater: a major source of environmental pollution and its biological treatment for environmental safety In: Singh, R., Kumar, S., (Eds.) Green technologies and environmental sustainability. Springer International, Switzerland, pp 409–435

Chowdhary P, Raj A, Bharagava RN (2018a) Environmental pollution and health hazards from distillery wastewater and treatment approaches to combat the environmental. Chemosphere 194:229–246

Chowdhary P, Yadav A, Singh R, Chandra R, Singh DP, Raj A, Bharagava RN (2018b) Stress response of *Triticum aestivum* L. and *Brassica juncea* L. against heavy metals growing at distillery and tannery wastewater contaminated site. Chemosphere 206:122–131

Chowdhary P, Mani S, Reyes IP, Bharagava RN (2018c) Effects of industrial wastewaters on soil sustainability and environment. In: Rakshit A, Sarkar B, Abhilash P (eds) Soil amendments for sustainability: challenges and perspectives. CRC, Boca Raton, FL

David SS (2010) Protease and phosphatase activity of soil contaminated with dairy wastewater. Asian J Microbiol Biotechnol Environ Sci 26:711–714

Deilek FB, Bese S (2001) Treatment of pulping effluents using alum and clay-color removal and sludge characteristics. Water SA 27(3):361–366

Dutton J, Fisher NS (2011) Bioaccumulation of As, Cd, Cr, Hg(II), and Me Hg in killifish (*Fundulus heteroclitus*) from amphipod and worm prey. Sci Total Environ 409:3438–3447

El-Bestawy E, El-Sokkary I, Hussein H, Abu-Keela AF (2008) Pollution control in pulp and paper industrial effluents using integrated chemical-biological sequences. J Ind Microbiol Biotechnol 35(11):1517–1529. https://doi.org/10.1007/s10295-008-0453-3

Feigin A, Ravina I, Shalhevet J (1991) Irrigation with treated sewage effluent: management for environmental protection. Springer, Berlin, p 224

Franzluebbers AJ, Wright SF, Stuedemann JA (2000) Soil aggregation and glomalin under pastures in the Southern Piedmont USA. Soil Sci Soc Am J 64:1018–1026

Friedel JK, Langer T, Siebe C, Stahr K (2000) Effects of long-term wastewater irrigation on soil organic matter, soil microbial biomass and its activities in central Mexico. Biol Fertil Soils 31:414–421

Garg VK, Kaushik P (2008) Influence of textile mill wastewater irrigation on the growth of sorghum cultivars. Appl Ecol Environ Res 6:1–12

Giller KE, McGrath SP, Hirsch PR (1989) Absence of nitrogen fixation in clover grown on soil subject to long-term contamination with heavy metals is due to survival of only ineffective Rhizobium. Soil Biol Biochem 21:841–848

Guo GX, Deng H, Qiao M, Yao HY, Zhu YG (2013) Effect of long-term wastewater irrigation on potential denitrification and denitrifying communities in soils at the watershed scale. Environ Sci Technol 47(7):3105–3113

Hare V, Chowdhary P, Baghel VS (2017) Influence of bacterial strains on *Oryza sativa* grown under arsenic tainted soil: accumulation and detoxification response. Plant Physiol Biochem 119:93–102. https://doi.org/10.1016/j.plaphy.2017.08.021

Heckman JR, Angle JS, Chaney RL (1987) Residual effects of sewage sludge on soybeans. I. Accumulation of heavy metals. II. Accumulation of soil and symbiotically fixed nitrogen. J Environ Qual 16:113–124

Hernandez MT, Moreno JI, Costa F, Gonzales-Vila FJ, Frund R (1990) Structural features of humic acid like substances from sewage sludge. Soil Sci 149:63–68

Huang CY (2000) Soil science. China Agriculture Press, Beijing

Jain RK, Kapur M, Labana S, Lal B, Sarma PM, Bhattacharya D, Thakur IS (2005) Microbial diversity: application of microorganisms for the biodegradation of xenobiotics. Curr Sci 89:101–112

Keren R, Bingham FT (1985) Boron in water, soils, and plants. Advances in soil science, vol 1. Springer, New York, pp 229–275

Kinkle BK, Angle JS, Keyser HH (1987) Long-term effects of metal rich sewage sludge application on soil populations of *Bradyrhizobium japonicum*. Appl Environ Microbiol 53:315–319

Koomen I, McGrath SP, Giller KE (1990) Mycorrhizal infection of clover is delayed in soils contaminated with heavy metals from past sewage sludge applications. Soil Biol Biochem 22:871–873

Kumar V, Chopra AK (2011) Alterations in physicochemical characteristics of soil after irrigation with paper mill effluent. J Chem Pharm Res 3(6):7–22

Kumar V, Chopra AK (2013a) Ferti-irrigational effect of paper mill effluent on agronomical characteristics of *Abelmoschus esculentus* L. (Okra). Pak J Biol Sci 16(22):1426–1437

Kumar V, Chopra AK (2013b) Distribution, enrichment and accumulation of heavy metals in soil and *Trigonella foenum-graecum* L. (Fenugreek) after fertilization with paper mill effluent. Open J Metals 3:8–20. https://doi.org/10.4236/ojmetal.2013.32A1002

Kumar V, Chopra AK (2014a) Ferti-irrigation effect of paper mill effluent on agronomical practices of *Phaseolus vulgaris* (L.) in two different seasons. Commun Soil Sci Plant Anal 45:2151–2217

Kumar V, Chopra AK (2014b) Ferti-irrigational impact of sugar mill effluent on agronomical characteristics of *Phaseolus vulgaris* (L.) in two seasons. Environ Monit Assess 186:7877–7892. https://doi.org/10.1007/s10661-0143974-4

Kumar V, Chopra AK (2014c) Pearl millet (*Pennisetum Glaucum* L.) response after ferti-irrigation with sugar mill effluent in two seasons. Int J Recycl Org Waste Agric 67. https://doi.org/10.1007/s40093-014-0067-x

Kumar V, Chopra AK (2015) Fertigation with agroresidue based paper mill effluent on a high yield spinach variety. Int J Veg Sci 21(1):69–97. https://doi.org/10.1080/19315260.2013.825690

Kumar V, Chopra AK, Pathak C, Pathak S (2010) Agro-potentiality of paper mill effluent on the characteristics of *Trigonella foenumgraecum* L. (Fenugreek). N Y Sci J 3(5):68–77

Kuske CR, Lawrence OT, Miller ME, Dunbar JM, Davis JA, Barns SM, Belnap J (2002) Comparison of soil bacterial communities in rhizosphere of three plant species and the interspace in an arid grassland. Appl Environ Microbiol 68:1854–1863

Lan MJ, Li MS, Zhao GJ, Rui L (2010) Effects of eutrophic sewage irrigation on soil-holding capacity. J Shihezi Univ 28:497–500

Li F, Benhur M, Keren R (2003) Effect of marginal water irrigation on soil salinity, sodicity and crop yield. Trans Chin Soc Agric Eng 19:63–66

Li F, Huang G, Ding Y, Peng C (2006) Effects of soil alkalinity, gypsum application, and filtration disposal on hydraulic conductivity under irrigation with domestic effluent water. Trans Chin Soc Agric Eng 22:48–52

Li C, Zhang Z, Li Y, Cao J (2015) Study on dyeing wastewater treatment at high temperature by MBBR and the thermotolerant mechanism based on its microbial analysis. Process Biochem 50:1934–1941

Logan TJ, Chaney RL (1983) Utilization of municipal wastewater and sludges on land-metals. In: Page AL, Gleason TL III, Smith JE Jr, Iskandar IK, Sommers LE (eds) Proceedings of the workshop on utilization of municipal wastewater and sludge on land. University of California, Riverside, CA, pp 235–326

Lorenz SE, McGrath SP, Giller KE (1992) Assessment of free-living nitrogen fixation activity as a biological indicator of heavy metal toxicity in soil. Soil Biol Biochem 24:601–601

Maas EV (1990) In: Tanji KK (ed) Crop salt tolerance of plants. Agricultural salinity assessment and management. American Society of Civil Engineers, New York, pp 262–304

Magesan GN, Williamson JC, Sparling GP, Schipper LA, Lloyd-Jones AR (1999) Hydraulic conductivity in soils irrigated with wastewaters of differing strengths: field and laboratory studies. Aust J Soil Res 37(2):391–401

Mani S, Bharagava RN (2016) Exposure to crystal violet, its toxic, genotoxic and carcinogenic effects on environmental and its degradation and detoxification for environmental safety. Rev Environ Contam Toxicol 237:71–104

Martens DA, Frankenberger WT Jr (1992) Modification of infiltration rates in organic-ammended irrigated soil. Agron J 84:707–717

Martensson AM, Witter E (1990) Influence of various soil amendments on nitrogen-fixing soil microorganism in a long-term field experiment, with special reference to sewage sludge. Soil Biol Biochem 22:977–982

McGrath SP, Brookes PC, Giller KE (1988) Effects of potentially toxic metals in soil derived from past applications of sewage sludge on nitrogen fixation by *Trifolium repens* L. Soil Biol Biochem 20:415–424

McGrath SP, Chang AC, Page AL, Witter E (1994) Land application of sewage sludge: scientific perspectives of heavy metal loading limits in Europe and the United States. Environ Rev 2:108–118

Meng WQ, Wang ZW, Hu BB, Wang ZL, Li HY, Goodman RC (2016) Heavy metals in soil and plants after long-term sewage irrigation at Tianjin China: a case study assessment. Agric Water Manag 171:153–161

Metzger L, Yaron B (1987) Influence of sludge organic matter on soil physical properties. Adv Soil Sci 7:141–163

Mullins GL, Reeves DW, Burmester CH, Bryant HH (1994) In-row subsoiling and potassium placement effects on root growth and potassium content of cotton. Agron J 86:136–139

Nagaraju LG, Narasimba B, Rangaswami V (2007) Impact of effluents of sugar industry on soil physico-chemical and biological properties. J Ind Pollut Contam 23:73–76

Narasimba GA, Sridevi A, Venkata SR, Rajasekhar B (1999) Effects of cotton gaining mill effluent on soil enzymatic activities and nitrogen mineralization in soil. J Chem Pharm Res 3:126–137

National Research Council (1977) Multimedium management of municipal sludge. National Academy Press, Washington, DC

Neves CSVJ, Feller C, Guimaraes MF, Medina CC, Filho JT, Fortier M (2003) Soil bulk density and porosity of homogeneous morphological units identified by the cropping profile method in clayey oxisols in Brazil. Soil Tillage Res 71:109–119

Nizamuddin SA, Sridevi A, Narasimba G (2008) Impact of dairy factory effluents on soil enzyme activities. Eco Environ Conserv 14:89–94

Oster JD, Rhodes JD (1985) Water management for salinity and sodicity control. In: Pettygrove GS, Asano T (eds) Irrigation with reclaimed municipal wastewater—a guidance manual. Lewis, Chelsea, MI, pp 1–20

Pagliai M, Guidi G, La Marea M, Giachetti M, Lucamante G (1981) Effects of sewage sludge and composts on soil porosity and aggregation. J Environ Qual 10:556–561

Pal S, Vimala Y (2012) Bioremediation and decolorization of Distillery effluent by novel Microbial Consortium. Eur J Exp Biol 2(3):496–504

Pearson RW, Adams F (1967) Soil acidity and lining. Agronomy monograph no. 12. Am Soc Agron, Madison, WI

Qin YD (2003) Soil. Physics Higher Education Press, Beijing, p 2003

Rattan RK, Datta SP, Chhonkar PK, Suribabu K, Singh AK (2005) Long-term impact of irrigation with sewage effluents on heavy metal content in soils, crops and groundwater-a case study. Agric Ecosyst Environ 109:310–322

Rusan MJM, Hinnawi S, Rousan L (2007) Long term effect of wastewater irrigation of forage crops on soil and plant quality parameters. Desalination 215:143–152

Sessitsch A, Weilharter A, Gerzabek MH, Kirchmann H, Kandeler E (2001) Microbial population structures in soil particle size fractions of a long term fertilizer field experiment. Appl Environ Microbiol 67:4215–4224

Shao MA, Wang QJ, Huang MB (2006) Soil physics. Higher Education Press, Beijing

Silva LVBD, Lima VLA, Pearson HW, Silva TTS, Maciel ACL, Sofiatti V (2016) Chemical properties of Haplustalf soil under irrigation with treated wastewater and nitrogen fertilization. Soil Water Plant Manag 20(4):308–315

Singh NK, Patel DB (2012) Microalgae for bioremediation of distillery effluent. Farm Food Water Secur 10:83–109

Smith SR (1991) Effects of sewage sludge application on soil microbial processes and soil fertility. In: Stewart BA (ed) Advances in Soil Science, vol 16. Springer, New York, pp 191–212

Takahashi CK, Turner A, Millward GE, Glegg GA (2012) Persistence and metallic composition of paint particles in sediments from a tidal inlet. Mar Pollut Bull 64:133–137

Thompson ML, Zhang H, Kazemi M, Sandor JA (1989) Contributions of organic matter to cation exchange capacity and specific surface area of fractionated soil materials. Soil Sci 148:250–257

Tripathi DM, Tripathi S, Tripathi BD (2011) Implications of secondary treated distillery effluent irrigation on soil cellulase and urease activities. J Environ Prot 2:655–661

Wallach R, Ben-Arie O, Graber ER (2005) Soil water repellency induced by long-term irrigation with treated sewage effluent. J Environ Qual 34:1910–1920

Wang GL, Lin WJ (2003) Contamination of soil from sewage irrigation and its remediation. J Agro-Environ Sci 22:163–166

Xia LJ, Wang HK (2001) Soil pollution and countermeasures. Huazhong University of Science Press, Wuhan

Xue ZJ (2012) Assessment of soil quality and pollution risk in main sewage-irrigated area of Hebei province. Agricultural University of Hebei Province, Baoding

Yang JF (2000) The problem on agricultural sewage irrigation and countermeasures. Water Resour Prot 2:4–8

Zhang JY, Wang GQ (2007) The impacts of climate change on hydrology and water resources. Science Press, Beijing

Zollinger H (1987) Colour chemistry-synthesis, properties of organic dyes and pigments. VCH Publishers, New York, pp 92–100

Chapter 5
Plant Metabolites Involved in Plant–Pathogen Interactions

Daraksha Parween, Binod Bihari Sahu, Maya Kumari, and Ramesh N. Pudake

Abstract Plants constantly confront different pathogens and undergo stress. To overcome such hurdles, plants produce primary and secondary metabolites. Primary metabolites are essential for the growth and development of plant and secondary metabolites are vital for plant survival by providing resistance against various pathogens and maintaining an elegant stability with the environment. Plants produce a huge number of metabolites, and many of such metabolites have yet to be identified. For the analysis of these wide range of highly complex metabolites synthesized by the plants, various tools and techniques are required for the study of metabolomics. Study of plant metabolomics comprises of sample preparation or extraction of bioactive molecules from the plants, detection and identification of the metabolites, and data processing and statistical analysis of the identified metabolites. Modern technologies used for the study of plant metabolomics includes metabolic fingerprinting, metabolite profiling and targeted and non-targeted detection analysis. Starting with the definition of primary and secondary metabolites, we aimed to focus on the behavior of different metabolites during plant–pathogen interaction and to finally concentrate on different tools and techniques, which are required for the identification and analysis of metabolites. With the help of current high-resolution mass spectrometers it has become quite feasible to identify low-molecular-mass metabolites. Efforts are made to develop computational tools for the identification of unknown metabolites and to develop mass spectral databases which will provide an authentic reference for the identified compounds.

D. Parween · B. B. Sahu
Department of Life Science, National Institute of Technology Rourkela, Rourkela, Odisha, India

M. Kumari
Directorate of Life Sciences, Defence Research & Development Organization (DRDO), New Delhi, India

R. N. Pudake (✉)
Amity Institute of Nanotechnology, Amity University Uttar Pradesh, Noida, Uttar Pradesh, India
e-mail: rnpudake@amity.edu

A. Varma et al. (eds.), *Plant Biotic Interactions*,
https://doi.org/10.1007/978-3-030-26657-8_5

5.1 Introduction

Under natural conditions, plants are surrounded by many probable enemies. To defend against pathogen bout, plants create defense strategies mainly through chemical and mechanical defenses (Olivoto et al. 2017). The former includes structures such as trichrome, thick cuticle, spines, and smooth, sticky, or hard surfaces, which avert pathogens from laying eggs or food selection. Chemical defenses comprise a variety of constituents which are repellent, toxic, or which make plant tissues difficult to digest for animals. These chemical substances retaliate against abiotic or biotic stimuli, traditionally referred to as secondary metabolites which play a major role in plant defense mechanism (Goyal et al. 2012).

Plants are an unlimited source of phytochemicals in the form of primary and secondary metabolites. However, secondary metabolites corrugate leading interest because of their multifunctional activities extending from antimicrobial, insecticidal, antibiotic properties, to extremely important pharmaceutical activities (Stöckigt et al. 1995). Studies on the functions of these compounds for plant defense has increased in the last two decades (Rhodes et al. 1994).

5.2 Primary Metabolites

Primary metabolites are limited molecules derived from living cells. Primary metabolites are the intermediary or final products of the metabolic pathways of transitional metabolism (e.g., lipid metabolism, amino acid metabolism, carbohydrate metabolism), and these are the composition units for vital macromolecules, or can be altered into coenzymes (Demain 2000). Primary metabolites such as organic acids, phytosterols, amino acids, nucleotides, and acyl lipids are biomolecules required to perform basic metabolic processes. These are found throughout the plant kingdom, required for basal metabolic roles that are usually noticeable and are highly useful to plants. These are produced in plentiful masses and can be effortlessly extracted from different species of plants. Primary metabolites are a portion of a cell's basic molecular structure (Croteau et al. 2000).

Primary metabolites are concentrated in vegetative storage organs and seeds in higher plants and are required for physiological development to perform basic cell metabolism. Generally, primary metabolites are obtained for commercial use, which is high volume–low value bulk chemicals. However, there are exemptions to this rule. For example, β-carotene and myoinositol are exclusive primary metabolites as their extraction, isolation, and purification are strenuous (Balandrin et al. 1985). Primary metabolites are mainly used in the food industry which incorporates fatty acids (utilized for manufacturing soaps and detergents), vegetable oils, flavor nucleotides (5′-inosinic acid, 5′-guanylic acid), alcohols (ethanol), polyols (mannitol, glycerol, erythritol, xylitol), amino acids (monosodium glutamate, lysine, threonine, phenylalanine, tryptophan), organic acids (acetic, propionic, succinic, fumaric, lactic), sugars (fructose, sorbose, ribose), vitamins [biotin, riboflavin (B2), cyanocobalamin (B12)], and polysaccharides (xanthan, gellan) (Demain 2000).

5.3 Secondary Metabolites

Secondary metabolites are derivatives of primary metabolites, synthesized by plants in a diverse array. Secondary metabolites are well known to act as chemical defenses that avert pests and pathogens; they have a key role in controlling plant growth and protecting plants from environmental pressures (Fraire-Velázquez and Balderas-Hernández 2013). They do not comprise basic molecular cell structure. These metabolites are produced in lesser amounts and their extraction from plants is tough. Classes of secondary metabolites are restricted to selected plant species or families; they are found at particular stages of development, with a limited role in the plant (Osbourn et al. 2003).

Plants collectively produce natural products of above 100,000 low molecular mass, i.e., secondary metabolites. These metabolites can be distinguished from the constituents of intermediary (primary) metabolic products in that they are generally nonessential for the chief metabolic processes of the plant. Most of them are derived from the phenylpropanoid, alkaloid or fatty acid/polyketide, and isoprenoid pathways. Such a rich diversity has made it difficult to apply conventional molecular and genetic techniques to know the functions of natural products, also to select the genotypes for improved defense against microbial infection or insect/animal predation in plant defense, or to increase plant disease resistance by using metabolic pathway engineering (Dixon 2001).

5.3.1 Major Classes of Secondary Metabolites

Secondary metabolites can be divided into two different chemical groups: nitrogen-containing compounds and nitrogen-free compounds.

5.3.1.1 Nitrogen-Containing Secondary Metabolites

Nitrogen-containing compounds include alkaloids, nonprotein amino acids, amines, glucosinolates, cyanogenic glycosides, protease inhibitors, and lectins.

5.3.1.1.1 Alkaloids

Alkaloids are biologically active compounds which contain a ring structure or a heterocyclic compound with a nitrogen atom connected to minimum two carbon atoms and have some role in biological, ecological, pharmacological, chemical, and medicinal activity. Alkaloids are special compounds which participate in various biological processes of microorganisms, plants, and animals at different cellular levels in altered environments (Aniszewski 2015). Alkaloids are classified depending upon their physical and biological attributes to help their morphological identification and understand their technical and applied uses. Alkaloids can be classified according to their chemical structure, ecological, and biological action; biosynthetic pathway; and relation with chemical and technological innovations.

According to biological activity, alkaloids are cleaved into neutral or weakly basic molecules (e.g., indicine, ricinine), animal-derived alkaloids (e.g., arthropod, anuran, and mammalian alkaloids), moss alkaloids, nonnatural alkaloids, marine alkaloids, and fungal and bacterial alkaloids (Pelletier 1983). Nonnatural alkaloids are a part of bioorganic and stereochemistry research, which are used in pharmacological research. Due to constant change in the species of the pathogens and their infection ability, it has become necessary for plants to become resistant to medicines and antibiotics.

Based on their relationship in the branches of chemistry and technology, alkaloids can be divided into three groups: (1) natural, (2) biomimic and bionic, and (3) synthetic. Natural alkaloids are synthesized by living organisms and are naturally synthesizing molecules which exist in nature because of the progression of life on Earth. Biomimetic alkaloids are structurally identical to natural alkaloids and are copied artificially in laboratories. Bionic alkaloids are those biomimetic compounds which are synthesized artificially but are not similar analogues to natural alkaloids. Synthetic alkaloids are molecules which are artificially synthesized using high-level techniques and planned models, having the chemical characteristics of alkaloids.

Based on their chemical structures at alkaloid base, it can be divided into various types: bisindoles, indolizidines, carbolines, purines, pyrolidines, pyrrolizidines, steroids, terpenoids, diterpenes, triterpenes, pyridines, quinolozidines, quinolones, and quinolizolines (Eftekhari-Sis et al. 2013). Based on shape, structure, and the biological pathway used to create the molecules, alkaloids can be of three main type: true alkaloids, protoalkaloids, and pseudoalkaloids (Hegnauer 1988). True alkaloids are derived from amino acid with nitrogen in their heterocyclic ring. These are extremely reactive compounds with biological activities even in low doses. Except nicotine, all true alkaloids are bitter in taste. The primary precursors of true alkaloids are L-tryptophan, L-tyrosine/phenylalanine, L-ornithine, L-lysine, and L-histidine. These alkaloids can be natural, bionic, or synthetic, and some examples of true alkaloids are dopamine, cocaine, quinine, and morphine. In protoalkaloids, the *N*-atom acquired from an amino acid is not part of the heterocyclic ring. These compounds are derived from L-typtophan and L-tyrosine. These can be bionic, natural, or synthetic alkaloids, and mescaline, hordenine, and yohimbine are some examples. Pseudoalkaloids are not derived from amino acids, but from the precursors of amino acids from the amination and transamination reactions. They can also be obtained from non-amino acid precursors (Aniszewski 2015).

5.3.1.1.2 Amines

Amines are ammonia derivatives where one, two, or all three hydrogens of ammonia are replaced by organic groups. They play a significant metabolic and physiologic role in living organisms. Biologically active amines are cyclic, heterocyclic, and aliphatic and most of them are named after their precursor amino acids, e.g., tryptamine from tryptophan, tyramine from tyrosine. Bioactive or biologically active amines can be classified based on the number of amine groups as mono- (phenylethylamine, tyramine), di- (cadaverine, histamine, tryptamine, serotonin, putrescine), or polyamines (spermine, agmatine, spermidine). According to their chemical structure, amines can be aromatic (phenylethylamine, tyramine), aliphatic (putrescine, spermine, spermidine, cadaverine, agmatine), or heterocyclic (histamine, serotonin, tryptamine). Based on their biosynthetic pathway, they are classified as natural or biogenic (Glória 2005).

5.3.1.1.3 Nonprotein Amino Acids (NPAAs)

In nature, more than thousands of nonprotein amino acids are extracted from microorganisms, plants, and other sources (Barrett 2012). These amino acids are not formed in the main chains of protein, but some times they do get added in protein by post-translational modification. For these amino acids an exact transfer RNA and codon triplet is absent (Hunt 1985). Many nonprotein amino acids are considered as structural analogs of protein amino acids. For example, S-aminoethylcysteine is an analog to L-azetidine-2-carboxylic acid to L-proline, L-lysine, 3-cyanoalanine to L-alanine and L-indospicine, or L-canavanine to L-arginine (Wink 2003).

5.3.1.1.4 Cyanogenic Glycosides (CNglcs)

CNglcs are the source of HCN that occur extensively in the plant kingdom (Conn 1969) and are specialized bioactive plant products derived from amino acids characterized by α-hydroxynitriles (cyanohydrins) and oximes as key intermediates. Cyanogenic glycosides release ketones and toxic hydrogen cyanide (HCN) when hydrolyzed by α-hydroxynitrilases and β-glycosidases in a process referred to as cyanogenesis. Cyanogenesis is an effective defense against herbivores but is not effective against fungal pathogens because many fungi convert HCN into carbon dioxide and ammonia (Gleadow and Møller 2014).

5.3.1.1.5 Glucosinolates (GSLs)

Glucosinolates (GSLs), the precursors of isothiocyanates, are organic anions containing β-thioglucoside-*N*-hydroxysulfates, which represents an important and unique class of secondary metabolites found in seeds, roots, stem, and leaves of plants (mainly in the Brassicaceae) (Fahey et al. 2001; Vig et al. 2009). Glucosinolates on hydrolyzation liberate D-glucose, sulfate, and an unstable aglycone, which converts to isothiocyanate (Mithen et al. 2000). There are more than 120 diverse glucosinolates identified till date, mainly belonging to the family Brassicaceae and other important crops. Glucosinolates represent a classical example of plant compounds which affect the plant–insect interactions (Hopkins et al. 2009). The defense activity of glucosinolates are increased upon hydrolysis by the enzyme myrosinase. In plants, myrosinase is stored in special myrosinase cells. Myrosinase is a thioglucosidase that transforms glucosinolates into toxic isothiocyanates (Rask et al. 2000). In damaged plant tissues, due to the myrosinase activity, the glucosinolates stowed in the vacuole come in contact with the myrosinase and result in the formation of various toxic products, such as nitriles, isothiocyanates, and oxazolidinethiones (Bones and Rossiter 2006).

5.3.1.2 Nitrogen-Free Secondary Metabolites

Nitrogen-free compounds are various terpenoids (mono-, di-, tri-, and tetraterpenes; saponins; and cardiac glycosides), polyketides (anthraquinones), polyacetylenes, and phenolics (phenolics acids, flavonoids, catechol tannins, anthocyanins, lignans, galloyl and lignins).

5.3.1.2.1 Terpenoids

Terpenoids are abundant in plants with more than 30,000 compounds (Aharoni et al. 2006). Among the myriad bioactive compounds produced by plants, terpenoids (isoprenoids) epitomize the largest and most varied group of chemicals. A majority of plant terpenoids are utilized for specialized chemical interactions and defense in abiotic and biotic stress environments (Tholl 2015). They play substantial roles in nature during plant–plant, plant–environment, plant–insect, and plant–animal interactions (Pichersky and Gershenzon 2002).

5.3.1.2.2 Phenolics

Plants phenolic compounds emerge as one of the main categories of secondary metabolites and are essential for the growth, development, resistance to pathogens, pigmentation, reproduction, and for many other functions in plants (Lattanzio et al. 2006). As stated by Harborne, the term “phenolic” embraces plant substances with an aromatic ring bearing one (phenol) or more (polyphenol) hydroxyl substituents in common (Harborne 1989). Phenolic substances are water soluble as they frequently occur homogenized with sugar as glycosides and are commonly located in the cell vacuole. Flavonoids form the largest group among the natural phenolic compounds whereas simple monocyclic phenols, phenolic quinones, and phenylpropanoids exist in considerable amounts (Harborne 1984). The participation of phenolic compounds in the defense role of plant–animal and/or plant–microorganism interaction, is related to their antinutritional and antibiotic properties (Wink 1997). Flavonoids help plants to reside in soils rich in noxious metals, such as aluminum (Barcelo and Poschenrieder 2002). Several flavonoids oozing out from plant roots act as signaling molecules, which induce bacterial gene transcription and production of proteins are required for the infection method (Cooper 2004; Hungria and Stacey 1997; Kobayashi et al. 2004). Several flavonoids oozing out from plant roots function as signals, which induce bacterial gene transcription, and protein products are required for the infection process (Cooper 2004; Hungria and Stacey 1997; Kobayashi et al. 2004). Based on their defense role, flavonoids can be split into two groups, i.e., preformed and induced compounds. Preformed flavonoids are innate compounds that may play a signaling and/or a direct role in defense. These are synthesized during the regular development of plant tissue. Induced flavonoid compounds are produced by plants during physical injury, stress, or infection (Treutter 2006).

5.3.2 Responses of Secondary Metabolites During Biotic Stress in Plants

Plants, being sessile organisms, live in persistently changing environments which are often unfavorable or stressful for their growth and development (Zhu 2016). Adverse environmental conditions result in affected plant growth; metabolism is enormously involved in physiological regulation, signaling, and defense responses (Fraire-Velázquez and Balderas-Hernández 2013). Very often the wild-type species is resistant to microorganisms, and abiotic or biotic stress. To strengthen the resistance of a specialized species will be of no use. So, an addition of secondary metabolites to increase the resistance of a plant in which the species is adapted could be the solution (Wink 1988).

Higher plants often persuade the synthesis and hoarding of defense-related secondary metabolites upon biotic stress (e.g., herbivore or pathogen attack), referred to as phytoalexins (Mithöfer et al. 2004). Plants possessing these biochemical defense mechanisms help them to prevent or reduce further damage from pathogens (Eder and Cosio 1994). Plants, during infection or stress, accumulate phytoalexins which are low-molecular-weight antimicrobial compounds (Kuc 1995).

Defensive secondary compounds or metabolites become associated in response to both abiotic and biotic stress conditions (Akula and Ravishankar 2011). PhA (Phenylamides) play an important role in plant growth, development, and stress defense. PhA involved in plant defense have an antimicrobial activity that can protect plants against both abiotic and biotic stresses (Edreva et al. 2007). It is also reported that the active production of reactive oxygen species (ROS) in plants controls several different physiological processes, such as pathogen defense, abiotic and biotic stress response and systemic signaling. However, cells are provided with outstanding antioxidant defensive machineries to detoxify the detrimental effects of ROS. These antioxidant defense mechanisms may be either nonenzymatic (e.g., carotenoids and flavonoids) or enzymatic (e.g., catalase glutathione peroxidase, superoxide dismutase) (Gill and Tuteja 2010).

5.3.3 Behavior of Secondary Metabolites During Plant Defense Mechanisms

5.3.3.1 Phenolic Compounds

Phenolic compounds are natural products which arise biogenetically from phenylalanine and tyrosine during shikimate, phenylpropanoids, and flavonoids pathways (Lattanzio et al. 2006). Tannins are a varied group of polyphenolics. Tannins may defend plant by reducing the digestibility of plants consumed by herbivores by binding with digestive enzymes, dietary proteins (Robbins et al. 1987). In 1970, it was reported by Feeny that tannins act as a defense compound to the Oak moth (*Opheropthera brumata*) larvae. He found that tannin level in the leaves increased significantly prior to cessation of feeding by the larvae. This limitation was assumed to be because of the reaction of tannins with digestive enzymes in the gut of the larvae and the complexation of tannins with host proteins (Feeny 1970). High tannin-containing 'bird-resistant' cultivars (sorghum plants) have also been recognized, which were compared to cultivars with low tannin which were severely damaged in field trials (Bennett and Wallsgrove 1994). Slugs are one of the major pests which attack several economically essential crops. *Deroceras reticulatum* is amongst the major slug pests of potato which leads to extensive crop damage. Some potato cultivars were found considerably less attractive to slugs owing to high levels of phenolics and polyphenoloxidase activities (Bennett and Wallsgrove 1994). Plants with high level of phenolics are very less palatable to herbivores, and polyphenolics like tannins are considered common antifeedants (Fahey and Jung 1989). More often, phenolics are noxious towards the fungal pathogens in vitro and are accumulated near the infection sites which lead to lignin deposition, necrosis, and resistance (Moran 1998). It was also reported that in sweet potato cultivars resistant to *Meloidogyne incognita* (root-knot nematode), the concentrations of soluble and wall-bound phenolics significantly increased after infection (Gapasin et al. 1988). It is well known that Rhizobia species utilizes phenolic acids as a carbon source (Irisarri et al. 1996). Plant phenolic compounds behave as potential candidates as signaling molecules in the establishment of arbuscular mycorrhizal symbioses, initiation of legume rhizobia symbioses, and can also act as a means in plant defense processes. Flavonoids are multifarious of polyphenolic compounds, act as signaling molecules in plant–microbe interactions. Flavonoids are released from different zones of the leguminous plant roots (Mandal et al. 2010). Flavonoids (luteolin) bind at the active transcriptional sites of Rhizobial *nod* genes that control root nodule organogenesis. This induction of the rhizobial *nod* genes leads to the production of

Nod (nodule-inducing) factors, lipochitooligosaccharides (LCOs), which are modified differently depending on the Rhizobium species (Schultze and Kondorosi 1998). Many phenolics and alkaloids released from roots or seeds function primarily as defense elements against soil-borne pathogens and root-feeding insects (Ndakidemi and Dakora 2003). Rapid accumulation of phenolic acids, mostly cinnamic, tannic, gallic and ferulic acids revealed the effectiveness of Rhizobia in inducing resistance in rice plants against the nectrotrophic soil-borne fungus *R. solani* (Mishra et al. 2006). Gallic acids provide antimicrobial activity (Binutu and Cordell 2000). Gallotannins are a derivative of gallic acid which get converted during accumulation and provide defense to plants against bacteria and fungi (Singh et al. 2002). Cinnamic and ferulic acids ascended from the shikimic acid pathway and are reported to be antioxidant and antifungal, respectively (Madhavi et al. 1997). Cinnamic acid serves as a precursor for the production of ferulic acid and is a key product of the phenylpropanoid pathway, which plays a crucial role in providing host resistance during pathogenic stress (Singh and Prithiviraj 1997).

5.3.3.2 Alkaloids

The majority of alkaloids are regarded as derivatives of certain amino acids, e.g., tryptophan, lysine, ornithine, phenylalanine, and tyrosine. Precursors of terpenoid and steroid are incorporated into the carbon skeletons of alkaloids. Till now, around 3000 different phytoalkaloids are known, which occur dissolved as cations in plant sap. They are mostly accumulated in the peripheral region of leaves, bark, or fruit, which can be shed (Levinson 1976). Common alkaloids can be found in the Liliaceae, the Leguminosae, the Amaryllidaceae, and the Solanaceae plant families which can be important resistance factors against herbivorous pests (Petterson et al. 1991). Because of their general toxicity and deterrence capability, alkaloids are believed to be key defensive elements against predators, especially mammals (Hartmann 1991; Robinson 1980). Death of a large number of animals in USA is reported due to the ingestion of plants containing alkaloids. A large number of grazing livestock is infected by consumption of alkaloids-containing plants such as lupines (*Lupinus*) and larkspur (*Delphinium*) (Keeler 1975).

5.4 Extraction and Isolation of Bioactive Compounds from Plant Extracts

It is an ancient thought that plant extracts have great healing power, and these phytocompounds have recently attracted interest because of their versatile applications (Bariş et al. 2006). Plant species contain various metabolites. However, only a small percentage of these phytochemicals have been investigated around the globe (Hostettmann and Wolfender 1997). Use of chemical pesticides induces environmental hazards in agricultural systems. So, the use of phytocompounds as an antimicrobial agent can be the best biorational alternative today (Tiwari et al. 2007). As large number of plant species are available for future studies, it is vital to have effective methods to evaluate the efficacy of plant origin antimicrobial agent and the identification of biologically active principles involved in it (Tanaka et al. 2006). Plants contain a number of metabolites which expose only a very narrow range of their constituents. Thus, the potential of higher plants is still largely unexplored as sources of new drugs (Hamburger and Hostettmann 1991). The selection of plant species for screening for biologically active constituents is a crucial factor in the investigation.

During searching for active phytocompounds, it is essential to verify the plant varieties for the success of the study. Targeted grouping of plant material is based on considering chemotaxonomic interaction and utilization of current ethnomedical information. The use of different technologies has allowed rich isolation of various fungicidal, larvicidal, and molluscicidal products (Hostettmann and Wolfender 1997).

Thousands of bioactive compounds from plants are found to be safe and have less adverse effects due to their beneficial biological activity, e.g., antimicrobial, antioxidant, and wound-healing activity. The leading steps to exploit the phytochemicals from plant resources include extraction, isolation, screening, identification, and characterization of bioactive compounds. Plant extracts contains multicomponent mixtures of bioactive compounds with different polarities, which still poses problems in identifying and characterizing phytochemicals and their separation. Purification of phytochemicals most commonly includes member of chromatographic techniques and other different purification methods to identify phytochemicals (Sasidharan et al. 2011). To extract the desired phytochemical from a plant, sample preparation is a crucial leading step in the analysis of plant or herbs for further separation and characterization of those isolated bioactive compounds (Huie 2002).

5.5 Metabolomics Tools and Their Application in Plants and Plant–Host Interactions

Metabolomics is now a briskly developing technology. With the help of specialized bioinformatics tools and data mining tools, metabolomics, like transcriptomics and proteomics, generates a huge amount of data. Metabolomics tools can possibly lead to identifying many of the compounds in plants undergoing stress (Shulaev et al. 2008). Current methodologies used in plant metabolomics comprise metabolic fingerprinting, metabolite profiling, and targeted and nontargeted detection analysis (Halket et al. 2004; Bajad and Shulaev 2007; Fiehn 2002); these are described below.

5.5.1 Metabolic Profiling and Fingerprinting

Research in metabolomics includes metabolite profiling and fingerprinting approaches. "Metabolic profiling" was coined during the 1970s (Horning and Horning 1971), and it is used to identify and quantify metabolites associated with their certain metabolic pathways or similarities in their compound classes. It involves common chromatographic separation techniques like liquid chromatography coupled with MS (LC-MS) or gas chromatography coupled with MS (GC-MS) to detect, quantify, and, if at all possible, identify the metabolites in an extract. In metabolite fingerprinting, metabolite profiles are obtained from simple cellular extracts or crude samples through rapid and high-throughput methods. Metabolite fingerprinting involves techniques like NMR (nuclear magnetic resonance spectroscopy) (Krishnan et al. 2004), Raman spectroscopy, Fourier-transform infrared spectroscopy (FTIR) (Johnson et al. 2003), MS (Goodacre et al. 2003), and electrospray ionization (ESI)-MS to detect all the metabolites present in a sample irrespective of their identification (Allwood et al. 2008).

Liquid chromatography mass spectrometry (LC-MS) is preferred for the analysis of flavonoids, phenylpropanoids, and alkaloids. Using GC-MS, fatty acids were found as the key component to provide resistance to gall midge rice varieties (Agarrwal et al. 2014). By using liquid chromatography tandem mass spectrometry (LC-MS/MS), the identification and quantification of more than 90 flavonoids were reported. It also studies the occurrence of their biosynthesis in various rice tissues during different developmental stages (Dong et al. 2014). As compared with the common cultivars, it was found that tomatoes contain 70-fold higher flavonoids by using LC/photodiode array detection along with liquid chromatography, quadrupole time of flight mass spectrometer (LC-QTOF-MS), and direct infusion QTOF-MS (Hall et al. 2002).

Capillary electrophoresis–mass spectrometry (CE-MS) as well as liquid chromatography–mass spectrometry (LC-MS) offers a better alternative for nonvolatile compounds. Capillary electrophoresis–mass spectrometry (CE-MS) has now come into view as a powerful tool for the analysis of charged molecules. CE-MS provides separation of metabolites based on charge and size and then it is detected using MS by observing over a large range of *m*/*z* values. CE-MS provides very high resolution, and nearly any charged species can be able to infuse into MS (Soga et al. 2003). The application of LC-MS in metabolomics is gradually growing after its recent acceptance of the ultra-performance liquid chromatography technology, which helps it to increase the efficiency of separation and decreases the analysis time of metabolites (Giri et al. 2007; Granger et al. 2007).

GC-MS is biased in contrast to nonvolatile high molecular weight metabolites and is functional towards polar nonvolatile metabolites (e.g., organic acids, amino acids, and sugars) that are volatile up to 250 °C through chemical derivatization (Allwood et al. 2008). In 2000, Roessner et al. (2000) described two stages of derivatization process for the analysis of plant extracts using GC-MS. Firstly, O-alkylhydroxylamine transforms the carbonyl group of the sample to oximes for thermal stabilization and then it is treated with a silylating compound, e.g., *N*-methyl-*N*-(trimethylsilyl) trifluoro-acetamide, which leads to the formation of volatile trimethyl-silyl esters (Roessner et al. 2000). Electron impact (EI), facilitating GC-MS ion formation, provides independent exclusion of the sample's solvent before vaporized sample being cleared into the ionization source which allows steady electron flow and thus ionizes the vaporized molecule (Gross 2006). Detection of mass is conducted by QTOF-MS or ion-trap-based mass analyzers. A single QTOF-mass analyzer requires an hour of chromatographic time to provide a standard separation of a complex metabolite (Dunn and Ellis 2005). Metabolites can be identified by tandem MS (MS-MS) which is responsible for metabolite fragmentation through collision with an inert gas like argon and causes collision-induced dissociation (Wysocki et al. 2005).

NMR (nuclear magnetic resonance) spectroscopy is a nondestructive technique which requires least sample preparation, and is presently also considered high throughput (hundreds of samples per day). NMR uses nuclei with odd mass or atomic numbers, which behave like magnets and intercommunicate with an external magnetic field by a method called nuclear spin (Kitayama and Hatada 2013). In particular, ^{1}H NMR has been extensively used for metabolites profiling in clinical samples (Holmes et al. 2000; Nicholson and Wilson 1989) and also has been functional towards complex compounds exuded out from the roots of cereals (Fan et al. 2001). Unlike GC-MS, which senses only volatilized compounds, ^{1}H NMR can instantaneously detect all compounds bearing proton in a sample. It covers mostly organic compounds, such as ethers, amino acids, fatty acids, carbohydrates, amines, and lipids esters, present in plant tissues. ^{1}H-NMR provide a nonbiased fingerprint in contrast with other metabolomics approaches, and therefore NMR is now evolving as one of the standard metabolic profiling platforms (Ward et al. 2003).

5.5.2 Targeted and Nontargeted Detection Analysis

Separation methods of numerous analytes from a particular sample have now been established. However, for effective application of such methods, it needs detectors which is accomplished of fast data-acquisition rates along with high specificity and sensitivity. Two major means of MS-based metabolomics are targeted and nontargeted detection analysis. Analytes in the targeted detection mode are predetermined, having a definite mass filter allotted to a specific analyte (Bajad and Shulaev 2007). Standard methods have been developed to identify specific members of a compound class while ignoring others. For example, polyamines thought to be involved in various important plant processes, e.g., drought stress, and quantification of polyamines in different plant species in response to various stimuli or environmental conditions has been developed (Bouchereau et al. 2000). Targeted analysis also results in comparative metabolite profiling of a huge number of identified metabolites. SRM (Selected reaction monitoring) provides high meticulosity and has been successfully used to quantitate a number of analytes at the same time. For example, SRM can analyze more than 100 metabolites in a single chromatographic run, based on highly parallel targeted assays (Bajad and Shulaev 2007; Bajad et al. 2006). Target analysis will continue to be the most wide-spread system in different areas of biological research. However, in case of functional genomics studies it has restricted use, because the levels of target analytes may be changed by unexpected reasons which cannot be understood without comprehensive approaches. Thus, a wider analysis of metabolic modifications is required to limit overinterpretation of data (Fiehn 2001).

The problem with targeted detection is that it is not an actual beneficial approach as it is not quite practical to assemble SRMs for all molecules of interest present in the sample containing numerous analytes. Thus, it cannot detect analytes with no SRMs. Also, it is difficult to predict the compositions of the sample. Therefore, in such cases, nontargeted detection analysis is used to cover a broad array of analytes and used to detect and find unknown or novel molecules (Hong et al. 2003; Tohge et al. 2005).

In nontargeted detection analysis, to detect common molecules of specific molecular mass range, scanning of mass spectrometer is done over a set m/z 100–1000 in both positive and negative ionization modes. For comparative profiling, full-scan mode acquired data along with low-resolution instruments are most commonly used. Data obtained in full-scan mode along with automated MS to MS/MS switching provide added information about the elemental composition and arrangement of fragments of the analytes, as well as the unknown components by using an accurate mass instrument (e.g., Fourier transform MS or qTOF). Obtained data are then subjected to library (e.g., the NIST library) search to identify the unknown compounds (Bobeldijk et al. 2001). Nontargeted analysis provides an unbiased detection method of the chemical nature of the sample which results in a holistic approach to detect and identify unexpected or unknown metabolites which can be important in environmental and pharmaceutical analysis (Bajad and Shulaev 2007). There are some examples of a nontargeted analysis in pharmaceutical and environmental analysis. Ibanez et al., using SPE-LC/QTOF-MS along with data processing, identified six unknown compounds from environmental waters (Ibáñez et al. 2005). A successful study of drug metabolites in pharmaceutical has also been outlined using application of nontargeted analysis (Idborg et al. 2004).

5.5.3 Data Processing and Analysis

Regardless of any analytical technique used, data analysis forms an essential part. The raw data must be preprocessed to transform them to a readable format. The modified data can be subjected to data reduction to facilitate the use of only appropriate input variables in the succeeding data analysis (Brown et al. 2005). Before analyzing data from most analytical instruments statistically, consequential preprocessing is required, and standardization of techniques is necessary. A lot of researchers put emphasis on requirement of post-sampling techniques such as deconvolution, noise reduction, internal standards reference, alignment of profile, and labeling of peak by using spectral libraries (Hall et al. 2002).

Large volumes of data produced from metabolome analysis are analyzed by the instruments to detect small signals with high resolution. For this metabolome analysis, programmed software is required to detect peaks of raw NMR or MS data to arrange the peaks in order amongst samples and to detect and measure the quantity of each metabolite (Fukushima et al. 2009; Fernie and Schauer 2009; Go 2010). Numerous statistical methods are used for metabolomics data. PCA (principal component analysis) is a multivariate analysis usually used in the study of metabolomics. PCA provides an outline of every sample or interpretation in a data and highlights the variance between the complex metabolites in each sample (Catchpole et al. 2005; Baker et al. 2006; Dixon 2003; Kim et al. 2006; Ku et al. 2009b). In addition to this, other statistical analytical methods are used to analyze metabolomic datasets, which are PLS-DA (partial least squares discriminant analysis) (Jonsson et al. 2004; Ku et al. 2009a; Kusano et al. 2007), HCA (hierarchical cluster analysis) (Grata et al. 2007; Parveen et al. 2007), and BL-SOM (batch-learning self-organizing map analysis) (Hirai et al. 2004).

Bioinformatics is the key supporter to gather information and make sense of the data. Currently, in the field of metabolomics, bioinformatics mainly focuses on the metabolic pathway simulation and construction of models (Fiehn et al. 2001). Schuster et al. (2000) tried to shed new light on the concept of identifying possible metabolic pathway leading to a given element. Databases like the Kyoto Encyclopedia of Genes and Genomes (KEGG) can serve as a model in providing information on the combination of simulated pathway with pathway databases. KEGG provides knowledge of systematic analysis of functions of gene based on the networks of genes and molecules. It develops and provides several computational tools for reforming biochemical pathways from a complete genomic sequence and predicting the regulatory networks of gene from the profiles of gene expression. The KEGG databases are updated on a daily basis and are available without restrictions (http://www.genome.ad.jp/kegg/) (Ogata et al. 1999).

Data handling and analysis of metabolomics using "omic" technologies have been improved dramatically in recent years. They help in the detection of specific metabolites in a biological sample in a nonbiased and nontargeted way. Compared to omics technology applied for the study of genomics, transcriptomics, and proteomics, metabolomics has numerous theoretical advantages over the other omics approaches (Horgan and Kenny 2011). The functional genomics databases include DOME (http://medicago.vbi.vt.edu), MetNetDB (http://www.metnetdb.org/MetNet_db.htm), data model for plant metabolomics research ArMet (http://www.armet.org/) (Shulaev et al. 2008), and pathway databases and pathway viewers like KEGG (http://www.genome.ad.jp/kegg/), KaPPA-View (http://kpv.kazusa.or.jp/kappa-view/) (Tokimatsu et al. 2005), MetaCyc (http://metacyc.org/) (Caspi et al. 2006), AraCyc (http://www.Arabidopsis.org/tools/aracyc/) (Zhang et al. 2005), BioCyc (http://biocyc.org) (Paley and Karp 2006), MapMan (http://gabi.rzpd.de/projects/MapMan/) (Thimm et al. 2004), BioPathAT (http://www.ibc.wsu.edu/research/lange/public%5Ffolder/) (Lange and Ghassemian 2005), and the Atomic Reconstruction of Metabolism database (http://www.metabolome.jp/) (Yamazaki et al. 2004).

Some of the selected open-access bioinformatics tools for multifarious LC-MS data analysis applied to environmental and pharmaceutical analysis are BL-SOM (http://prime.psc.riken.jp/?action=blsom_index) (Kanaya et al. 2001), MZmine (http://mzmine.sourceforge.net/) (Katajamaa and Orešič 2005), XCMS (http://metlin.scripps.edu/download/) (Smith 2013), MSFACTs (http://noble.org/) (Duran et al. 2003), MeMo (http://dbkgroup.org/memo/) (Spasić et al. 2006), and MET-IDEA (http://noble.org/) (Broeckling et al. 2006).

LC-MS is broadly used in both proteomics and metabolomics. Integrated non-target metabolomics (LC-MS/MS) and proteomics (2D gel electrophoresis) have been applied in wheat genotype Nyubai. Gunnaiah et al. (2012) found that Fusarium head blight resistance locus, *Fhb1*, provides resistance against the spread of *F. graminearum* within the spikes. The involvement of *Fhb1* in providing resistance in wheat, in response to *F. graminearum*, is mainly recognized due to the triggering of fatty acid, terpenoid, and phenylpropanoid metabolic pathways. This study used NILs (near isogenic lines) which set a good example to demonstrate that proteometabolomic studies are not delimited up to the genetics of a given QTL (Gunnaiah et al. 2012).

Biotechnological approaches require genetic modification to govern the assembly of specific metabolites in plants, to progress food quality, to increase their adaptation against environmental stress, and to increase crop yield. Unluckily, these approaches do not essentially lead to an estimated result because of the complex mechanisms required for the plants metabolic regulation. Metabolites such as inositol, salicylic acid, ethylene, and jasmonic acid have been linked to plant defense signaling pathways against biotic stress (Kushalappa and Gunnaiah 2013). Many resistances related (RR) metabolites which were identified based upon nontargeted analysis to possess antimicrobial properties (Ahuja et al. 2012; Ballester et al. 2013) are discussed in Table 5.1. Many bioinformatic tools are accessible for LC-MS data processing for mass spectral output processing and compound annotation mentioned in Table 5.2.

5.6 Conclusion and Future Aspects

Currently, plant–pathogen interactions exemplify the utmost biochemically complex and thought-provoking scenarios being evaluated by metabolomics approaches. For example, there is complication in identifying which metabolites are procured from the plant and which metabolites are interacting from the pathogen side. Phytocompounds are involved in resistance mechanisms of plant. Until now, there is much information on the mechanism of resistance of plants against invading pathogens, but very less is known about the pathogenicity of invading pathogens. Toxins certainly play a role in the pathogenicity factor during plant–pathogen interaction. Plant components which have a negative effect upon the growth and development or survival of another organism can be considered as toxins. The way plants store their toxins are often critical for their effectiveness. Some plant species

Table 5.1 List of resistance related (RR) metabolites identified in plants against biotic stresses, following nontarget metabolomics approaches

Plants	Disease/pathogens	Resistance related biochemicals	Chemical groups/pathways	Disease resistance mechanism	References
Wheat	*F. graminearum*	Sinapic acid, deoxyphodophyllotoxin, phyllanthusmin, linoleic acid, 13(S)hydroperoxylinolenic acid, deoxyloganate	Phenylpropanoids Fatty acids	Antimicrobial	Gunnaiah et al. (2012)
Arabidopsis	*Golovinomyces orontii*	Camalexin Indol-3-yl-methylamine	Indole group	Phytoanticipin/ phytoalexin	Consonni et al. (2010)
Citrus	Huanglongbing (*Candidatus* spp., *Liberibacter* spp.)	L-glycine Mannose	Amino acid Sugar	Antimicrobial	Cevallos-Cevallos et al. (2012)
Grape	Fungi	Caffeic acid Inositol Alanine, glutamine, and glutamate	Phenylpropanoid sugar Amino acids	Antimicrobial Signal transduction Ammonia recycling liberated by PAL	Figueiredo et al. (2008)
Barley	*Fusarium graminearum*	*p*-coumaric acid, sinapate Kaempferol-3-*O*-β-d-glucopyranosyl, kaempferol-*O*-rutinoside, Flavonoid glucosides, Capric acid, lauric acid	Phenylpropanoids Flavonoids Fatty acids	Signaling molecules Antimicrobial Cell wall strengthening	Bollina et al. (2010)

Referred from Kushalappa and Gunnaiah (2013)

Table 5.2 Open access databases for metabolite search and compound annotation

Database	Features	Weblink
PRIMe	Provide web-based service for metabolomics and transcriptomics tools. It measures standard metabolites through GC/MS, CE/MS, LC/MS, and multi-dimensional NMR spectroscopy, unique tools for transcriptomics, metabolomics, and integrated analysis of omics data	http://prime.psc.riken.jp/
PlantCyc	Plant metabolic pathway database for compounds, enzymes, genes, and pathways intricated in primary and secondary metabolism. Tools for BLAST, user input pathway generation and comparative analysis Downloadable reference pathways for rice, Arabidopsis, cassava corn, papaya, grape, poplar, potato	http://www.plantcyc.org/
METLIN	Over 64,000 structures, tandem mass spectra of more than about 10,000 metabolites, external link to other databases, batch search	http://metlin.scripps.edu/
ReSpect	Provides phytochemicals tandem mass spectral database	http://spectra.psc.riken.jp/
KNAPSACK	Database for metabolites-species relation, search options; organism name, organism taxonomic tree, metabolite name, molecular weight, formula, batch search	http://kanaya.aist-nara.ac.jp/KNApSAcK/
McGill-MD	Metabolites related to plant biotic stress resistance, *in planta* fragmentations using LC-LTQ-orbitrap and annotated with in silico fragmentation	http://metabolomics.mcgill.ca
ChEBI	A database, dictionary, and ontology of manually annotated small molecules	http://www.ebi.ac.uk/chebi/init.do

Referred from Kushalappa and Gunnaiah (2013)

store toxins in resin ducts, laticifers (Dussourd and Hoyle 2000), or glandular trichomes (Hallahan 2000) from where the toxins are released in huge amounts as soon as these structures are broken by pathogens. For instance, hydrogen cyanide released from cyanogenic glycosides inhibits cellular respiration (Jones et al. 2000), saponins disrupt cellular membranes (Osbourn 1996), and cardenolides are specific Na^+/K^+-ATPase inhibitors (Bramer et al. 2015). One substantial challenge for the study of plant metabolomics is it lacks fully defined and interpreted metabolome for any plant species. It is estimated that the plant kingdom produces around 90,000–200,000 diverse metabolites. Still, the definite number of metabolites present in independent species of a plant is unknown (Fiehn et al. 2001). A reiterate theme in all aspects of spheres of plant–pathogen interaction is the ability of each participant to recognize and respond to cues generated by the other. Still, understanding of molecular recognition and response systems, receptors involved in plant perception of pathogens in its infancy, and many other important questions remain unanswered. In future, research focusing on the identification of effector molecules from pathogens and their mechanism of action is likely to set a new stage in plant–pathogen interaction.

Acknowledgments We thank the research funding support from both SERB, Govt. of India (sanction number: YSS2014/000142), and Science and Technology, Govt. of Odisha (Sanction number:27552800232014/202808), to BBS.

References

Agarrwal R, Bentur JS, Nair S (2014) Gas chromatography mass spectrometry based metabolic profiling reveals biomarkers involved in rice-gall midge interactions. J Integr Plant Biol 56:837–848

Aharoni A, Jongsma MA, Kim T-Y, Ri M-B, Giri AP, Verstappen FW, Schwab W, Bouwmeester HJ (2006) Metabolic engineering of terpenoid biosynthesis in plants. Phytochem Rev 5:49–58

Ahuja I, Kissen R, Bones AM (2012) Phytoalexins in defense against pathogens. Trends Plant Sci 17:73–90

Akula R, Ravishankar GA (2011) Influence of abiotic stress signals on secondary metabolites in plants. Plant Signal Behav 6:1720–1731

Allwood JW, Ellis DI, Goodacre R (2008) Metabolomic technologies and their application to the study of plants and plant–host interactions. Physiol Plant 132:117–135

Aniszewski T (2015) Alkaloids: chemistry, biology, ecology, and applications. Elsevier Science, Amsterdam

Bajad S, Shulaev V (2007) Highly-parallel metabolomics approaches using LC-MS2 for pharmaceutical and environmental analysis. TrAC Trends Anal Chem 26:625–636

Bajad SU, Lu W, Kimball EH, Yuan J, Peterson C, Rabinowitz JD (2006) Separation and quantitation of water soluble cellular metabolites by hydrophilic interaction chromatography-tandem mass spectrometry. J Chromatogr A 1125:76–88

Baker JM, Hawkins ND, Ward JL, Lovegrove A, Napier JA, Shewry PR, Beale MH (2006) A metabolomic study of substantial equivalence of field-grown genetically modified wheat. Plant Biotechnol J 4:381–392

Balandrin MF, Klocke JA, Wurtele ES, Bollinger WH (1985) Natural plant chemicals: sources of industrial and medicinal materials. Science 228:1154–1160

Ballester A-R, Lafuente MT, de Vos RC, Bovy AG, González-Candelas L (2013) Citrus phenylpropanoids and defence against pathogens. Part I: metabolic profiling in elicited fruits. Food Chem 136:178–185

Barcelo J, Poschenrieder C (2002) Fast root growth responses, root exudates, and internal detoxification as clues to the mechanisms of aluminium toxicity and resistance: a review. Environ Exp Bot 48:75–92

Bariş Ö, Güllüce M, ŞAHİN F, Özer H, Kiliç H, Özkan H, Sökmen M, Özbek T (2006) Biological activities of the essential oil and methanol extract of Achillea biebersteinii Afan. (Asteraceae). Turk J Biol 30:65–73

Barrett G (2012) Chemistry and biochemistry of the amino acids. Springer, Dordrecht

Bennett RN, Wallsgrove RM (1994) Secondary metabolites in plant defence mechanisms. New Phytol 127:617–633

Binutu OA, Cordell GA (2000) Gallic acid derivatives from mezoneuron benthamianum leaves. Pharm Biol 38:284–286

Bobeldijk I, Vissers J, Kearney G, Major H, Van Leerdam J (2001) Screening and identification of unknown contaminants in water with liquid chromatography and quadrupole-orthogonal acceleration-time-of-flight tandem mass spectrometry. J Chromatogr A 929:63–74

Bollina V, Kumaraswamy GK, Kushalappa AC, Choo TM, Dion Y, Rioux S, Faubert D, Hamzehzarghani H (2010) Mass spectrometry-based metabolomics application to identify quantitative resistance-related metabolites in barley against Fusarium head blight. Mol Plant Pathol 11:769–782

Bones AM, Rossiter JT (2006) The enzymic and chemically induced decomposition of glucosinolates. Phytochemistry 67:1053–1067

Bouchereau A, Guénot P, Larher F (2000) Analysis of amines in plant materials. J Chromatogr B Biomed Sci Appl 747:49–67

Bramer C, Dobler S, Deckert J, Stemmer M, Petschenka G (2015) Na+/K+-ATPase resistance and cardenolide sequestration: basal adaptations to host plant toxins in the milkweed bugs (Hemiptera: Lygaeidae: Lygaeinae). Proc R Soc B 282:20142346

Broeckling CD, Reddy IR, Duran AL, Zhao X, Sumner LW (2006) MET-IDEA: data extraction tool for mass spectrometry-based metabolomics. Anal Chem 78:4334–4341

Brown M, Dunn WB, Ellis DI, Goodacre R, Handl J, Knowles JD, O'Hagan S, Spasić I, Kell DB (2005) A metabolome pipeline: from concept to data to knowledge. Metabolomics 1:39–51

Caspi R, Foerster H, Fulcher CA, Hopkinson R, Ingraham J, Kaipa P, Krummenacker M, Paley S, Pick J, Rhee SY (2006) MetaCyc: a multiorganism database of metabolic pathways and enzymes. Nucleic Acids Res 34:D511–D516

Catchpole GS, Beckmann M, Enot DP, Mondhe M, Zywicki B, Taylor J, Hardy N, Smith A, King RD, Kell DB (2005) Hierarchical metabolomics demonstrates substantial compositional similarity between genetically modified and conventional potato crops. Proc Natl Acad Sci 102:14458–14462

Cevallos-Cevallos JM, Futch DB, Shilts T, Folimonova SY, Reyes-De-Corcuera JI (2012) GC–MS metabolomic differentiation of selected citrus varieties with different sensitivity to citrus huanglongbing. Plant Physiol Biochem 53:69–76

Conn EE (1969) Cyanogenic glycosides. J Agric Food Chem 17(3):519–526

Consonni C, Bednarek P, Humphry M, Francocci F, Ferrari S, Harzen A, van Themaat EVL, Panstruga R (2010) Tryptophan-derived metabolites are required for antifungal defense in the arabidopsis. Plant Physiol 152(3):1544–1561

Cooper J (2004) Dominant role of flavonoids among signal molecules involved in the formation of legume-rhizobia symbiosis. Polyphenol Commun 87

Croteau R, Kutchan TM, Lewis NG (2000) Natural products (secondary metabolites). Biochem Mol Biol Plants 24:1250–1319

Demain AL (2000) Microbial biotechnology. Trends Biotechnol 18:26–31

Dixon RA (2001) Natural products and plant disease resistance. Nature 411:843

Dixon RA (2003) Phytochemistry meets genome analysis, and beyond. Phytochemistry 62:815–816

Dong X, Chen W, Wang W, Zhang H, Liu X, Luo J (2014) Comprehensive profiling and natural variation of flavonoids in rice. J Integr Plant Biol 56:876–886

Dunn WB, Ellis DI (2005) Metabolomics: current analytical platforms and methodologies. TrAC Trends Anal Chem 24:285–294

Duran AL, Yang J, Wang L, Sumner LW (2003) Metabolomics spectral formatting, alignment and conversion tools (MSFACTs). Bioinformatics 19:2283–2293

Dussourd DE, Hoyle AM (2000) Poisoned plusiines: toxicity of milkweed latex and cardenolides to some generalist caterpillars. Chemoecology 10:11–16

Eder J, Cosio EG (1994) Elicitors of plant defense responses. In: International review of cytology. Elsevier, New York, pp 1–36

Edreva A, Velikova V, Tsonev T (2007) Phenylamides in plants. Russ J Plant Physiol 54:287–301

Eftekhari-Sis B, Zirak M, Akbari A (2013) Arylglyoxals in synthesis of heterocyclic compounds. Chem Rev 113:2958–3043

Fahey G Jr, Jung H (1989) Phenolic compounds in forages and fibrous feedstuffs. In: Toxicants of plant origin, vol vol 4. CRC, Boca Raton, pp 123–190

Fahey JW, Zalcmann AT, Talalay P (2001) The chemical diversity and distribution of glucosinolates and isothiocyanates among plants. Phytochemistry 56:5–51

Fan TW-M, Lane AN, Shenker M, Bartley JP, Crowley D, Higashi RM (2001) Comprehensive chemical profiling of gramineous plant root exudates using high-resolution NMR and MS. Phytochemistry 57:209–221

Feeny P (1970) Seasonal changes in oak leaf tannins and nutrients as a cause of spring feeding by winter moth caterpillars. Ecology 51:565–581

Fernie AR, Schauer N (2009) Metabolomics-assisted breeding: a viable option for crop improvement? Trends Genet 25:39–48

Fiehn O (2001) Combining genomics, metabolome analysis, and biochemical modelling to understand metabolic networks. Int J Genomics 2:155–168

Fiehn O (2002) Metabolomics—the link between genotypes and phenotypes. In: Functional genomics. Springer, Berlin, pp 155–171

Fiehn O, Kloska S, Altmann T (2001) Integrated studies on plant biology using multiparallel techniques. Curr Opin Biotechnol 12:82–86

Figueiredo A, Fortes AM, Ferreira S, Sebastiana M, Choi YH, Sousa L, Acioli-Santos B, Pessoa F, Verpoorte R, Pais MS (2008) Transcriptional and metabolic profiling of grape (Vitis vinifera L.) leaves unravel possible innate resistance against pathogenic fungi. J Exp Bot 59:3371–3381

Fraire-Velázquez S, Balderas-Hernández VE (2013) Abiotic stress in plants and metabolic responses. In: Abiotic stress-plant responses and applications in agriculture. InTech. https://doi.org/10.5772/54859. Available from https://www.intechopen.com/books/abiotic-stress-plant-responses-and-applications-in-agriculture/abiotic-stress-in-plants-and-metabolic-responses

Fukushima A, Kusano M, Redestig H, Arita M, Saito K (2009) Integrated omics approaches in plant systems biology. Curr Opin Chem Biol 13:532–538

Gapasin R, Valdez R, Mendoza E (1988) Phenolic involvement in sweet potato resistance to Meloidogyne incognita and M. javanica. Ann Trop Res 10:63–73

Gill SS, Tuteja N (2010) Reactive oxygen species and antioxidant machinery in abiotic stress tolerance in crop plants. Plant Physiol Biochem 48:909–930

Giri S, Krausz KW, Idle JR, Gonzalez FJ (2007) The metabolomics of (±)-arecoline 1-oxide in the mouse and its formation by human flavin-containing monooxygenases. Biochem Pharmacol 73:561–573

Gleadow RM, Møller BL (2014) Cyanogenic glycosides: synthesis, physiology, and phenotypic plasticity. Annu Rev Plant Biol 65:155–185

Glória MBA (2005) Bioactive amines. In: Handbook of food science, technology, and engineering, vol vol 4. Set. 15. CRC, Boca Raton

Go EP (2010) Database resources in metabolomics: an overview. J Neuroimmune Pharmacol 5:18–30

Goodacre R, York EV, Heald JK, Scott IM (2003) Chemometric discrimination of unfractionated plant extracts analyzed by electrospray mass spectrometry. Phytochemistry 62:859–863

Goyal S, Lambert C, Cluzet S, Mérillon J, Ramawat KG (2012) Secondary metabolites and plant defence. In: Plant defence, biological control. Springer, Dordrecht, pp 109–138

Granger JH, Williams R, Lenz EM, Plumb RS, Stumpf CL, Wilson ID (2007) A metabonomic study of strain-and age-related differences in the Zucker rat. In: Rapid communications in mass spectrometry: an international journal devoted to the rapid dissemination of up-to-the-minute research in mass spectrometry, vol 21. Wiley, Chichester, pp 2039–2045

Grata E, Boccard J, Glauser G, Carrupt PA, Farmer EE, Wolfender JL, Rudaz S (2007) Development of a two-step screening ESI-TOF-MS method for rapid determination of significant stress-induced metabolome modifications in plant leaf extracts: the wound response in Arabidopsis thaliana as a case study. J Sep Sci 30:2268–2278

Gross JH (2006) Mass spectrometry: a textbook. Springer, Berlin

Gunnaiah R, Kushalappa AC, Duggavathi R, Fox S, Somers DJ (2012) Integrated metabolo-proteomic approach to decipher the mechanisms by which wheat QTL (Fhb1) contributes to resistance against Fusarium graminearum. PLoS One 7:e40695

Halket JM, Waterman D, Przyborowska AM, Patel RK, Fraser PD, Bramley PM (2004) Chemical derivatization and mass spectral libraries in metabolic profiling by GC/MS and LC/MS/MS. J Exp Bot 56:219–243

Hall R, Beale M, Fiehn O, Hardy N, Sumner L, Bino R (2002) Plant metabolomics: the missing link in functional genomics strategies. American Society of Plant Biologists, Rockville, MD

Hallahan D (2000) Monoterpenoid biosynthesis in glandular trichomes of labiate plants. Adv Bot Res 31:77–120

Hamburger M, Hostettmann K (1991) 7. Bioactivity in plants: the link between phytochemistry and medicine. Phytochemistry 30:3864–3874

Harborne JB (1984) Phenolic compounds. In: Phytochemical methods. Springer, Dordrecht, pp 37–99

Harborne J (1989) General procedures and measurement of total phenolics. Methods Plant Biochem 1:1–28

Hartmann T (1991) Alkaloids in herbivores; their interaction with secondary plant metabolites. In: Rosenthal GA, Berenbaum MR (eds) The chemical participants, vol I, 2 edn. Academic, San Diego

Hegnauer R (1988) Biochemistry, distribution and taxonomic relevance of higher plant alkaloids. Phytochemistry 27:2423–2427

Hirai MY, Yano M, Goodenowe DB, Kanaya S, Kimura T, Awazuhara M, Arita M, Fujiwara T, Saito K (2004) Integration of transcriptomics and metabolomics for understanding of global responses to nutritional stresses in Arabidopsis thaliana. Proc Natl Acad Sci 101:10205–10210

Holmes E, Nicholls AW, Lindon JC, Connor SC, Connelly JC, Haselden JN, Damment SJ, Spraul M, Neidig P, Nicholson JK (2000) Chemometric models for toxicity classification based on NMR spectra of biofluids. Chem Res Toxicol 13:471–478

Hong S, Gronert K, Devchand PR, Moussignac R-L, Serhan CN (2003) Novel docosatrienes and 17S-resolvins generated from docosahexaenoic acid in murine brain, human blood, and glial cells autacoids in anti-inflammation. J Biol Chem 278:14677–14687

Hopkins RJ, van Dam NM, van Loon JJ (2009) Role of glucosinolates in insect-plant relationships and multitrophic interactions. Annu Rev Entomol 54:57–83

Horgan RP, Kenny LC (2011) 'Omic'technologies: genomics, transcriptomics, proteomics and metabolomics. Obstet Gynaecol 13:189–195

Horning E, Horning M (1971) Human metabolic profiles obtained by GC and GC/MS. J Chromatogr Sci 9:129–140

Hostettmann K, Wolfender JL (1997) The search for biologically active secondary metabolites. Pestic Sci 51:471–482

Huie CW (2002) A review of modern sample-preparation techniques for the extraction and analysis of medicinal plants. Anal Bioanal Chem 373:23–30

Hungria M, Stacey G (1997) Molecular signals exchanged between host plants and rhizobia: basic aspects and potential application in agriculture. Soil Biol Biochem 29:819–830

Hunt S (1985) The non-protein amino acids. In chemistry and biochemistry of the amino acids. Springer, Wien, pp 55–138

Ibáñez M, Sancho JV, Pozo ÓJ, Niessen W, Hernández F (2005) Use of quadrupole time-of-flight mass spectrometry in the elucidation of unknown compounds present in environmental water. In: Rapid communications in mass spectrometry: an international journal devoted to the rapid dissemination of up-to-the-minute research in mass spectrometry, vol 19. Wiley, Chichester, pp 169–178

Idborg H, Edlund PO, Jacobsson SP (2004) Multivariate approaches for efficient detection of potential metabolites from liquid chromatography/mass spectrometry data. Rapid Commun Mass Spectrom 18:944–954

Irisarri P, Milnitsky F, Monza J, Bedmar E (1996) Characterization of rhizobia nodulating Lotus subbiflorus from Uruguayan soils. Plant Soil 180:39–47

Johnson HE, Broadhurst D, Goodacre R, Smith AR (2003) Metabolic fingerprinting of salt-stressed tomatoes. Phytochemistry 62:919–928

Jones PR, Andersen MD, Nielsen JS, Høj PB, Møller BL (2000) The biosynthesis, degradation, transport and possible function of cyanogenic. Evol Metab Pathways 34:191

Jonsson P, Gullberg J, Nordström A, Kusano M, Kowalczyk M, Sjöström M, Moritz T (2004) A strategy for identifying differences in large series of metabolomic samples analyzed by GC/MS. Anal Chem 76:1738–1745

Kanaya S, Kinouchi M, Abe T, Kudo Y, Yamada Y, Nishi T, Mori H, Ikemura T (2001) Analysis of codon usage diversity of bacterial genes with a self-organizing map (SOM): characterization of horizontally transferred genes with emphasis on the E. coli O157 genome. Gene 276:89–99
Katajamaa M, Orešič M (2005) Processing methods for differential analysis of LC/MS profile data. BMC Bioinforma 6(1):179
Keeler R (1975) Toxins and teratogens of higher plants. Lloydia 38(1):56–86
Kim JK, Bamba T, Harada K, Fukusaki E, Kobayashi A (2006) Time-course metabolic profiling in Arabidopsis thaliana cell cultures after salt stress treatment. J Exp Bot 58:415–424
Kitayama T, Hatada K (2013) NMR spectroscopy of polymers. Springer, Berlin, Heidelberg
Kobayashi H, Graven YN, Broughton WJ, Perret X (2004) Flavonoids induce temporal shifts in gene-expression of nod-box controlled loci in Rhizobium sp. NGR234. Mol Microbiol 51:335–347
Krishnan P, Kruger N, Ratcliffe R (2004) Metabolite fingerprinting and profiling in plants using NMR. J Exp Bot 56:255–265
Ku KM, Choi JN, Kim J, Kim JK, Yoo LG, Lee SJ, Hong Y-S, Lee CH (2009a) Metabolomics analysis reveals the compositional differences of shade grown tea (Camellia sinensis L.). J Agric Food Chem 58:418–426
Ku KM, Kim J, Park H-J, Liu K-H, Lee CH (2009b) Application of metabolomics in the analysis of manufacturing type of pu-erh tea and composition changes with different postfermentation year. J Agric Food Chem 58:345–352
Kuc J (1995) Phytoalexins, stress metabolism, and disease resistance in plants. Annu Rev Phytopathol 33:275–297
Kusano M, Fukushima A, Kobayashi M, Hayashi N, Jonsson P, Moritz T, Ebana K, Saito K (2007) Application of a metabolomic method combining one-dimensional and two-dimensional gas chromatography-time-of-flight/mass spectrometry to metabolic phenotyping of natural variants in rice. J Chromatogr B 855:71–79
Kushalappa AC, Gunnaiah R (2013) Metabolo-proteomics to discover plant biotic stress resistance genes. Trends Plant Sci 18:522–531
Lange BM, Ghassemian M (2005) Comprehensive post-genomic data analysis approaches integrating biochemical pathway maps. Phytochemistry 66:413–451
Lattanzio V, Lattanzio VM, Cardinali A (2006) Role of phenolics in the resistance mechanisms of plants against fungal pathogens and insects. Phytochem Adv Res 661:23–67
Levinson HZ (1976) The defensive role of alkaloids in insects and plants. Experientia 32:408–411
Madhavi D, Smith M, Linas A, Mitiku G (1997) Accumulation of ferulic acid in cell cultures of Ajuga pyramidalis metallica crispa. J Agric Food Chem 45:1506–1508
Mandal SM, Chakraborty D, Dey S (2010) Phenolic acids act as signaling molecules in plant-microbe symbioses. Plant Signal Behav 5:359–368
Mithen RF, Dekker M, Verkerk R, Rabot S, Johnson IT (2000) The nutritional significance, biosynthesis and bioavailability of glucosinolates in human foods. J Sci Food Agric 80 (7):967–984
Mishra RP, Singh RK, Jaiswal HK, Kumar V, Maurya S (2006) Rhizobium-mediated induction of phenolics and plant growth promotion in rice (Oryza sativa L.). Curr Microbiol 52:383–389
Mithöfer A, Schulze B, Boland W (2004) Biotic and heavy metal stress response in plants: evidence for common signals. FEBS Lett 566:1–5
Moran PJ (1998) Plant-mediated interactions between insects and a fungal plant pathogen and the role of plant chemical responses to infection. Oecologia 115:523–530
Ndakidemi PA, Dakora FD (2003) Legume seed flavonoids and nitrogenous metabolites as signals and protectants in early seedling development. Funct Plant Biol 30:729–745
Nicholson JK, Wilson ID (1989) High resolution proton magnetic resonance spectroscopy of biological fluids. Prog Nucl Magn Reson Spectrosc 21:449–501
Ogata H, Goto S, Sato K, Fujibuchi W, Bono H, Kanehisa M (1999) KEGG: Kyoto encyclopedia of genes and genomes. Nucleic Acids Res 27:29–34

Olivoto T, Nardino M, Carvalho IR, Follmann DN, Szareski VJ, Ferrari M, de Pelegrin AJ, de Souza VQ (2017) Plant secondary metabolites and its dynamical systems of induction in response to environmental factors: a review. Afr J Agric Res 12:71–84

Osbourn A (1996) Saponins and plant defence—a soap story. Trends Plant Sci 1:4–9

Osbourn AE, Qi X, Townsend B, Qin B (2003) Dissecting plant secondary metabolism–constitutive chemical defences in cereals. New Phytol 159:101–108

Paley SM, Karp PD (2006) The pathway tools cellular overview diagram and omics viewer. Nucleic Acids Res 34:3771–3778

Parveen I, Moorby JM, Fraser MD, Allison GG, Kopka J (2007) Application of gas chromatography−mass spectrometry metabolite profiling techniques to the analysis of heathland plant diets of sheep. J Agric Food Chem 55:1129–1138

Pelletier SW (1983) The nature and definition of an alkaloid. In: Alkaloids: chemical and biological perspectives, vol 1. Pergamon, Oxford, pp 1–31

Petterson DS, Harris DJ, Allen DG (1991) Alkaloids. In: Toxic substances in crop plants. Elsevier, Amsterdam, pp 148–179

Pichersky E, Gershenzon J (2002) The formation and function of plant volatiles: perfumes for pollinator attraction and defense. Curr Opin Plant Biol 5:237–243

Rask L, Andréasson E, Ekbom B, Eriksson S, Pontoppidan B, Meijer J (2000) Myrosinase: gene family evolution and herbivore defense in Brassicaceae. Plant Mol Biol 42:93–114

Rhodes M, Parr A, Giulietti A, Aird E (1994) Influence of exogenous hormones on the growth and secondary metabolite formation in transformed root cultures. Plant Cell Tissue Organ Cult 38:143–151

Robbins C, Mole S, Hagerman A, Hanley T (1987) Role of tannins in defending plants against ruminants: reduction in dry matter digestion? Ecology 68:1606–1615

Robinson, T. (1980). The organic constituents of higher plants: their chemistry and interrelationships. Cordus Press

Roessner U, Wagner C, Kopka J, Trethewey RN, Willmitzer L (2000) Simultaneous analysis of metabolites in potato tuber by gas chromatography–mass spectrometry. Plant J 23:131–142

Sasidharan S, Chen Y, Saravanan D, Sundram K, Latha LY (2011) Extraction, isolation and characterization of bioactive compounds from plants' extracts. Afr J Tradit Complement Altern Med 8:1–10

Schultze M, Kondorosi A (1998) Regulation of symbiotic root nodule development. Annu Rev Genet 32:33–57

Schuster S, Fell DA, Dandekar T (2000) A general definition of metabolic pathways useful for systematic organization and analysis of complex metabolic networks. Nat Biotechnol 18:326

Shulaev V, Cortes D, Miller G, Mittler R (2008) Metabolomics for plant stress response. Physiol Plant 132:199–208

Singh U, Prithiviraj B (1997) Neemazal, a product of neem (Azadirachta indica), induces resistance in pea (Pisum sativum) against Erysiphe pisi. Physiol Mol Plant Pathol 51:181–194

Singh U, Sarma B, Singh D, Bahadur A (2002) Plant growth-promoting rhizobacteria-mediated induction of phenolics in pea (Pisum sativum) after infection with Erysiphe pisi. Curr Microbiol 44:396–400

Smith CA (2013) LC/MS preprocessing and analysis with xcms. Available from http://www.bioconductor.org/packages//2.11/bioc/vignettes/xcms/inst/doc/xcmsPreprocess.pdf

Soga T, Ohashi Y, Ueno Y, Naraoka H, Tomita M, Nishioka T (2003) Quantitative metabolome analysis using capillary electrophoresis mass spectrometry. J Proteome Res 2:488–494

Spasić I, Dunn WB, Velarde G, Tseng A, Jenkins H, Hardy N, Oliver SG, Kell DB (2006) MeMo: a hybrid SQL/XML approach to metabolomic data management for functional genomics. BMC Bioinf 7:281

Stöckigt J, Obitz P, Falkenhagen H, Lutterbach R, Endress S (1995) Natural products and enzymes from plant cell cultures. Plant Cell Tissue Organ Cult 43:97–109

Tanaka J, Da Silva C, De Oliveira A, Nakamura C, Dias Filho B (2006) Antibacterial activity of indole alkaloids from Aspidosperma ramiflorum. Braz J Med Biol Res 39:387–391

Thimm O, Bläsing O, Gibon Y, Nagel A, Meyer S, Krüger P, Selbig J, Müller LA, Rhee SY, Stitt M (2004) MAPMAN: a user-driven tool to display genomics data sets onto diagrams of metabolic pathways and other biological processes. Plant J 37:914–939

Tholl D (2015) Biosynthesis and biological functions of terpenoids in plants. In: Biotechnology of isoprenoids. Springer, Cham, pp 63–106

Tiwari R, Singh A, Das K, Sinha A (2007) Efficacy of extracts of medicinal plants against Rhizoctonia solani. Ann Plant Protect Sci 15:460–539

Tohge T, Nishiyama Y, Hirai MY, Yano M, Nakajima Ji, Awazuhara M, Inoue E, Takahashi H, Goodenowe DB, Kitayama M (2005) Functional genomics by integrated analysis of metabolome and transcriptome of Arabidopsis plants over-expressing an MYB transcription factor. Plant J 42:218–235

Tokimatsu T, Sakurai N, Suzuki H, Ohta H, Nishitani K, Koyama T, Umezawa T, Misawa N, Saito K, Shibata D (2005) KaPPA-view. A web-based analysis tool for integration of transcript and metabolite data on plant metabolic pathway maps. Plant Physiol 138:1289–1300

Treutter D (2006) Significance of flavonoids in plant resistance: a review. Environ Chem Lett 4:147

Vig AP, Rampal G, Thind TS, Arora S (2009) Bio-protective effects of glucosinolates–a review. LWT-Food Sci Technol 42:1561–1572

Ward JL, Harris C, Lewis J, Beale MH (2003) Assessment of 1H NMR spectroscopy and multivariate analysis as a technique for metabolite fingerprinting of Arabidopsis thaliana. Phytochemistry 62:949–957

Wink M (1988) Plant breeding: importance of plant secondary metabolites for protection against pathogens and herbivores. TAG Theor Appl Genet 75:225–233

Wink M (1997) Compartmentation of secondary metabolites and xenobiotics in plant vacuoles. In: Advances in botanical research. Elsevier, Amsterdam, pp 141–169

Wink M (2003) Evolution of secondary metabolites from an ecological and molecular phylogenetic perspective. Phytochemistry 64:3–19

Wysocki VH, Resing KA, Zhang Q, Cheng G (2005) Mass spectrometry of peptides and proteins. Methods 35:211–222

Yamazaki Y, Kitajima M, Arita M, Takayama H, Sudo H, Yamazaki M, Aimi N, Saito K (2004) Biosynthesis of camptothecin. In silico and in vivo tracer study from [1-13C] glucose. Plant Physiol 134:161–170

Zhang P, Foerster H, Tissier CP, Mueller L, Paley S, Karp PD, Rhee SY (2005) MetaCyc and AraCyc. Metabolic pathway databases for plant research. Plant Physiol 138:27–37

Zhu J-K (2016) Abiotic stress signaling and responses in plants. Cell 167:313–324

Chapter 6
Management of Root-Knot Nematode in Different Crops Using Microorganisms

Aastha Singh, Pankaj Sharma, Anju Kumari, Rakesh Kumar, and D. V. Pathak

Abstract Root-knot nematodes are severe pests that affect a wide variety of crops. These nematodes are controlled by various types of nematicides that have hazardous effects on the environment. Hence, identification of new approaches alternate to harmful chemical nematicides could be effective in controlling root knot nematodes. Conventional control measures use soil fumigants, but they have certain limitations. Being chemical in nature and cost-effective, they pose severe economic concerns. Besides, they have a broad-spectrum range by harming nontarget species. This raised a question regarding the biosafety of sustainable environment posed by hazardous chemicals. Application of plant-derived products is an effective eco-friendly approach to mitigate the infestations caused by nematodes in different crops.

6.1 Introduction

Root-knot nematodes are austere pests of various food and industrial crops and are globally distributed. They are polyphagous and have short life cycle, which allows them to multiply rapidly in numerous agroecosystems, ranging from vicious monocultures to subsistence crops. It is indeed an arduous task for the growers to avert crop losses due to root-knot nematode problems, susceptibility of the current crop varieties (Stirling 2006), and absence of control measures which are not economically effective. These nematodes are obligate plant parasites and invade live plants for their growth, development, and reproduction. They cause an annual global loss of around

A. Singh · P. Sharma · R. Kumar (✉)
Department of Microbiology, CCS Haryana Agricultural University, Hisar, Haryana, India
e-mail: sehrawatrk@hau.ernet.in

A. Kumari
Center of Food Science and Technology, CCS Haryana Agricultural University, Hisar, Haryana, India

D. V. Pathak
Department of Microbiology, CCS Haryana Agricultural University, Hisar, Haryana, India

Regional Research Station, CCS Haryana Agricultural University, Bawal (Rewadi), Haryana, India

A. Varma et al. (eds.), *Plant Biotic Interactions*,
https://doi.org/10.1007/978-3-030-26657-8_6

US$100 billion in agriculture. The damage caused by root-knot nematodes is not always apprehended because their ecological niche is present below the earth surface, which makes their management task hellacious. The plant parts such as roots, bulbs, and tubers present in the soil are usually attacked by plant parasitic nematodes, thus playing a key role in the interruption of nutrients and water uptake. The impairment of water and mineral transport pathways leads to the appearance of symptoms resembling water or nutrition deficiency like small, chlorotic, and less vigorous plants (Oka et al. 2000). The root tips are primarily invaded, and damage to plants is done by degradation of the cell wall followed by migration toward the vascular cylinder and formation of giant cells which are also known as "galls." The development of root galls siphons the nutrients and photosynthates of the plant. The root-knot nematode affects both young and mature plants. The infection process can be fatal for the young plants, whereas infected mature plants experience a considerable loss in the yield. The damages caused by root-knot nematode are as follows: stunt growth, loss of crop quality and yield, and a reduced resistance to several other stresses (e.g., drought, other diseases) (Borah et al. 2018; Kepenekci et al. 2018). They cause rigorous damage, resulting in significant yield losses in different crops like tomato, cotton, rice, carrot, pepper, potato, eggplant, watermelon, cucumber, ashwagandha etc. Root-knot nematode infection induces several biochemical changes such as alterations in the levels of amino acids and organic acids and reduced chlorophyll content in the plants (Saikia et al. 2013).

There are around 80 different infective species, but three (*Meloidogyne arenaria*, *M. incognita*, and *M. javanica*) of them are of utmost agronomic importance because they are the only ones responsible for at least 90% of the crop damage, which contributes to 5% of the global crop loss. These species are highly polyphagous as they infect more than 3000 plant species. Furthermore, their global distribution makes them eminently successful plant parasites (Castagnone-Sereno 2002).

The conventional control measures of root-knot nematode include use of soil fumigants. The use of these chemicals poses severe economic concerns due to their high cost. Moreover, the use of these chemical nematicides hinders sustainable environmental management by posing potential risks to nontarget microorganisms. Thus, the dire threats posed by chemical control measures have raised the concern for the development of nonchemical and eco-friendly management strategies for controlling root-knot nematode (Huang et al. 2016). Therefore, nonchemical and eco-friendly alternatives such as biological control are being sought.

6.2 Chemical Control of Root-Knot Nematodes

The chemical control of nematodes is mainly mediated by the use of soil fumigants. Farmers mainly relied on a broad-spectrum soil fumigant like methyl bromide (MeBr) against various soilborne diseases, weeds, and nematodes. However, the use of methyl bromide was banned in 2005 and it was completely phased out. Thus, farmers were left with only limited options against the root-knot nematodes. Pic-Clor

60 [1,3-dichloropropene plus chloropicrin (40:60, w/w)] (Agrian, Inc., Fresno, CA) was one of the main fumigants which was used as an alternative to methyl bromide for growing tomato in Florida (Castillo et al. 2017). Some nonfumigant nematicides are available also which are currently registered for use in a number of crops like tomatoes, peppers, and eggplants. Such nematicides, except Vydate, are usually applied on soil, which can also be applied foliarly. These compounds need to be applied in a uniform manner to the soil, so that the future rooting zone of the plant can be targeted, where the nematodes will come in contact, or they can be applied in the areas where they can be easily absorbed and taken up into the plant (Noling 2016).

6.3 Root-Knot Management in Vegetables and Horticulture

6.3.1 Root-Knot Nematode Management in Tomato

Tomato (*Lycopersicon esculentum*) has got a unique place among the most popular vegetables in the world. It suffers a great number of biotic stresses during its growth. The root-knot nematode is liable for inducing the most scourging and widespread stress among the many biotic stresses. These nematodes not only affect the crop yield in a direct manner but also make the plant more susceptible to bacterial and fungal attacks (Zhou et al. 2016). The management of root-knot nematodes in tomato includes various strategies like crop rotation, use of resistant cultivars, and treatment of soils with chemical nematicides. Plant's resistance toward root-knot nematodes is unstable and often results in decreased yields (Williamson and Robert 2009). They cause a considerable loss in annual yield of vegetables, which may range from about 10 to 30% depending on the severity of infection (Radwan et al. 2012). There are various control measures that can be employed for controlling root-knot nematodes, including conventional methods such as chemical control in infested areas. Use of such chemicals has potential adverse impacts on environment. Furthermore, their prolonged use has also made them inefficient, resulting in a total ban or restricted use, thus directing the acute need for safer and more effective alternatives (Zukerman and Esnard 1994). Biological control appears to be one of the most promising alternatives. The nematicidal activity of the biocontrol agents *Bacillus megaterium*, *Trichoderma album*, *Trichoderma harzianum*, and *Ascophyllum nodosum* against the root-knot nematode, *Meloidogyne incognita*, that infects tomato, was found to be comparatively much superior than the control. There was a substantial decline in the count of root galls which was observed much higher in case of oxamyl or carbofuran (Radwan et al. 2012). Actinomycetes can secrete a variety of antimicrobial substances that can alter the microbial variety and population of the root zone. *Streptomyces*, an important group of actinomycetes, are mainly reported for the secretion of insecticidal substances (Ruanpanun et al. 2010). Treatment of soil with actinomycetes strongly affects the culturable microorganisms present in the root zone of the plants, root knots, and root systems, along with marked changes in the population of nematodes. The disease index of plant is significantly reduced on treatment with

Streptomyces. A decrease of 37% in disease index has been reported in tomato when treated with *Streptomyces* sp. Furthermore, there was a sharp decline of 14% in the number of nematodes feeding on bacteria present in the root zone. The fresh weight of root and shoot also increased significantly. The culturable microflora of the root knots is also altered in a significant manner. Moreover, the populations of nematicidal bacteria as well as of plant growth-promoting rhizobacteria are greatly enhanced, whereas the numbers of plant pathogenic bacteria are sharply declined. The significant reduction in root-knot disease in tomato could be due the activation of systemic resistance and defensive mechanism in plant against nematode infection (Ma et al. 2017).

Arbuscular mycorrhiza is a highly infective species and greatly enhances the growth of plants. They reduce pathogenicity in the root system and induce tolerance in plants against disease. The preestablishment of arbuscular mycorrhizal fungi significantly reduces the reproduction of *Meloidogyne incognita*, which in turn reduces the disease severity in infected soil. The inoculation with AM can improves the growth of plants and also protects the plants against soilborne pathogens.. It also contributes toward increasing the nutrient uptake and also acts as a potential biocontrol agent improving plant growth by direct or indirect mechanisms, thus compensating the damages caused by root-knot nematodes (Sharma and Sharma 2017). The application of plant growth-promoting rhizobacteria (PGPR) belonging to *Pseudomonas* spp. and *Bacillus* spp. is also helpful in the efficient management of Root knot nematode (RKN) infestation in tomato. PGPR protects the plants by inducing systemic resistance against the pathogens. A diverse array of microbes having varied physiological requirements is advantageous than using a single biocontrol agent such as *Piriformospora indica*, an endophytic fungus, in combination with two plant growth-promoting rhizobacteria (*Bacillus pumilus* and *Pseudomonas fluorescens*) successfully suppress root-knot nematode infection (Varkey et al. 2018). Cyanobacteria (blue–green algae), a primitive group of organisms that contains at least 40 toxicogenic species, can be used as biocontrol agents. The endospores of the species of *Microcoleus* and *Oscillatoria* have been reported for their ability to kill nematodes. They inhibited the hatching of second-stage juveniles and also killed the hatched juveniles, and therefore, they can be used as potent biocontrol agents (Khan et al. 2007).

6.3.2 Root-Knot Nematode Management in Carrot

Carrot is one of the important root crops. It is among the main ten vegetables grown globally in subtropical and tropical areas. The roots of this plant find uses in a variety of food items including pickle, salad, and juice. Its seeds are also a good source of essential oils that are used for various purposes. Its roots are rich in a number of vitamins like vitamin B6, vitamin K, and provitamin A. This food crop finds an area of 1.59 million hectares under its cultivation across the world. It is grown in over 130 countries with an annual global production of 49.35 million tonnes. This crop is also cultivated in India under the area of 0.32 million hectares and the annual production is 0.49 million tonnes, thus making India a major producer of carrot (Nagachandrabose 2018). The major hurdle in the commercial cultivation of carrot is the root-knot nematode that parasitizes plant roots. This results in severe qualitative and quantitative loss of yield. The cultivation of carrot is provoked by six different varieties of root-knot nematodes, namely, *Meloidogyne incognita*, *Meloidogyne hapla*, *Meloidogyne javanica*, *Meloidogyne chitwoodi*, *Meloidogyne fallax*, and *Meloidogyne polycephannulata* (Charchar et al. 2009). In India, carrots are mainly affected by northern root-knot nematode, *M. hapla*, which is a sedentary endoparasitic nematode. Infection of nematodes followed by penetration of juveniles through growing root tips leads to the formation of various giant cells containing multiple nuclei in the vascular tissues. The activity of nematode further results in the formation of galls, digitation, hairiness, and compression in taproots, thereby resulting in the formation of defective and forked carrots. Low infestations usually deteriorate the quality of carrots by disfigurement (forking), whereas high infestations lead to the arrest of taproot formation. The root infection caused by *M. hapla* in carrots results in an overall 24–55% quantitative loss and 13–77% qualitative loss (Nagachandrabose 2018). Furthermore, the wounds created by the nematodes enable the easy entry of different kinds of soilborne pathogens particularly *Pectobacterium carotovorum*, the soft rot bacterium, and together these organisms result in the formation of a disease complex resulting in enormous crop loss (Sowmya et al. 2012).

Crop damages caused by root-knot nematode infection can be controlled by using biological control agents like fungus or bacteria that can antagonize the nematodes effectively. The application of biocontrol agents for controlling nematodes is gaining popularity among farmers. If the introduction of biocontrol agent leads to its establishment with nematode populations, the invasion may last for a long time, ranging from complete season to a period even greater than the required period (Jacobs et al. 2003).

Liquid formulations of various biocontrol agents exhibit more efficiency than the solid carriers such as talc- and peat-based formulations. These liquid formulations are advantageous in a number of ways than the solid formulations: they increase the viable cell count, reduce the risk of contamination, show more virulence, and have high shelf life. Moreover, the cells in liquid formulations maintain their dormancy, which become active again on application to the soil rhizosphere.

In a study by Nagachandrabose (2018), the field efficiency of liquid formulations of *Pseudomonas fluorescens*, *Trichoderma viride*, and *Purpureocillium lilacinum* was tested against the populations of nematode, *Meloidogyne hapla*. Seed treatment with *P. fluorescens* caused a paramount decline in the populations by around 68%, whereas treatment with *P. lilacinum* reduced the population by 64–67%. The treatment of plant with *P. fluorescens* and *P. lilacinum* increased 36% height of the plants. Furthermore, the plants were also reported to have comparatively higher leaf count and improved leaf size than the untreated plants. The plants experienced 28–36% increase in the number of leaves and 27–30% longer leaves were observed. Surprisingly, treatment with *P. fluorescens* resulted in 20–21% higher yield of root tuber, whereas treatment with *P. lilacinum* resulted in 19% higher yield.

Bacillus subtilis is a widely used bacterium which has been commercialized as a potent biocontrol agent against a number of diseases and infectious nematodes parasitizing an extensive range of plants. In addition to its ability to form spores, it possesses numerous characteristics which strongly enhance its survival in rhizosphere. It exhibits various modes of action like competition, antibiosis, and induction of systemic resistance, and it also exhibits various traits of plant growth promotion. The success of any biocontrol agent strongly depends on its delivery mechanism into the field.

B. subtilis produces different types of antimicrobial compounds like subtillin, bacillin, bacitracin, bacillomycin, and subtenolin (Killani et al. 2011). In addition, it is also reported to produce various volatile compounds such as 2-nonanone, 2-undecanone, benzene acetaldehyde, dimethyl disulfide, and decanal, which show nematicidal activity; these compounds exert antagonistic activity toward the egg hatching of root-knot nematode and its second-stage juveniles (Huang et al. 2010). The seed treatment together with soil application of *B. subtilis*-enriched vermicompost increased the yield of carrot to 28.8%, decreased the total nematode population by 69.3%, and decreased the disease incidence by 70.2% (Rao et al. 2017). There are various natural enemies of nematodes, and fungus *Pochonia chlamydosporia* is one among them. This fungus parasitizes the eggs and the exposed females of the root-knot nematodes, so it can tremendously manage the root-knot nematodes. Moreover, it produces a large amount of resistant structures called chlamydospores, which allows its survival during adverse conditions. The incorporation of *Pochonia chlamydosporia* into the soil increased the total weight of taproots by approximately 55% in a study by Bontempo et al. (2014).

Aspergillus is another most effective biocontrol agent. They inhibit the hatching of nematode eggs and also promote enzymatic degradation by disintegrating chitin and vitelline layers of the nematode eggshell, which in turn leads to the disintegration of egg contents. The effect of this species is mainly exogenous. The coinoculation of *P. lilacinus* with *A. niger* significantly reduces nematode multiplication and also improves plant growth more compared to the inoculation of single organism due to the combined mechanisms of action of both organisms (Nesha and Siddiqui 2017).

6.3.3 Root-Knot Nematode Management in Chili

Chili (*Capsicum annuum*) is one of the important commercial crops grown worldwide and is a rich source of proteins, vitamin, ascorbic acid, and other nutrients. It belongs to the family Solanaceae and grows well in hot regions of the world. Several crop losses caused by plant parasitic nematode have been reported in chili (Khan et al. 2012). Chili crops infested with root-knot nematode *Meloidogyne incognita* appears to be stunted in growth, produces less flowers, and gives less yield. *Pseudo-*

monas fluorescens has proved its efficacy as a biocontrol agent for nematode management; in addition, compared to pesticides, it has also provided a prolonged resistance against nematode attack. The root-knot nematode infects the plant roots, when it becomes second-stage juveniles, and establishes a feeding site within the pericycles and vascular tissues of the plant cells and forms giant cells. Galls are formed due to hyperplasia of root cells around giant cells. Root-knot nematode destroys the root system, which creates competition for food and nutrition among the developing nematodes within the root system. During infection, root-knot nematode induces cell enlargement accompanied by nuclear division without cytokinesis. It results in the formation of multinucleate giant cells which confer resistance to root-knot nematode, characterized by hypersensitive reactions. The juveniles surrounded by necrotic cells fail to develop and die, which prevent nematode penetration and migration at early infection stages.

In a study of disease management in chili pepper infected with the root-knot nematode *Meloidogyne incognita*, Moon et al. (2010) reported 39 chili pepper cultivars/lines, out of which 6 were found to be resistant and 33 susceptible. The resistant cultivars/lines showed enhanced resistance against gall formation as compared to susceptible lines. These results revealed that disease resistance in chili pepper may be related to post-inflectional defense mechanisms.

Nowadays, plant products and their derivatives are also gaining importance in the management of plant parasitic nematodes and have increased the awareness of environmental hazards associated with nematicidal chemicals. Several indigenous plants and plants parts have been identified as promising sources of biopesticides. Plant products such as terpenoid, triterpenoid, limonoid, flavanoid, azadirachtins, meliantriol, salannin, nimbin, and nimbidin can be explored for their nematicidal potential. The neem-based formulations used in dressing and seedling bare root dip showed a significant decrease in the population of *M. incognita* and an increase in the yield of green chili.

6.3.4 Root-Knot Nematode Management in Banana

Banana (*Musa* sp.) is one of the chief economic tropical fruit crops grown worldwide and is rich in carbohydrates, proteins, minerals, and vitamins. It grows in a wide variety of soils. A total of 132 species of nematodes belonging to 54 genera are associated with rhizosphere of banana. Although crop yield is affected by a number of crop pests, the productivity of banana orchards is highly limited due to root-knot nematode infestation (Eissa et al. 2005). Infestation of plant roots by the pests results in toppling, reduces plant longevity, increases crop cycle duration, and results in yield losses. *Meloidogyne* spp. are the predominant nematode agents in bananas with an occurrence of about 76% (Mokbel et al. 2006). The management of root-knot nematodes by using hazardous chemicals is not a useful practice because these are highly prone to the humans and the environment. Thus, alternative control strategies such as biocontrol agents are needed. The biological, chemical, and physical properties of soil are also improved with the amendments of organic and biological products. During the decomposition of these organic amendments, some volatile compounds such as formic, propionic, and butyric acids and phenols are released, which proved to be toxic to nematodes. Biocontrol agents such as *Penicillium* spp., *Glomus fasciculatum*, *Bacillus subtilis*, *Trichoderma viride*, and *Paecilomyces lilacinus* also showed effective nematicidal activities against *M. incognita* (Esnard et al. 1998). *P. lilacinus*-treated plants showed significant reduction in root galls index and eggs per egg mass in soil but showed an increase in the number of leaves, root length, shoot length, and height of plant. Recent studies have reported that alga is also an effective biocontrol agent in treating plants affected by root-knot nematodes. Few algae such as *Ulva lactuca*, *Jania rubens*, *Sargassum vulgare*, and *Laurencia obtusa* can be successively used for controlling root-knot nematode and for promoting plant growth.

El-Nagdi et al. (2015) studied the biological control of root-knot nematode and *Fusarium* root rot fungus in banana by using two commercial biocontrol agent products (CBAP)— Fornem x5® (contains *Rhodotorula pustula*, *S. marcescens*, *Serratia entomophila*, *P. putida*, and *P. fluorescens*) and Micronema® (contains *Azotobacter* sp., *Pseudomonas* sp., *Bacillus thuringiensis*, *Serratia* sp., and *Bacillus circulans*)—compared to chemical nematicide Nemacur® against *Meloidogyne incognita* and *Fusarium* root rot (*Fusarium solani*) fungus. Micronema® and Fornem x5® significantly reduced the numbers of nematode parameters after 2, 4, and 6 months of treatment. The highest reduction of juveniles in stage three (J2), stage two (J3), eggs and females in roots were achieved by application of Fornem x5® followed by Micronema® at 30 ml/plant after 6 months. Results showed that the treatments increased the frequency of *Aspergillus* spp. and *Penicillium* spp., while they decreased the frequency of *Fusarium* spp., moderately inhibitory effect on *Fusarium* root rot was obtained by CBAP. Yield parameters were increased with Micronema® and Fornem x5® at 30 ml/plant as compared to Nemacur® and untreated control.

6.4 Root-Knot Nematode Management in Agricultural Crops

6.4.1 Root-Knot Nematode Management in Sugarcane

Sugarcane is a very popular sugar crop and is also a regenerative energy crop. The soils which are not suitable for conventional agricultural crops are used for sugarcane plantation. This crop has occupied a significant area under cultivation in more than 80 countries across the globe. The crop is mainly grown for fiber, crude sugar, and bioethanol production (Chirchir et al. 2008). This crop is susceptible to a number of threats including nematode diseases, which affect its production significantly and also cause austere economic losses globally (Stirling et al. 2001). The susceptibility of current varieties of sugarcane toward the attack of root-knot nematode makes it a very challenging problem for the farmers as the preclusion of crop losses resulting from nematode infections appears to be a strenuous exercise (Stirling 2006) due to the absence of control measures which are economically effective. The nematodes are responsible for approximately 20% reduction in the production. In addition to the presence of nematodes also poses a threat to the cultivation of new crops thus making the process of growing new crops an uneconomical operation in infested areas (Morgado et al. 2015).

The yields of sugarcane affected with nematodes are significantly increased when treated with functional biofertilizers as compared to chemical fertilizers. There is a significant decrease in the population of plant parasites along with an increase in the number of beneficial nematodes. The yield of sugarcane is increased by comparatively high number of bacterivore and a reduced population of plant parasites (Zhang et al. 2017).

The introduction of spore-forming bacterium *B. subtilis* in the soils has been reported to cause a reduction in the number of plant parasitic nematodes. Furthermore, its ability to form spores ensures its survival over long periods. Moreover, suppression of nematodes by treating with *B. subtilis* is found to be equivalent to the use of conventional chemical control measures (Morgado et al. 2015).

When alternative methods of nematode control are taken into consideration, *Pasteuria penetrans* also proves to be an effective biocontrol agent that is commendable for investigation. Its host specificity toward the root-knot nematode can be explored for a significant reduction in the populations of root-knot nematode. The elevated number of endospores can even lead to the elimination of 99% of the nematode population (Bhuiyan et al. 2018). The rigorousness of root galls and the amount of nematode eggs can be knocked down by amplifying the concentration of endospores. Thus, the perpetual sustentation of altitudinous concentration of endospores in the vicinity of roots can strikingly reduce the infections induced by root-knot nematode, a dire pest of sugarcane.

An integrated use of nematophagous fungi along with the application of arbuscular mycorrhizal fungi also proves to be an effective biocontrol measure. Arbuscular mycorrhizal fungi like *Glomus mosseae* and *Glomus fasciculatum* have the potential to act as effective biocontrol agents when used with nematophagous fungi like *Arthrobotrys oligospora*, *Pochonia chlamydosporia*, and *Paecilomyces lilacinus*. A maximum reduction of 47% in the population of *M. javanica* has been reported with the integrated use of *G. fasciculatum* and *P. lilacinus*. This combination proved to be very effective in increasing the shoot weight of the plant (Sankaranarayanan and Hari 2013). The plant parasitic root-knot nematode can also be controlled by amendments of organic matter in the soil. The organic amendments of filter mud and sugarcane bagasse can significantly lower the numbers of parasitic nematodes (Chirchir et al. 2008).

6.4.2 Root-Knot Nematode Management in Rice

Rice is the most important staple crop consumed globally, which finds an area of 162 mha under annual cultivation with an overall global production of 464 mmt annually. Rice crop is very much susceptible to the attack of root-knot nematodes and is infected by *Meloidogyne triticoryzae*, *M. arenaria*, *M. incognita*, *M. graminicola*, *M. oryzae*, and *M. javanica*. The primary pest that attacks rice is *M. graminicola*, which poses an extensive threat to the cultivation of rice particularly in Southeast Asia where around 90% of the global rice is cultivated as well as consumed (Khan et al. 2014). Rice undergoes attack by nematodes in nursery as well as in the main field, which results in considerable loss of crop yield. This nematode causes an annual yield loss of around 16–32% in upland and rainfed rice in India (Haque et al. 2018). The chemical control of root-knot nematode proves to be much effective, and organophosphates are the widely used pesticides for nematode control. Although chemical treatment is very effective, the persistence of saturated conditions in rice fields due to irrigation leads to the percolation of pesticides from the root zones, which further results in the persistence of root-knot nematode (Khan et al. 2014). The use of biocontrol agents proves to be the best alternative for root-knot nematode management in rice.

There are a number of biocontrol agents such as *Pseudomonas putida*, *Trichoderma harzianum*, *P. fluorescens*, *Purpureocillium lilacinum*, and *Bacillus subtilis*, which provide better and effective control of root-knot disease. *P. putida* is found to be much effective than *P. fluorescens* and *P. lilacinum*, and *T. harzianum* effectively suppresses the nematode as an endophyte (Haque et al. 2018). Biological control employing different endophytic microorganisms is found to be highly effective in mitigating endoparasites which complete their life cycle inside the host. Various endophytes are capable of targeting various life stages of the parasitic nematodes by delaying development, reducing penetration, and diminishing reproductive capacity. *Bacillus megaterium*, a bacterial endophyte, is capable of reducing the attraction and penetration of rice roots by *M. graminicola* and also of diminishing the rate of hatching of nematode eggs. The inoculation of rice roots with *Fusarium* isolates and *Trichoderma* species has been found to decrease the severity of infection. These fungal isolates have also been found to decrease the severity of root galling by 29–42% and increase root weight by 33% (Le et al. 2009).

There are various biocontrol agents which have been reported to control the infections of plant parasitic nematodes. Chitinolytic microorganisms—for instance, the plant-growth-promoting rhizobacterium, *Pseudomonas fluorescens*, and the egg parasitic fungus, *Paecilomyces lilacinus*—seem to be the ideal agents for controlling rice nematodes due to their enhanced survival under clay soils which are meant for cultivation of irrigated rice. The talc-based formulations of *Paecilomyces lilacinus* and *Pseudomonas fluorescens* have been reported to produce comparable results with nematicide carbofuran application (Seenivasan 2011).

Dactylaria brochopaga is also supposed to be an effective biocontrol agent of root-knot nematode. In a study by Kumar and Singh (2008), it was observed that the application of *D. brochopaga* at 1% significantly reduced the parasitic nematode population and number of galls. The application of spore suspension of *D. brochopaga* also significantly controlled the disease along with significant upsurge in shoot and root weight.

6.4.3 Root-Knot Nematode Management in Chickpea

Chickpea (*Cicer arietinum* L.), a pulse crop, is a chief source of vitamins, minerals, and nutritional protein. It is widely used as a source of protein. Parasitic nematodes cause a great loss to chickpea, estimated to 13.7% loss in its production globally with an approximate total of 7620 hg/ha yield. It is more prone to be attacked by numerous ectoparasitic and endoparasitic nematodes like *M. javanica*, *Helicotylenchus* spp. (Ali and Sharma 2002), *Heterodera swarupi*, *Meloidogyne incognita*, and *Pratylenchus thornei*; therefore, management of chickpea is very complex. A few reports in India stated that root-knot nematodes reduce the yield of chickpea from 17 to 60% depending on the soil types and nematode inoculum density, and this has been reported in many states of India (Khan and Siddiqui 2005). *Meloidogyne* sp. is one of the most common root-knot nematodes in both tropical and subtropical regions where it has highly reduced the crop production and caused extensive economic loss globally (Sikora and Fernandez 2005).

Nematode alters plants metabolic processes such as nitrogen fixation and nodule formation, which ultimately affect the total yield of plant. To combat this problem, nematicides are currently being used to control nematode population. But this is not an effective practice; as nematicides are toxic, they may accumulate in plant, which may cause environmental pollution and there is reduction in amount stratosphere layer (Wheeler et al. 1979). Therefore, a substitute for nematode management is urgently required. Substances that occur naturally may also possess nematicidal activity. Yadav et al. (2006) reported that oil cakes are an important organic substitute in the management of root-knot nematodes that affect the yield of chickpea.

Rehman et al. (2012) studied the effect of *M. incognita*, a root-knot nematode, in reducing the yield of chickpea. They observed that application of chickpea with high concentration of *M. azedarach* leaf extracts showed reduction of *M. incognita* infection as compared to other doses. Chickpea plants were inoculated with second-stage juveniles of *M. incognita*, and a highest reduction in plant growth was observed with a reduction in root and shoot length (cm), total chlorophyll content, nitrate reductase activity, fresh weight (g), dry weight (g), and flower number per pods as root-knot index increased. A significant reduction in infection was observed in chickpea plants (control) inoculated with lesser concentration of leaf extract of *M. azedarach*. This study supports the use of organic substitutes in the control management of nematodes as an effective approach to manage environmental pollution. This study was supported by Mojumder and Mittal (2000); the results showed that chickpea transplanted with neem seedlings showed significant reduction in *M. incognita* infection. Therefore, these studies provide evidence that the infestation of *Meloidogyne* spp. may be reduced by the use of organic compounds obtained from plant in view of eco-friendly environment to avoid expensive and hazardous chemical nematicides for eco-friendly environment.

6.5 Conclusion

Root-knot nematodes are serious threat to global food industries and other industrial sectors. They cause reduction in crop yield due to the formation of galls that siphon the photosynthetic machinery of plants. Application of soil fumigants as a control measure of root-knot nematode poses severe economic concerns and also raises a question on the sustainable management of agroecosystems. To cope with this, best alternatives such as organic substitutes of microbial origin possessing nematicidal activity are to be opted for root-knot nematode management.

References

Ali SS, Sharma SB (2002) Distribution and importance of plant parasitic nematodes associated with chickpea in Rajasthan state. Indian J Pulses Res 15:57–65

Bhuiyan SA, Garlick K, Anderson JM et al (2018) Biological control of root-knot nematode on sugarcane in soil naturally or artificially infested with *Pasteuria penetrans*. Australas Plant Pathol 47:45–52. https://doi.org/10.1007/s13313-017-0530-z

Bontempo AF, Fernandes RH, Lopes J, Freitas LG, Lopes EA (2014) *Pochonia chlamydosporia* controls *Meloidogyne incognita* on carrot. Aust Plant Pathol 43:421–424

Borah B, Ahmed R, Hussain M et al (2018) Suppression of root-knot disease in *Pogostemon cablin* caused by *Meloidogyne incognita* in a rhizobacteria mediated activation of phenylpropanoid pathway. Biol Control 119:43–50

Castagnone-Sereno P (2002) Genetic variability in parthenogenetic root-knot nematodes, *Meloidogyne* spp., and their ability to overcome plant resistance genes. Nematology 4:605–608. https://doi.org/10.1163/15685410260438872

Castillo GX, Ozores-Hampton M, Navia Gine PA (2017) Effects of fluensulfone combined with soil fumigation on root-knot nematodes and fruit yield of drip-irrigated fresh-market tomatoes. Crop Prot 98:166–171. https://doi.org/10.1016/j.cropro.2017.03.029

Charchar JM, Eisenback JD, Vieira JV, Fonseca-Boiteux MEDN, Boiteux LS (2009) *Meloidogyne polycephannulata* n. sp.(Nematoda: Meloidogynidae), a root-knot nematode parasitizing carrot in Brazil. J Nematol 41:174–186

Chirchir AK, Kimenju JW, Olubayo FM et al (2008) Abundance and distribution of plant parasitic nematodes associated with sugarcane in Western Kenya. Asian J Plant Pathol 2:48–53

Eissa MFM, El-Gindi AY, Abd-Elgawad MM, Ismail AE, El-Nagdi WA (2005) Application of some bioagents and oxamyl in controlling *Meloidogyne incognita*, *Helicotylenchus exallus* and *Criconemoides* spp. infesting banana cv. Williams. Pak J Biotechnol 2(1–2):70–79

El-Nagdi WMA, Abd El-Khair H, El-Ghonaimy AM (2015) Field application of biological control on root-knot nematode and Fusarium root rot fungus in banana cv. grand naine. Middle East J Agr Res 04(3):545–554

Esnard J, Marban MN, Zuckerman BM (1998) Effects of three microbial broth cultures and an organic amendment on growth and populations of free living and plant-parasitic nematodes on banana. Eur J Plant Pathol 104:457–463

Haque Z, Khan MR, Ahamad F (2018) Relative antagonistic potential of some rhizosphere biocontrol agents for the management of rice root-knot nematode, *Meloidogyne graminicola*. Biol Control 126:109–116. https://doi.org/10.1016/j.biocontrol.2018.07.018

Huang Y, Xu C, Ma L, Zhang K, Duan C, Mo M (2010) Characterization of volatiles produced from *Bacillus megaterium* YFM 3.25 and their nematicidal activity against *Meloidogyne incognita*. Eur J Plant Pathol 26:417–422

Huang WK, Cui JK, Liu SM, Kong LA, Wu QS, Peng H, He WT, Sun JH, Peng DL (2016) Testing various biocontrol agents against the root-knot nematode (*Meloidogyne incognita*) in cucumber plants identifies a combination of *Syncephalastrum racemosum* and *Paecilomyces lilacinus* as being most effective. Biol Control 92:31–37. https://doi.org/10.1016/j.biocontrol.2015.09.008

Jacobs H, Gray SN, Crump DH (2003) Interactions between nematophagous fungi and consequences for their potential as biological agents for the control of potato cyst nematodes. Mycol Res 107:47–56

Kepenekci I, Hazir S, Oksal E, Lewis E (2018) Application methods of *Steinernema feltiae*, *Xenorhabdus bovienii* and *Purpureocillium lilacinum* to control root-knot nematodes in greenhouse tomato systems. Crop Prot 108:31–38. https://doi.org/10.1016/j.cropro.2018.02.009

Khan AA, Siddiqui MA (2005) Status of root- knot nematodes in U.P. India. In: Nehra S (ed) Plant microbes and biotechnology. Pointer Pub, Jaipur, pp 209–226

Khan Z, Kim YH, Kim SG, Kim HW (2007) Observations on the suppression of root-knot nematode (*Meloidogyne arenaria*) on tomato by incorporation of cyanobacterial powder (*Oscillatoria chlorina*) into potting field soil. Bioresour Technol 98:69–73. https://doi.org/10.1016/j.biortech.2005.11.029

Khan Z, Tiyagi SA, Mahmood I, Rizvi R (2012) Effect of N fertilisation, organic matter and biofertilisers on growth and yield of chilli in relation to management of plant parasitic nematodes. Turk J Bot 36:73–81

Khan MR, Haque Z, Kausar N (2014) Management of the root-knot nematode *Meloidogyne graminicola* infesting rice in the nursery and crop field by integrating seed primingand soil application treatments of pesticides. Crop Prot 63:15–25. https://doi.org/10.1016/j.cropro.2014.04.024

Killani AS, Abaidoo RC, Akintokun AK, Abiala MA (2011) Antagonistic effect of indigenous *Bacillus subtilis* on root soil borne fungal pathogens of cowpea. Researcher 3:11–18

Kumar N, Singh KP (2008) Assessment of predacity and efficacy of *Dactylaria brochopaga* for biological control of root knot disease of rice *(Oryza sativa* L.). J Biotech 136:S23. https://doi.org/10.1016/j.jbiotec.2008.07.039

Le HTT, Padgham JL, Sikora RA (2009) Biological control of the rice root-knot nematode *Meloidogyne graminicola* on rice, using endophytic and rhizosphere fungi. Int J Pest Manage 55:31–36

Ma YY, Li YL, Lai HX et al (2017) Effects of two strains of Streptomyces on root-zone microbes and nematodes for biocontrol of root-knot nematode disease in tomato. Appl Soil Ecol 112:34–41. https://doi.org/10.1016/j.apsoil.2017.01.004

Mojumder V, Mittal A (2000) Effect of neem seedlings on infestations of *Meloidogyne incognita* in chickpea. Legum Res 3:195–196

Mokbel AA, Ibrahim IKA, El-Saedy MAM, Hammad SE (2006) Plant parasitic nematodes associated with some fruit trees and vegetable crops in northern Egypt. Egyptian J Phytopathol 34 (2):43–51

Moon HS, Khan Z, Kim SG, Son SH, Kim YH (2010) Biological and structural mechanisms of disease development and resistance in chili pepper infected with the root-knot nematode. Plant Pathol J 26:149–153

Morgado TDT, Guerra JT, Araujo FFD et al (2015) Effectiveness and persistence of biological control of nematodes in sugarcane. Afr J Agr Res 10:4490–4495

Nagachandrabose S (2018) Liquid bioformulations for the management of root-knot nematode, *Meloidogyne hapla* that infects carrot. Crop Prot 114:155–161. https://doi.org/10.1016/j.cropro.2018.08.022

Nesha R, Siddiqui ZA (2017) Effects of *Paecilomyces lilacinus* and *Aspergillus niger* alone and in combination on the growth, chlorophyll contents and soft rot disease complex of carrot. Sci Hortic 218:258–264

Noling JW (2016) Nematode management in tomatoes, peppers, and eggplant. UF IFAS Ext:1–14

Oka Y, Koltai H, Bar-Eyal M et al (2000) New strategies for the control of plant-parasitic nematodes. Pest Manag Sci 56:983–988

Radwan M, Farrag S, Abu-Elamayem M, Ahmed N (2012) Biological control of the root-knot nematode, *Meloidogyne incognita* on tomato using bioproducts of microbial origin. Appl Soil Ecol 56:58–62. https://doi.org/10.1016/j.apsoil.2012.02.008

Rao M, Kamalnath M, Umamaheswari R et al (2017) *Bacillus subtilis* IIHR BS-2 enriched vermicompost controls root knot nematode and soft rot disease complex in carrot. Sci Hortic 218:56–62

Rehman B, Ganai MA, Parihar K et al (2012) Management of root knot nematode, meloidogyne incognita affecting chickpea, *Cicer arietinum* for sustainable production. Biosci Int 1:1–5

Ruanpanun P, Tangchitsomkid N, Hyde KD, Lumyong S (2010) Actinomycetes and fungi isolated from plant-parasitic nematode infested soils: screening of the effective biocontrol potential, indole-3-acetic acid and siderophore production. World J Microb Biot 26:1569–1578

Saikia SK, Tiwari S, Pandey R (2013) Rhizospheric biological weapons for growth enhancement and *Meloidogyne incognita* management in *Withania somnifera* cv. Poshita. Biol Control 65:225–234. https://doi.org/10.1016/j.biocontrol.2013.01.014

Sankaranarayanan C, Hari K (2013) Bio-management of root knot nematode *Meloidogyne javanica* in sugarcane by combined application of arbuscular mycorrhizal fungi and nematophagous fungi. J Sugar Res 3:62–70

Seenivasan N (2011) Efficacy of *Pseudomonas fluorescens* and *Paecilomyces lilacinus* against *meloidogyne graminicola* infecting rice under system of rice intensification. Arch Phytopathol Plant Prot 44:1467–1482. https://doi.org/10.1080/03235408.2010.505788

Sharma IP, Sharma AK (2017) Co-inoculation of tomato with an arbuscular mycorrhizal fungus improves plant immunity and reduces root-knot nematode infection. Rhizosphere 4:25–28. https://doi.org/10.1016/j.rhisph.2017.05.008

Sikora RA, Fernandez E (2005) Nematode parasites of vegetables. In: Luc M, Sikora RA, Bridge J (eds) Plant parasitic nematodes in subtropical and tropical agriculture, 2nd edn. CABI, Wallingford, pp 319–392

Sowmya DS, Rao MS, Manojkumar R, Gavaskar J, Priti K (2012) Bio-management of *Meloidogyne incognita* and *Erwinia carotovora* in carrot (*Daucus carota* L.) using *Pseudomonas putida* and *Paecilomyces lilacinus*. Nematol Mediterr 40:189–194

Stirling GR (2006) Susceptibility of sugarcane varieties to two species of root-knot nematode (*Meloidogyne javanica* and *M. incognita*), and implications for crops grown in rotation with sugarcane. Proc Aust Soc Sugarcane Technol 28:345–350

Stirling GR, Blair BL, Pattemore JA, Garside AL, Bell MJ (2001) Changes in nematode populations following fallow, fumigation and crop rotation, and implications for the role of nematodes in yield decline. Australas Plant Pathol 30:323–335

Varkey S, Anith KN, Narayana R, Aswini S (2018) A consortium of rhizobacteria and fungal endophyte suppress the root-knot nematode parasite in tomato. Rhizosphere 5:38–42. https://doi.org/10.1016/j.rhisph.2017.11.005

Wheeler WB, Thompson MP, Edelstein RL, Krause RT (1979) Ultrasonic extraction of carbon furon residues from radishes. Bull of Environ Contam Toxicol 21:238–242

Williamson VM, Robert PA (2009) Mechanisms and genetics of resistance. In: Perry RN, Moens M, Starr JL (eds) Root-knot nematodes. CABI, Wallingford, pp 301–325

Yadav YS, Siddqui AU, Parihar A (2006) Efficacy of oil cakes as seed dressing treatment for the management of root-knot nematode Meloidogyne incognita affecting chickpea. Indian J Nematol 36:151–152

Zhang F, Gao C, Wang J et al (2017) Coupling sugarcane yield to soil nematodes: implications from different fertilization regimes and growth stages. Agric Ecosyst Environ 247:157–165. https://doi.org/10.1016/j.agee.2017.06.020

Zhou L, Yuen G, Wang Y et al (2016) Evaluation of bacterial biological control agents for control of root-knot nematode disease on tomato. Crop Prot 84:8–13. https://doi.org/10.1016/j.cropro.2015.12.009

Zukerman BM, Esnard J (1994) Biological control of plant nematodes current status and hypothesis. Jpn J Nematol 24:1–13

Chapter 7
Plant Growth-Promoting Bacterial Life at High Salt Concentrations: Genetic Variability

Ritika Kapoor and S. S. Kanwar

Abstract Abiotic stresses are emerging environmental factors limiting agricultural productivity around the world. Among these stresses, salt stress is a serious threat affecting crop production especially in arid and semiarid regions of the world. Development of strategies to ameliorate deleterious effects of salt stress on plants has received considerable attention. In this scenario, the use of salt-tolerant plant growth-promoting microorganisms to enhance salinity resilience in crops is encouraged due to their vital interactions with crop plants. Bacteria are widely used to diminish deleterious impacts of high salinity on crop plants because they possess various direct and indirect plant growth-promoting characteristics. This chapter focuses on the effect of salt stress on plants, plant growth-promoting bacterial survival in saline conditions, and their mechanisms to mitigate salt stress at genetic level.

7.1 Introduction

Salinity is one of the major abiotic stresses which negatively affects crop growth and yields and puts down crop production. The presence of high sodium chloride concentration has been reported to cause reduction in microbial flora in the soil (Ibekwe et al. 2010). Most of the world's plateaus confined to the tropics and Mediterranean regions have potential salinity problems (Cordovilla et al. 1994). It is accounted for the presence of naturally high salt levels, salt accumulation during irrigation, or the application of chemical fertilizers. High salinity owing to its toxic effects inhibits the growth of plants by affecting cellular growth and entry of ions across the root system by slowing down the water uptake of plants. Sodium chloride is the most disparaging salt that affects the growth of plants. Saline habitats are frequently inhabited by an abundance of microbial communities adapted to these ecosystems (Zahran et al. 1992). Halophilic bacteria which flourish in hypersaline habitats may retain their potential to express various types of plant growth-

R. Kapoor · S. S. Kanwar (✉)
Department of Microbiology, CSK Himachal Pradesh Agricultural University, Palampur, Himachal Pradesh, India

A. Varma et al. (eds.), *Plant Biotic Interactions*,
https://doi.org/10.1007/978-3-030-26657-8_7

promoting activities such as phosphate solubilization, nitrogen fixation, or phytohormone production. These PGPRs offer promise as potential biofertilizers for improvement of plant growth under stress conditions by reducing the impact of salinity on plant growth and its productivity. Salt-tolerant bacterial species of *Bacillus*, *Pseudomonas*, *Azotobacter*, and *Enterobacter* have been isolated from salt-affected soil and were found to be efficient plant growth promoters (Gopalkrishnan et al. 2012; Allam et al. 2018; Nakbanpote et al. 2014; Kapoor et al. 2017). It is well drafted that indigenous strains have better potential to multiply under stress conditions as compared to the exotic strains. These facts are important while selecting the microbial inoculants for a specific environment. This chapter emphasizes on the assessment of plant growth-promoting rhizobacteria (PGPR) approaches for the alleviation of salinity stress with a brief overview of adaptation mechanism and genetic variability of salt-tolerant strains facilitating them to grow in saline environments.

7.2 Diversity of Salt-Tolerant Bacteria

Salinity affects the structure and species composition of the rhizospheric communities. Saline environments harbor taxonomically diverse bacterial groups such as *Enterobacter*, *Pseudomonas*, *Vibrio*, and a few Gram-positive bacterial species, e.g., *Bacillus*, *Micrococcus*, and *Salinicoccus*, which exhibit modified physiological and structural characteristics under the prevailing saline conditions (DasSarma and DasSarma 2012). Salt-tolerant bacteria have been isolated from different sources such as salt lakes (Hedi et al. 2009), river water (Tiquia et al. 2007), rhizosphere (Hasnain and Taskeen 1989), root nodules (Gal and Choi 2003), and soil samples (Takashina et al. 1994). Gram-negative bacteria including nodulating bacteria have been reported to colonize the saline soil (Zahran et al. 1992). Nodulating bacteria such as *Rhizobium* have been reported in association with the salt marsh grass (Whiting et al. 1986). Among free-living bacteria, those belonging to genus *Azospirillum*, *Bacillus*, *Enterobacter*, and *Azotobacter* play a crucial role in different stressed conditions (Sahoo et al. 2014). In fact, inoculation with *Azotobacter* has been found to exert several beneficial effects on plant yields as possess various plant growth promoting traits and also found to produce exopolysaccharides under saline conditions (De la Vega et al. 1991; Mrkovacki et al. 1996). Consequently, it has been implicit that isolating bacteria with PGP traits from naturally saline environments would give indigenous isolates to improve the effect of salt stress on plants.

7.3 Plant Growth-Promoting Activities of Salt-Tolerant Bacteria

7.3.1 Nitrogen Fixation Under Salt Stress

Nitrogen is the essential macronutrient required for plant growth. Bacteria inhabiting under saline conditions alter some of their activities and pathways to adapt themselves. One of the sensitive activities is the nitrogenase activity, which is affected by extreme saline conditions. Nitrogen fixation was found to be decreased in saline soils as salt stress adversely affects the nitrogenase enzyme activity (Gao et al. 2014). The extent of effects of salinity on denitrification process is dependent on the type of nitrogen compound present in the soil (El-Shinnawi et al. 1982). These stressed conditions disrupt the nitrogen cycle and lead to the disappearance of nitrate (NO^{3-}) from saline soil through denitrification process, resulting in alteration of enzymatic processes (Azhar et al. 1989). Biological nitrogen fixation (BNF) involves the enzymatic reduction of nitrogen to ammonia (NH_3), which acts as the precursor molecule for the biosynthesis of amino acids and other nitrogen-containing biomolecules. Islam et al. (2010) studied the free-living culturable diazotrophic bacteria of paddy soils under salt stress conditions and found that 32 bacteria were positive for acetylene reduction assay (ARA) and the values ranged from 1.8 to 2844.7 nmol ethylene h^{-1} mg^{-1} protein. The study carried out by Chowdhury et al. (2007) on diazotrophic bacterial isolates showed that the predominance of Gram-negative bacteria from the surface-sterilized roots of *Lasiurus scindicus* were capable of fixing nitrogen. Nitrogen-fixing *Bacillus* strains were also obtained from saline lands of Egypt; these strains reduced acetylene in pure culture at 5% NaCl (Zahran et al. 1992).

7.3.2 Phosphate Solubilization

Phosphorus is one of the key nutrients for plants, but a major portion of it is available in insoluble form. Microorganisms play a vital role in solubilizing phosphorous and in increasing the availability of phosphorous to plants. Phosphate-solubilizing microorganisms belonging to genera *Klebsiella*, *Erwinia*, *Rhizobium*, *Achromobacter*, *Aerobacter*, *Enterobacter*, *Pseudomonas*, *Micrococcus*, and *Bacillus* have been reported earlier. However, strains belonging to *Pseudomonads* and *Bacillus* are deliberated as the most proficient phosphate solubilizers (Villegas and Fortin 2002), whereas fungal species such as *Aspergillus*, *Penicillium*, and *Curvularia* and yeast are widely reported to solubilize various forms of inorganic phosphates (Das et al. 2013).

Several researchers have isolated phosphate-solubilizing microorganisms from various niches of saline soils (Sharan et al. 2008; Park et al. 2010; Srinivasan et al. 2012). A salt-tolerant, nitrogen-fixing, and phosphate-solubilizing species *Swaminathania salitolerans* has been isolated from the rhizosphere, roots, and stems of salt-tolerant mangrove associated with wild rice (Loganathan and Nair 2004). In another study, phosphate-solubilizing bacteria *Alteromonas* sp. and *Pseudomonas aeruginosa* have been isolated from salt-affected soils. These isolates were found to solubilize phosphate under saline conditions, i.e., up to 2M NaCl concentration (Srinivasan et al. 2012). Rosado et al. (1998) and Nautiyal (1999) observed increased phosphate-solubilizing activity of bacteria in the presence of 10% NaCl, but the solubilizing activity decreased with increase in NaCl concentration.

7.3.3 Siderophore Production

Iron is the fourth abundant and essential growth element for all living organisms and perhaps the most important micronutrient used by bacteria for their metabolism. To confiscate and solubilize ferric iron, many microorganisms utilize low-molecular-weight (<1000 Da) compounds with high iron affinity known as "siderophores." Siderophores are produced by rhizospheric bacteria to enhance the growth and development of plants by increasing the availability of iron. Siderophore-producing microorganisms prevailing in the rhizosphere suggest that plants would all become iron deficient in the absence of iron-chelating siderophores (Kloepper et al. 1980). Nine halophilic archaea were isolated from marine salterns for siderophore production (Dave and Desai 2006). Ramadoss et al. (2013) found that *Bacillus halodenitrificans* and *Halobacillus* sp. isolated from saline habitats exhibited siderophore-producing activity.

7.3.4 Indole Acetic Acid (IAA) Production

Some PGPR strains enhance plants' growth and development by modulating the concentration of known phytohormones. Among plant hormones, auxins and ethylene play an essential role in root system development and crop yield. Indole-3-acetic acid (IAA) is the common natural auxin that extensively affects plant physiology. Diverse microbial groups are capable of producing physiological active auxins, which exert pronounced effects on plant growth and its establishment. In order to produce auxin, bacteria use tryptophan as a precursor molecule and convert it into IAA (Etesami et al. 2009). In plants, saline stress often affects the production of IAA and makes them imbalance. Thus, it is important to study IAA-producing rhizobacteria in saline conditions which could facilitate plant growth under salt stress. It has been reported that pre-sowing seeds with phytohormones alleviated the growth-restricted effect of salt stress (Ramadoss et al. 2013). Zahir et al. (2010) isolated IAA-producing halophilic *Rhizobium phaseoli* strains from the mung bean nodules and evaluated their growth parameters in the presence and absence of tryptophan under salt stress conditions. Growth promotion effects were observed, and this might be due to higher auxin production and mineral uptake in rhizosphere, which reduced the adverse effect of salinity.

7.3.5 *Lytic Enzyme Production*

Lytic enzyme production is one of the indirect approaches for plant growth promotion. A wide array of organisms have been obtained from harsh environments that produce many active and stable enzymes including proteases (Durham et al. 1987), amylases (Horikoshi 1971), lipases (Watanable et al. 1977), etc. Lytic enzymes produced by biocontrol organisms mediate defense against the pathogens and improve plant growth (Vivekananthan et al. 2004). Enzymes that are stable and active at extreme saline conditions are very much in demand for various industrial processes. Shaheen et al. (2008) reported the protease enzyme production by *Bacillus subtilis* at different concentrations of salt (0–6% NaCl). Sivaprakasam et al. (2011) obtained salt-tolerant alkaline protease from *P. aeroginosa* that was capable of enzymatic degradation.

7.4 Mechanism of Salt Tolerance

Salt stress reduces microbial population in the rhizosphere. Microbes that inhabit hypersaline environments experience intense osmotic pressure and thus use "compatible solute strategy" or the "salt-in strategy" to resist salt stress (Etesami and Beattie 2017). Bacteria accumulate compatible solutes and other amino acids under saline conditions (Brown 1976). Some salt-tolerant bacteria can use salt in strategy mechanism and accumulate electrolytes, e.g., K^+ glutamate. Furthermore, enzymes, ribosomes, and transport proteins of these bacteria require high level of potassium for stability and activity. Organic solutes increase the intracellular osmotic strength and stabilize the cellular macromolecules (Lippert and Galinski 1992).

Specific genetic induction is required to accumulate compatible organic solutes in salt-tolerant bacteria (Plemenitas et al. 2014). Intracellular proline was found to increase rapidly in *Bacillus* in response to osmotic stress by NaCl, and the corresponding genes were detected as proB, proA, and proC encoding γ-glutamyl kinase (γ-GK), γ-glutamyl-phosphate reductase (γ-GPR), and pyrroline-5-carboxylate (P5C) reductase, respectively (Chen et al. 2007). Various genes encoding L-aspartokinase (Ask), L-2,4-diaminobutyric acid transaminase (EctB), L-2,4-diaminobutyric acid acetyltransferase (EctA), and L-ectoine synthase (EctC) have been located and found to be involved in the biosynthesis of ectoine in *Halobacillus dabanensis* (Nada et al. 2011). Four genes, viz., *betI*, *betC*, *betB*, and *beta*, were found to be essential for oxidation of choline or choline-*O*-sulfate to glycine betaine organized in one operon (Sevin and Sauer 2014). Other antiporter genes that have been reported in salt-tolerant bacteria are also essential for maintaining the balance of Na^+ and K^+ ions in the cell in order to attain an osmotic equilibrium. This mechanism is accompanied by certain physiological modifications which are required to protect all the metabolic and regulatory functions at high salinity (Eisenberg and Wachtel 1987). Na^+/H^+ antiporters are membrane proteins involved in pH and Na^+ homeostasis in cells that exchange Na^+ for H^+ (Inaba et al. 2001). The genes that are proved to be involved in halotolerance in bacteria either through knockout studies or through overexpression studies are given in Table 7.1.

7.4.1 Genetic Variations Based on Nha

Sodium hydrogen antiporters transport Na^+ or Li^+ in exchange for H^+ across the cytoplasmic membrane of cell (Alkoby et al. 2014) and maintain intracellular pH homeostasis, detoxification of cells from Na^+ ions, regulation of cell volume, and establishment of an electrochemical potential of Na^+ ions (Padan 2014). Various Na^+/H^+ antiporters such as *nhaA*, *nhaB*, *nhaC*, *nhaD*, *nhaP*, *chaA*, *tetA(L)*, and *napA* have been identified in Gram-positive and Gram-negative bacteria (Padan et al. 2001; Majernik et al. 2001). *NhaA* gene responsible for salt tolerance in

Table 7.1 Genes conferring salt-tolerance response in selected bacteria

Strains	Genes	Product	Source
Pseudomonas aeruginosa	nhaP	Na^+/H^+ antiporter	Inaba et al. (2001)
Sinorhizobium meliloti	relA bet genes betS gene greA Kup	(p)ppGpp synthetase Glycine Betaine/proline Betaine transporter Transcription cleavage factor Potassium uptake protein	Wei et al. (2004) Pocard et al. (1997) Nogales et al. (2002)
Rhizobium tropici	ntrY, ndvA and ndvB (synthetic gene) nhaA, nhaB, nhaC	Histidine kinase Na^+/H^+ antiporter	Miller and Wood (1996)
Azotobacter vinelandii	ggpPS	Glucosyl glycerol biosynthesis	Klahn et al. (2009)
Enterobacter ludwigii	nhaA	Na^+/H^+ antiporter	Kapoor et al. (2017)
Synechocystis	nhaS1, nhaS2, nhaS3, nhaS4, and nhaS5	Na^+/H^+ antiporter	http://www.ncbi.nlm.nih.gov/nuccore/NC_016514
Aphanothece halophytica	napA	Na^+/H^+ antiporter	Inaba et al. (2001)
Bacillus subtilis	proA, proB, proC	γ-Glutamyl kinase	Chen et al. (2007)
Enterobacter cloacae	nhaA	Na^+/H^+ antiporter	Kapoor (2014)
Bacillus aquimaris	nhaA	Na^+/H^+ antiporter	Kapoor (2014)

Enterobacter sp. has been reported previously (Kapoor et al. 2017). The primary structure of all the abovementioned genes exhibits very weak or no significant homology. This indicates that different transport systems coupling H^+ and Na^+ circulation have developed during evolution. Several genes encoding Na^+/H^+antiporters from different microorganisms have been shown variability by replacing *nhaA* of *Escherichia coli* e.g., *nhaA* of *Vibrio alginolyticus* (Nakamura et al. 1994), *Vibrio parahaemolyticus* (Kuroda et al. 1994), *Bacillus aquimaris* and *Enterobacter cloacae* (Kapoor 2014), *Enterobacter ludwigii* (Kapoor et al. 2017), as well as *nhaB* of *V. parahaemolyticus* (Nozaki et al. 1996), *nhaD* of *V. parahaemolyticus* (Nozaki et al. 1998), *nhaP* of *Pseudomonas aeruginosa* (Utsugi et al. 1998), *nhaC* of *Bacillus pseudofirmus* OF4 (Ivey et al. 1993), *napA* of *Enterococcus hirae* (Strausak et al. 1993), and *mnh* of *Staphylococcus aureus* (Hiramatsu et al. 1998).

Amino acid residues Asp-133, Asp-163, and Asp-164 (Inoue et al. 1995) and His-225 were proposed to be involved in pH sensitivity in *E. coli* for binding of sodium ions (Gerchman et al. 1993). Furthermore, amino acid residues Gly-14, Gly-166, Phe-267, Leu-302, Gly-303, Cys-335, Ser-342, and Ser-369 located in the cell membrane were identified by Nuomi et al. (1997) and found to be essential for the activity of *nhaA* in *E. coli*. In general, 111 amino acid residues were found to be fully conserved in the *nhaA* gene products from different bacteria (Inoue et al. 1995; Vimont and Berche 2000). Our previous study showed that specific insertions/deletions caused major variations of amino acids in salt-tolerant strains (Kapoor et al. 2017). However, these types of alleles mined the rare mutation among the salt- and non-salt-tolerant strains, and little information is available on allele mining of genes responsible for salt tolerance.

7.5 Conclusions

The salt tolerance mechanism of plants modulated by rhizosphere bacteria opens up new prospects to understand plant–microbe interaction. PGPR strains have been conventionally used as biofertilizers to augment the growth and yield of different crops under salt stress conditions. The variability in salt tolerance behavior of bacteria can be explored by targeting the genes involved in salt tolerance mechanism. Genetic and genomic studies used to determine allele mining in gene sequences among salt- and non-salt-tolerant strains are yet to be explored. Gene silencing approach can be used to study the precise function of specific gene in salt-tolerant strains.

References

Alkoby D, Rimon A, Burdak M, Patino-Ruiz M, Calinescu O (2014) NhaA Na+/H+ antiporter mutants that hardly react to the membrane potential. PLoS One 9:e93200

Allam NG, Kinany R, El-Refai E, Ali WY (2018) Potential use of beneficial salt tolerant bacteria for improving wheat productivity grown in salinized soil. J Microbiol Res 8:43–53. https://doi.org/10.5923/j.microbiology.20180802.03

Azhar E, Van Cleemput O, Verstraete W (1989) The effect of sodium chlorate and nitrapyrin on the nitrification mediated nitrosation process in soils. Plant Soil 116:133–139

Brown AD (1976) Microbial water stress. Bacteriol Rev 40:803–846

Chen M, Wei H, Cao J, Liu R, Wang Y, Zheng C (2007) Expression of *Bacillus subtilis* proBA genes and reduction of feedback inhibition of proline synthesis increases proline production and confers osmotolerance in transgenic Arabidopsis. J Biochem Mol Biol 40:396–403

Chowdhury SP, Schmid M, Hartmann A, Tripathi AK (2007) Identification of diazotrophs in the culturable bacterial community associated with roots of *Lasiurus sindicus*, a perennial grass of Thar Desert, India. Microb Ecol 54:82–90

Cordovilla MP, Ligero F, Lluch C (1994) The effect of salinity on N fixation and assimilation in *Vicia faba*. J Exp Bot 45:1483–1488

Das A, Dutta BK, Barooah AK (2013) In vitro solubilization of inorganic phosphate by phosphate solubilizing fungi isolated from tea agroecosystem soil of Barak Valley, Southern Assam. Int J Microbiol Res 4:336–341

DasSarma S, DasSarma P (2012) Halophiles. In: eLS. Wiley, Chichester

Dave SR, Desai HB (2006) Microbial diversity at marine salterns near Bhavnagar, Gujarat, India. Curr Sci 90:497–500

De la Vega MG, Cejudo FJ, Paneque A (1991) Production of exocellular polysaccharide by *Azotobacter chroococcum*. Appl Biochem Biotechnol 30:273–284

Durham DR, Stewart DB, Stellwag EJ (1987) Novel alkaline and heat stable proteases from alkalophilic Bacillus sp. Strain GX 6638. J Bacteriol 169:2262–2768

Eisenberg H, Wachtel EJ (1987) Structural studies of halophilic proteins, ribosomes, and organelles of bacteria adapted to extreme salt concentrations. Annu Rev Biophys Biophys Chem 16:69–92

El-Shinnawi MM, Omran MS, Abo El-Naga SA (1982) Denitrification in soil saturated with saline water. Appl Microbiol Biotechnol 43:1099–1106

Etesami H, Beattie GA (2017) Plant-microbe interactions in adaptation of agricultural crops to abiotic stress conditions. In: Kumar V, Kumar M, Sharma S, Prasad R (eds) Probiotics and plant health. Springer, Singapore, pp 163–200

Etesami H, Alikhani HA, Akbari A (2009) Evaluation of plant growth hormones production (IAA) ability by Iranian soils rhizobial strains and effects of superior strains application on wheat growth indexes. World Appl Sci J 6:1576–1584

Gal SW, Choi YJ (2003) Isolation and characterization of salt tolerance Rhizobia from *Acacia* root nodules. Agric Chem Biotechnol 46:58–62

Gao H, Bail J, He X, Zhao Q, Lu Q, Wang J (2014) High temperature and salinity enhance soil nitrogen mineralization in a tidal freshwater marsh. PLoS One 9:e95011

Gerchman Y, Olami Y, Rimon A, Taglicht D, Schuldiner S, Padan E (1993) Histidine-226 is part of the pH sensor of NhaA, a Na^+/H^+ antiporter in *Escherichia coli*. PNAS 90:1212–1216

Gopalkrishnan S, Upadhyaya HD, Vadlamudi S, Humayun P, Vidya MS, Alekhya G, Singh A, Vijyabharathi Bhimineni RK, Seema M, Rathore RO (2012) Plant growth promoting traits of biocontrol potential bacteria isolated from rice rhizosphere. Springerplus 1:71–76

Hasnain S, Taskeen N (1989) Characterization of salt tolerant bacteria isolated from the rhizosphere of *Leptochloa fusca* and *Atritplex rhocodoidaes*. Pak J Pharm Sci 2:55–57

Hedi A, Sadfi N, Fardeau M-L, Rebib H, Cayol J-L, Ollivier B, Boudabous A (2009) Studies on the biodiversity of halophilic microorganisms isolated from El-Djerid salt lake (Tunisia) under aerobic conditions. Int Microbiol 9:731786

Hiramatsu T, Kodama K, Kuroda T, Mizushima T, Tsuchiya T (1998) A putative multisubunit Na^+/H^+ antiporter from *Staphylococcus aureus*. J Bacteriol 180:6642–6648

Horikoshi K (1971) Production of alkaline enzyme by alkalophilic microorganisms. Part II. Alkaline amylase produced by Bacillus No. A-40-2. Agric Biol Chem 35:1783–1791

Ibekwe AM, Poss JA, Grattan SR, Grieve CM, Suarez D (2010) Bacterial diversity in cucumber (*Cucumis sativus*) rhizosphere in response to salinity, soil pH, and boron. Soil Biol Biochem 42:567–575

Inaba M, Sakamoto A, Murata N (2001) Functional expression in *Escherichia coli* of low-affinity and high-affinity Na $^+(Li^+)/H^+$ antiporters of *Synechocystis*. J Bacteriol 183:1376–1384

Inoue H, Nuomi T, Tsuchiya T, Kanzawa H (1995) Essential aspartic acid residues, Asp-133, Asp-163 and Asp-164, in the transmembrane helices of a Na^+/H^+antiporter (NhaA) from *Escherichia coli*. FEBS Lett 363:264–268

Islam R, Trivedi P, Madhaiyan M, seshadre S, Lee G, Yang J, Kim Y, Kim M, Han G, Chauhan PS, Sa T (2010) Isolation, enumeration, and characterization of diazotrophic bacteria from paddy soil sample under long term fertilizer management experiment. Biol Fertil Soils 46:261–269

Ivey DM, Guffanti AA, Zemsky J (1993) Cloning and characterization of a putative Ca^{2+}/H^+ antiporter gene from *Escherichia coli* upon functional complementation of Na^+/H^+ antiporter-deficient strains by the overexpressed gene. J Biol Chem 268:11296–11303

Kapoor R (2014) Bacterial diversity of salt tolerant nitrogen fixers around the salt mines of Himachal Pradesh. PhD Thesis. CSKHPKV, Palampur (HP)

Kapoor R, Gupta MK, kumar N, Kanwar SS (2017) Analysis of *nhaA* gene from salt tolerant and plant growth promoting *Enterobacter ludwigii*. Rhizosphere 4:62–69. https://doi.org/10.1016/j.rhisph.2017.07.002

Klahn S, Marquardt DM, Rollwitz I, Hagemann M (2009) Expression of the *ggpPS* gene for glucosylglycerol biosynthesis from *Azotobacter vinelandii* improves the salt tolerance of *Arabidopsis thaliana*. J Exp Bot 60(6):1679–1689

Kloepper JW, Leong J, Teintze M, Schroth MN (1980) Enhanced plant growth by siderophores produced by plant growth promoting rhizobacteria. Nature 286:885–886

Kuroda T, Shimamoto T, Inaba K, Tsuda M, Tsuchiya T (1994) Properties and sequence of the NhaA Na1/H1 antiporter of *Vibrio parahaemolyticus*. J Biochem 116:1030–1038

Lippert K, Galinski EA (1992) Enzyme stabilization by ectoine-type compatible solutes: protection against heating, freezing and drying. Appl Microbiol Biotechnol 37:61–65

Loganathan P, Nair S (2004) *Swaminathania salitolerans* gen. nov. sp. nov., a salt-tolerant nitrogen-fixing and phosphate solubilizing bacterium from wild rice (*Porteresia coarctata* Tateoka). Int J Syst Evol Microbiol 54:1185–1190

Majernik A, Gottschalk G, Daniel R (2001) Screening of environmental DNA libraries for the presence of genes conferring $Na^+(Li^+)/H^+$antiporter activity on *Escherichia coli*: characterization of the recovered genes and the corresponding gene products. J Bacteriol 183:6645–6653

Miller KJ, Wood JM (1996) Osmoadaptation by rhizosphere bacteria. Annu Rev Microbiol 50:101–136

Mrkovacki N, Mezei S, Kovacev L (1996) Effect of Azotobacter inoculation on dry matter mass and nitrogen content in the hybrid varieties of sugar beet. A Periodical of Scientific Research on Field and Vegetable Crops 25:107–113

Nada AMK, Refaat MH, Abdel-Sabour MS, Hassan AM, Abd El Kader MM (2011) Molecular studies on EctC gene (Ectoine) in some halophilic bacterial isolates. Researcher 3:34–42

Nakamura T, Komano Y, Itaya E, Tsukamoto K, Tsuchiya T, Unemoto T (1994) Cloning and sequencing of an Na^+/H^+ antiporter gene from the marine bacterium *Vibrio alginolyticus*. Biochim Biophys Acta 1190:465–468

Nakbanpote W, Panitlurtumpai N, Sangdee A, Sakulpone N, Sirisom P, Pimthong A (2014) Salt-tolerant and plant growth-promoting bacteria isolated from Zn/Cd contaminated soil: identification and effect on rice under saline conditions. J Plant Interact 9:379–387

Nautiyal CS (1999) An efficient microbiological growth medium for screening phosphate solubilizing microorganisms. FEMS Microbiol Lett 170:265–270

Nogales J, Campos R, BenAbdelkhalek H, Olivares J, Lluch C, Sanjuan J (2002) *Rhizobium tropici* genes involved in free-living salt tolerance are required for the establishing of efficient nitrogen-fixing symbiosis with *Phaseolus vulgaris*. Mol Plant-Microbe Interact 15:225–232

Nozaki K, Inaba K, Kuroda T, Tsuda M, Tsuchiya T (1996) Cloning and sequencing of the gene for Na^+/H^+ antiporter of *Vibrio parahaemolyticus*. Biochem Biophys Res Commun 222:774–779

Nozaki K, Kuroda T, Mizuschima T, Tsuchiya T (1998) A new Na^+/H^+ antiporter, NhaD, of *Vibrio parahaemolyticus*. Biochim Biophys Acta 1369:213–220

Nuomi TH, Inoue T, Tsuchiya ST, Kanzawa H (1997) Identification and characterization of functional residues in a Na^+/H^+ antiporter (NhaA) from *Escherichia coli* by random mutagenesis. J Biochem 121:661–670

Padan E (2014) Functional and structural dynamics of NhaA, a prototype for Na^+ and H^+ antiporters, which are responsible for Na^+ and H^+ homeostasis in cells. Biochim Biophys Acta 1837:1047–1062

Padan E, Venturi M, Gerchman Y, Dover N (2001) Na(+)/H(+) antiporters. Biochim Biophys Acta 1505:144–157

Park KH, Lee OM, Jung HI, Jeong JH, Jeon YD, Hwang DY, Lee CY, Son HJ (2010) Rapid solubilization of insoluble phosphate by a novel environmental stress-tolerant *Burkholderia vietnamiensis* M6 isolated from ginseng rhizospheric soil. Appl Microbiol Biotechnol 86:947–955

Plemenitas A, Lenassi M, Konte T, Kejzar A, Zajc J, Gostincar C, Cimerman NG (2014) Adaptation to high salt concentrations in halotolerant/Halophilic fungi: a molecular prospective. Front Microbiol 5:199

Pocard JA, Vincent N, Boncompagni E, Smith LT, Poggi MC, DLe R (1997) Molecular characterization of the *bet* genes encoding glycine betaine synthesis in *Sinorhizobium meliloti* 102F34. Microbiology 143:1369–1379

Ramadoss D, Lakkineni VK, Bose P, Ali S, Annapurna K (2013) Mitigation of salt stress in wheat seedlings by halotolerant bacteria isolated from saline habitats. Springerplus 2:6

Rosado AS, de Azevedo FS, da Cruz DW, van Elsasand JD, Seldin L (1998) Phenotypic and genetic diversity of *Paenibacillus azotofixans* strains isolated from the rhizoplane or rhizosphere soil of different grasses. J Appl Microbiol 84:216–226

Sahoo RM, Ansari MW, Pradhan M, Dangar TK, Mihanty S, Tuteja M (2014) A novel *Azotobacter vinellandii* (SRIAz3) functions in salinity stress tolerance in rice. Plant Signal Behav 9:e29377

Sevin DC, Sauer U (2014) Ubiquinone accumulation improves osmotic-stress tolerance in *Escherichia coli*. Nat Chem Biol 10:266–227

Shaheen M, Shah AA, Hameed A, Hasan F (2008) Influence of culture conditions on production and activity of protease from *Bacillus subtilis* bs1. Pak J Bot 40:2161–2169

Sharan A, Shikha DNS, Gaur R (2008) *Xanthomonas compestris*, a novel stress tolerant, phosphate solubilizing bacterial strain from saline –alkali soils. World J Microbiol Biotechnol 24:753–759

Sivaprakasam S, Dhandapani B, Mahadevan S (2011) Optimization studies on production of a salt-tolerant protease from *Pseudomonas aeruginosa* strain bc1 and its application on tannery saline wastewater treatment. Braz J Microbiol 42:1506–1515

Srinivasan R, Alagawadi AR, Mahesh S, Meena KK, Saxena AK (2012) Characterization of phosphate solubilizing microorganisms from salt-affected soils of India and their effect on growth of sorghum plants *Sorghum bicolor* (L.). Moench. Ann Microbiol 62:93–105

Strausak D, Waser M, Solioz M (1993) Functional expression of the *Enterococcus hirae* NaH-antiporter in *Escherichia coli*. J Biol Chem 268:26334–26337

Takashina T, Otozati K, Hamamoto T, Horikoshi K (1994) Isolation of halophilic and halotolerant bacteria from a Japanese salt field and comparison of the partial 16S rRNA gene sequence of an extremely halophilic isolate with those of other extreme halophiles. Biodivers Conserv 3:632–642

Tiquia SM, Davis D, Hadid H, Kasparian S, Ismail M, Sahly R, Shim J, Singh S, Murray KS (2007) Halophilic and halotolerant bacteria from river waters and shallow groundwater along the Rouge river of southeastern Michigan. Environ Technol 28:297–230

Utsugi J, Inaba K, Kuroda T, Tsuda M, Tsuchiya T (1998) Cloning and sequencing of a novel Na^+/H^+ antiporter gene from *Pseudomonas aeruginosa*. Biochim Biophys Acta 1398:330–334

Villegas J, Fortin JA (2002) Phosphorous solubilization and pH changes as a result of the interactions between soil bacteria and arbuscular mycorrhizal fungi on a medium containing NO_3^- as nitrogen source. Can J Bot 80:571–576

Vimont S, Berche P (2000) NhaA, an Na1/H1 antiporter involved in environmental survival of *Vibrio cholera*. J Bacteriol 182:2937–2944

Vivekananthan R, Ravi M, Ramanathan A, Samiyappan R (2004) Lytic enzymes induced by *Pseudomonas fluorescene* and other biocontrol organisms mediated defence against the anthracnose pathogen in mango. World J Microbiol Bioltechnol 20:235–244

Watanable N, Ota Y, Minoda Y, Yomada K (1977) Isolation and identification of alkaline lipase producing microorganisms, cultural conditions and some properties of crude enzymes. Agric Biol Chem 41:1353–1358

Wei W, Jiang J, Yang SS (2004) Mutagenesis and complementation of *relA* from *Sinorhizobium meliloti* 042BM as a salt tolerance involvement gene. Ann Microbiol 54:317–324

Whiting GJ, Gandy EL, Yoch DC (1986) Tight coupling of root-associated nitrogen fixation and plant photosynthesis in the salt marsh grass *Spartina alterniflora* and carbon dioxide enhancement of nitrogenase activity. Appl Environ Microbiol 52:108–113

Zahir ZA, Shah KM, Naveed M, Akhter JM (2010) Substrate dependent auxin production by *Rhizobium phaseoli* improves the growth and yield of *Vignaradiata L.* under salt stress conditions. J Microbiol Biotechnol 20:1288–1294

Zahran HH, Moharram AM, Mohammad HA (1992) Some ecological and physiological studies on bacteria isolated from salt affected soils of Egypt. J Basic Microbiol 32:405–413

Chapter 8
Rhizosphere: A Home for Human Pathogens

Richa Sharma, V. S. Bisaria, and Shilpi Sharma

Abstract Rhizosphere is the zone where the microbe-mediated processes are influenced by root exudates. Owing to its high nutrient content due to root exudates, and ability to provide a safe home, it acts as a natural reservoir to not only beneficial bacteria but also to those which can be potential threat for humans, and hence acts as a 'microbial hot spot'. There has been an increase in incidences of human infections by opportunistic human pathogens residing in the rhizosphere. Many bacterial species are known to have dual interactions with both plants and humans. These bacterial species share similar colonization mechanisms for the rhizosphere and human organs. Other phenomena of common occurrence in rhizosphere are the higher rate of horizontal gene transfer, enhanced competition, and presence of various antibiotics resulting in greater level of natural resistances. The present chapter highlights the prevalence and concern of human pathogens residing in the rhizosphere.

8.1 Introduction

Rhizosphere (term coined by Hiltner) is the area surrounding the roots of a plant where processes mediated by microorganisms are influenced by the root system. Root exudates result in high nutrient content in the rhizosphere, providing a hot spot for different types of microorganisms, like bacteria, archaea, fungi, viruses, nematodes, protozoa, algae, and arthropods (Lynch 1990; Meeting 1992; Sørensen 1997; Bonkowski et al. 2009; Raaijmakers et al. 2009) which are much higher than the plant cells in terms of cell density (Mendes et al. 2013). Rhizosphere consists of the most structurally and functionally diverse niches owing to these rhizodeposits and microbial interactions (Hinsinger and Marschner 2006; Pierret et al. 2007; Raaijmakers et al. 2009; Hinsinger et al. 2009) and, hence, provides a platform for extensive scientific research. Several factors, viz. physical and chemical, alter the

R. Sharma · V. S. Bisaria · S. Sharma (✉)
Department of Biochemical Engineering and Biotechnology, Indian Institute of Technology Delhi, Hauz Khas, New Delhi, India
e-mail: shilpi@dbeb.iitd.ac.in

A. Varma et al. (eds.), *Plant Biotic Interactions*,
https://doi.org/10.1007/978-3-030-26657-8_8

structural and functional diversity of the resident microbial communities in soil. Whether root exudates actively participate in recruiting a specific microbial population (Cook et al. 1995; Berg 2009) or they are passively released because of the overflow of the plant products (Jones et al. 2009; Hartmann et al. 2009; Dennis et al. 2010) still needs to be addressed. As the rhizosphere is one of the most diverse microecosystems, harbouring a variety of species, it provides a home for the microbes beneficial for plants too. These microorganisms can be broadly classified as biocontrol agents (to combat the plant pathogens), biofertilizers [plant growth-promoting rhizobacteria (PGPR)] and phytostimulants (Whipps 2001; Berg 2009). For the biotechnological application of these agricultural amendments, studying their overall impact on plants and resident microbes is crucial.

Apart from these plant beneficial organisms, plant pathogens and other microbes also reside in rhizosphere, some of which can even be opportunistic human pathogens. Opportunistic human pathogenic bacteria cause diseases in patients with weakened immune systems or breached integumentary barriers (Parke and Gurian-Sherman 2001; Steinkamp et al. 2005). In recent times, an increase in the influence of opportunistic infections on human health has been reported. Hence, it is important to gain complete understating of the group. It is, therefore, important to assess their interactions with eukaryotes (Berg et al. 2010) and their pathogenic potential. In this chapter, we will throw light on (1) the diversity of bacterial species in rhizosphere which are potentially pathogenic to humans, and (2) the factors influencing eukaryote–microbe interaction.

8.2 Rhizosphere as a Favourable System for Potential Human Pathogens

The composition of rhizosphere is determined by a wide variety of biotic and abiotic factors. When compared to the bulk/root-free soil, the rhizospheric soil being nutrient rich attracts an astounding number of microorganisms resulting in increased biomass and activity. This phenomenon is known as the rhizospheric effect. Many cultivation-independent techniques showed that the microbial diversity in the soil is highly underestimated. The diversity of resident microbial community depends on various elements such as soil type, plant developmental stages, and different treatments of soil (Bulgarelli et al. 2012; Lundberg et al. 2012; Sharma et al. 2017). The roots of the plants release a range of compounds such as amino acids, vitamins, sugars and organic acids rendering rhizosphere to be a hot spot for a variety of microbes (Neumann and Römheld 2001). Sloughing-off from the root cap cells also contributes to rhizodeposition. In addition to the analysis of the abundance of the microbes present in soil, it is essential to go a step further to explore their activities during different developmental stages of plant growth. The combination of rhizosphere and its inhabiting microbes serve many ecological functions such as involvement in various nutrient cycles and plant growth promotion (Sørensen 1997).

Several advanced molecular approaches have been used to study gene expression in the rhizosphere (Mark et al. 2005; Barret et al. 2011). It was shown in the study conducted by Mark et al. (2005) that root exudates significantly altered about 104 genes. The detailed description of what happens in the rhizosphere can be revealed by employing 'omics' approach which enables the identification of genes, transcripts, proteins and metabolites. Delmotte et al. (2009) were the pioneers in adopting metaproteogenomic approach to explore the phyllosphere bacterial communities of Arabidopsis, soybean and clover plants. Complex plant–microbes interactions in the rhizosphere have been explored using MALDI-TOF/TOF-MS and T-RFLP (Knief et al. 2011; Wang et al. 2011; Wu et al. 2011).

Being nutrient rich, protective against UV radiations, and provider for water films for dispersal, rhizosphere serves as a favourable environment to not only beneficial bacteria but also to those that can be potential threat for humans (Berg et al. 2005; Tyler and Triplett 2008). The resident soil microbial communities (both plant beneficial and pant pathogens) display diverse interactions with each other (Whipps 2001), but there is no well-defined boundary between these organisms, and the interactions are host-dependent (Berg et al. 2005). Owing to the nutrient-rich conditions and diverse interactions within resident microbial community, the competition level is high in this 'microenvironment'. To compete well with their contenders, many microbes release antibiotics, like 2, 4-diacetylphloroglucinol, pyrrolnitrin, herbicolin A, pyoluteorin, etc. (Thomashow et al. 1997), which results in their establishment in the rhizosphere (Berg et al. 2002, 2005). Intense microbial interactions in the rhizosphere, together with the prevalence of a vast diversity of antibiotics, direct the resident microbes to acquire multiple antibiotic resistance genes (Riesenfeld et al. 2004). These bacterial species not only compete with other bacterial or fungal species, but also have antagonistic activities towards other eukaryotic organisms. This in turn suggests that these organisms can act against humans too. A number of opportunistic human pathogens have been reported to be more in the rhizospheric soil as compared to the bulk soil, because rhizosphere not only provides a perfect environment for a wide variety of interactions among resident microbial community but also facilitates suitable condition for horizontal gene transfer (Knudsen et al. 1988). There are higher chances of horizontal gene transfer within cross-domain species, which inhabit varied ecosystems but come into physical contact at some point (Overbeek et al. 2014). Reports are available for increased frequency of horizontal gene transfers in the rhizosphere compared to the bulk soil. Troxler et al. (1997) reported the affirmation of conjugative chromosomal gene transfer of the soil-resident strain of *Pseudomonas fluorescens* CHA0 with its clinical counterpart *Pseudomonas aeruginosa* PAO1 in the wheat rhizosphere, whereas they could not observe it in the bulk soil. Hence, the rhizosphere provides a suitable environment for opportunistic human pathogens as they are immensely competitive in terms of nutrient acquisition and production of vast varieties of antibiotics.

Pathogens include 'true' human pathogens like *Listeria monocytogenes*, *Salmonella enterica* serovar Typhimurium and *E. coli* O157:H7 and also opportunistic pathogens that cause disease only in humans with compromised immune system. The latter have been an intense topic of study as their impact on human health has shot up significantly in the past two decades in Europe and Northern America (Berg et al. 2005). Opportunistic human pathogens, such as *Stenotrophomonas maltophilia*, *Pseudomonas aeruginosa*, *Burkholderia cenocepacia*, *Bacillus cereus* and *Proteus vulgaris* have been hosted by rhizosphere of various plant species (cultivated and wild). These pathogens are responsible for skin, wound and urinary tract infection (Berg et al. 2005).

Disease outbreaks due to contaminated plant products have been one of the major concerns all over the world (Berg et al. 2005; Tyler and Triplett 2008; Teplitski et al. 2009; Critzer and Doyle 2010). In addition, many studies have reported that there are chances of contamination, not only from fresh food, but it can occur throughout the plant development stages, with the probable sources of contamination being contaminated manure and water used for irrigation. Their capability to persist in soil, and invade plants, suggests that these opportunistic human pathogens have the ability to inhabit the animal gastrointestinal tracts along with equal capability to survive in soil, which leads to their cross-kingdom leap as cited in various studies (van Baarleen et al. 2007; Kaestli et al. 2012). Plants, therefore, can be used as an alternative host for the survival of opportunistic human pathogens, and as a channel to re-colonize their animal hosts when ingested (Tyler and Triplett 2008).

Despite the fact that several studies have reported the occurrence of opportunistic human pathogens, knowledge of their virulence, as compared to their clinical counterparts, is still in its infancy. Kumar et al. (2013) studied the relatedness of PaBP35 strain of *P. aeruginosa* (isolated from aerial shoots of *Piper nigrum*) to its clinical counterpart by using polyphasic approaches like sequencing of *recN* gene, hybridization of comparative genome, and multilocus sequence typing. They found that the strain PaBP35 formed separate cluster from the typical clinical isolates. However, it was interesting to note that the strain PaBP35 exhibited resistance to multiple antibiotics, and cytotoxicity to mammalian cells. Moreover, it was able to cause infection in murine airway infection model. Contrasting results were obtained by Wu et al. (2011) who claimed that plant-related strain of *P. aeruginosa* M18 was sensitive to various antibiotics, and it was easier to eradicate the strain in an acute lung infection model of mouse when compared to its clinical counterpart, strain LESB58. Such unpredictable behaviour of different strains can be attributed to the fact that different strains might differ in genome expression profiles, virulence factors, and antibiotic resistance and susceptibility. Though rhizosphere provides a suitable habitat for human pathogens, many studies have argued that the resident microbial community imparts an obstacle against the intrusion of these human pathogens. For example, colonization of *S. enterica* and *E. coli* O157:H7 on the roots of *Arabidopsis thaliana* was impeded by *Enterobacter asburiae* (Cooley et al. 2003). On the other hand, Troxler et al. (1997) showed that many of these pathogens displayed high competitiveness with respect to nutrient acquisition, hence were able to colonize and subsequently proliferate in the presence of resident microbial community, as exhibited by *P. aeruginosa* in wheat rhizosphere. Similarly, *Stenotrophomonas* strains were not only able to inhabit strawberry rhizosphere but also stimulate root growth (Suckstorff and Berg 2003). The mechanism responsible for root colonization happens to be similar to those associated with human tissue colonization (Berg et al.

2005; Holden et al. 2009). On the other hand, several strains of *B. cepacia* were found to be virulent both in lung infection model and in alfalfa (Bernier et al. 2003). Klerks et al. (2007) hypothesized that the genes for adherence in *S. enterica* was turned on under the influence of root exudates. Apart from chemotaxis, the human pathogens get attached to the plants with the help of fimbriae, capsules and adhesins, which are regulated by several other genes. A list of several opportunistic human pathogens occurring in rhizosphere has been included in Table 8.1.

8.2.1 Pathogens as PGPR

Several genera (*Staphylococcus*, *Burkholderia*, *Pseudomonas*, *Enterobacter*, *Cronobacter*, *Herbaspirillum*, *Azospirillum*, *Ochrobactrum*, *Pantoea*, *Ralstonia*, *Serratia* and *Stenotrophomonas*) show bivalent interaction with both plants and humans exhibiting several plant growth-promoting (PGP) properties in addition to having antagonistic properties against the plant pathogens (Whipps 2001; Govan et al. 2000; Parke and Gurian-Sherman 2001).

The genus *Burkholderia* consists of many species, several of which are major pathogens of animals, humans and plants. Many other species have appeared as key opportunistic pathogens as well. There has been a marked increase in the incidences of diseases caused by *Burkholderia* in the last two decades (Govan et al. 1996; Coenye and Vandamme 2003). *Burkholderia* spp. have often been found in the rhizosphere of many grasses and crops (Tabacchioni et al. 2002). They have been reported to be of substantial interest as biocontrol agents, repressing plant pathogens and, in turn, improving crop yields (Govan et al. 1996; Hebbar et al. 1998; Holmes et al. 1998). These biocontrol properties are due to the production of a variety of compounds such as antifungal metabolites, and siderophores, and their ability to degrade a range of organic compounds (Kang et al. 1998). Hence, despite having immense biotechnological applications in agricultural fields, their commercial usage is impeded by probable danger to human health. *Ochrobacter* species inhabit a vast variety of niches, which includes soil, rhizosphere, activated sludge and sediments. These species being versatile are capable of denitrification and degrade recalcitrant chemicals (Goris et al. 2003). They have also been reported to be associated with various clinical issues (Jelveh and Cunha 1999; Moller et al. 1999). At both genotypic and phenotypic level, *Stenotrophomonas* species have exhibited a vast diversity (Berg et al. 1999; Hauben et al. 1999). They play a crucial role in biogeochemical cycles (Ikemoto et al. 1980), plant protection against pathogens (Berg et al. 1994; Kobayashi et al. 1995; Nakayama et al. 1999), and bioremediation and degradation of xenobiotic components (Binks et al. 1995; Lee et al. 2002). During the last decade, multidrug-resistant species have gained attraction, causing high fatality ratios in immuno-compromised patients (Denton and Kerr 1998). Though the pathogenicity of *Stenotrophomonas* is not clear because of the unavailability of literature on their virulence factors, antibiotics produced by them are utilized in controlling plant pathogens (Nakayama et al. 1999).

Table 8.1 Various opportunistic human pathogens associated with the rhizosphere of plants

S. no.	Plant	Opportunistic human pathogens	References
1.	*Brassica napus*	*Enterobacter intermedius, Burkholderia cepacia, Salmonella typhimurium, Cytophaga johnsonnae, Pseudomonas aeruginosa, Aeromonas salmonicida, Chromobacterium violaceum, Pantoea agglomerans, S. proteamaculans, S. rubidaea, S. liquefaciens, Stenotrophomonas maltophilia, Sphingomonas paucimobilis, Bacillus cereus, Chryseomonas luteola, Serratia grimesii, Proteus vulgaris, Chryseobacterium indologenes*	Berg et al. (1996, 2002), Graner et al. (2003)
2.	*Solanum tuberosum*	*Staphylococcus epidermidis, S. xylosus, S. pasteuri, Kluyvera cryocrescens, Proteus vulgaris, Flavimonas oryzihabitans, Bacillus cereus, Serratia grimesii, Achromobacter xylosoxidans, Cytophaga johnsonae, Stenotrophomonas maltophilia, Enterobacter amnigenus, E. intermedius, E. cloacae, Janthinobacterium lividum, Pantoea agglomerans, Pseudomonas aeruginosa, Alcaligenes faecalis, Chromobacterium violaceum, Sphingomonas paucimobilis, Burkholderia cepacia, Francisella philomiragia, Ochrobactrum anthropi, Sphingobacterium spiritivorum*	Berg et al. (2002, 2005), Gupta et al. (2001), Krechel et al. (2002), Lottmann et al. (1999), Lottmann and Berg (2001), Reiter et al. (2002), Sessitsch et al. (2004)
3.	*Fragaria X ananassa*	*Proteus vulgaris, Salmonella typhimurium, Pantoea agglomerans, Burkholderia cepacia, Serratia proteamaculans, Acinetobacter baumannii, Staphylococcus epidermidis, Acinetobacter calcoaceticus, Serratia grimesii*	Berg et al. (2002, 2005)
4.	*Medicago sativa* (target plant) *Chenopodium album* (non-target plant)	*Flavobacterium johnsoniae, Stenotrophomonas maltophilia*	Schwieger and Tebbe (2000)

(continued)

Table 8.1 (continued)

S. no.	Plant	Opportunistic human pathogens	References
5.	*Helianthus annuus*	*Stenotrophomonas maltophilia, Burkholderia cepacia, Flavobacterium odoratum*	Hebbar et al. (1998)
6.	*Zea mays*	*Klebsiella pneumoniae, Stenotrophomonas maltophilia, Sphingomonas paucimobilis, Burkholderia cepacia, Serratia liquefaciens*	Chelius and Triplett 2000), Dalmastri et al. (1999), Lambert et al. (1987)
7.	*Oryza sativa*	*Serratia marcescens, Pseudomonas aeruginosa, Ochrobactrum anthropi, Alcaligenes xylosoxidans, Aeromonas veronii, Enterobacter cloacae*	Gyaneshwar et al. (2001), Mehnaz et al. (2001), Tripathi et al. (2002)
8.	*Triticum aestivum*	*Ochrobactrum tritici, Burkholderia cepacia, Streptococcus pyogenes, Enterobacter agglomerans, Stenotrophomonas maltophilia Ochrobactrum anthropi, Staphylococcus aureus, Pseudomonas aeruginosa, Salmonella typhimurium*	Germida and Siciliano (2001), Morales et al. (1996)
9.	Hay grass	*Listeria monocytogenes, Salmonella* and Shiga toxin-producing *Escherichia coli* (STEC)	Strawn et al. (2013)
10.	*Daucus carota*	*Listeria monocytogenes*	Dowe et al. (1997)

8.2.2 Factors Responsible for Host–Pathogen Interaction

The antagonistic mechanism of interaction of rhizobacteria with its human host includes various means such as antibiosis, competition for niche and nutrients, and production of extracellular enzymes (Fravel 1988; Raaijmakers et al. 2002; De Souza et al. 2003). Rhizobacteria utilize fimbriae and cell surface proteins for their early interaction with the plant host. It then utilizes root exudates and seeds as the carbon source (Lugtenberg and Dekkers 1999). Under stress like changing osmolarity, these bacteria synthesize solutes to combat the stress (Miller and Wood 1996). The mechanisms of action such as invasion, colonization, growth, and approach used to establish virulence are similar for pathogenicity of bacteria in both plants and humans (Rahme et al. 1995; Cao et al. 2001). To survive in humans, the additional crucial feature is the potential to grow at 37 °C. To counter the infection, plants stimulate their innate immunity against bacterial components (flagella, lipopolysaccharides), which is very similar to the reaction of mammalian innate immune system in response to pathogens (Van Loon et al. 1998). Hartmann et al. (2004) reported the induction of innate immune system in tomato plants against N-acyl homoserine lactone secreted by *Serratia liquefaciens* for quorum sensing. Many studies showed that the factors responsible for favourable interaction with plants and pathogenicity in humans are almost similar or even identical, e.g. production of siderophores and extracellular enzymes (Tan et al. 1999). The study conducted by Dörr et al. (1998) showed that type IV pili of *Azoarcus* sp. BH72, which is used for the adhesion on plants, was also used for adhesion on fungus. Moreover, the sequence of amino acids of the pilus represents high resemblance to that of human pathogens such as *P. aeruginosa* and *Neisseria gonorrhoeae*. When in contact with different host niche or host, there are chances that harmless bacteria might turn into pathogens because of the display of their virulence potential to its full extent. Other factors include structural changes in bacterial genome due to horizontal gene transfer, recombination and mutations (Hacker et al. 2003). However, sequence information based on 16S rRNA is not at all a solution to draw conclusion regarding the pathogenicity, and therefore it is necessary to perform analysis at protein level.

8.2.3 Bacterial Genera with Unknown/Varied Pathogenicity

Many rhizospheric bacterial species have been reported to have varied pathogenicity on plants, animals and humans. *Bacillus thuringiensis* has been used for decades in fields to keep a check on insects and mosquitoes. While it is a ubiquitous soil bacteria, its ecological behaviour was not found to be related with *B. anthracis* and *B. cereus* (both of which are human pathogens) (Jensen et al. 2003). *B. thuringiesis*-based biopesticide preparations fed to sheep did not exert any harmful effect on these animals (Hadley et al. 1987). Therefore, it is possible to postulate that the strains incapable of growing at 37 °C can be regarded as safe for humans and biotechnological applications.

Burkholderia species, on the other hand, is considered to be one of the most debatable in terms of its agricultural applications. The reason for this is the difficulty in discriminating between its beneficial and clinical strains. Yabuuchi et al. (1992) defined the *Burkholderia* genus. Initially, it comprised of seven species, consisting of animal pathogens, human pathogens, plant pathogens, and opportunistic human pathogens. The emerging strains of *Burkholderia* possess beneficial properties with rare clinical reports and, therefore, were considered to be safe for application. These *Burkholderia* strains exhibit immense potential with respect to biotechnological applications by releasing a list of hydrolytic enzymes, and several bioactive components. Despite these properties, their application in industries and agriculture is extremely restricted because of any potential threat they might impose on humans (Eberl and Vandamme 2016).

Cross-domain pathogens include the species under the genera *Pseudomonas*. Differentiation of clinical type from non-pathogenic types (includes PGPR) is generally hard to accomplish. Representatives of the species *Pseudomonas fluorescens* have also been found to be associated with Crohn's diseases, apart from their common occurrence in rhizosphere (Wei et al. 2002; Eckburg and Relman 2007). An additional rhizosphere-borne representative of *Pseudomonas veronii* was reported to be linked with the formation of intestinal pseudotumour in humans (Cheuk et al. 2000).

Rhizosphere also provides a suitable home for those which cause diseases not only in plants but also in humans, e.g. *Erwinia* and *Pantoea* (Cruz et al. 2007; Coutinho and Venter 2009). Nithya and Babu (2017) demonstrated that human pathogens and opportunistic human pathogens coexisted with the plant beneficial bacteria as endophytes in salad vegetables. It was demonstrated that out of 19 genera and 46 species, 16.25% were human pathogens, whereas 26.5% were opportunistic human pathogens. Higher abundance of plant beneficial bacteria resulted in a negative impact on both plant and human pathogens.

8.3 Horizontal Gene Transfer as a Mode for Cross-Domain Talk

There are several modes of horizontal gene transfer, such as plasmids, bacteriophages and transposons (Juhas et al. 2009). When a human pathogen is not residing inside a human host, there are chances that novel phenotypic traits are acquired by it. Those human pathogens which reside in the rhizosphere, may acquire additional characteristics from the resident microbial community through horizontal gene transfer (Overbeek et al. 2014). The key helpers in the transmission of virulent factors between different bacterial strains are found to be bacteriophages, by either transduction or lysogenic conversion. This process results in bacterial evolution. Non-pathogenic variety of *E. coli* can be transformed into virulent strains post-infection with the temperate bacteriophages possessing *stx*1 and *stx*2 genes encoding Shiga toxins. This type of bacterial conversion from avirulent to virulent strains is more common in *Corynebacterium diphtheria* and several *Salmonella* species (Saunders et al. 2001). Conjugal plasmids are another mode of horizontal gene transfer. As there are commonalities in virulent factors among *Enterobacteriaceae* of plants and animals (Toth et al. 2006), conjugal plasmids might be responsible for the horizontal transfer of these factors in plant–soil ecosystem. Genes that are accountable for causing diseases in humans can be found among soil and plant resident bacterial species having human pathogens in close proximity. This stipulates that the rhizosphere is a hub of genes which might enhance robustness of human pathogens residing in the soil. Hence, rhizosphere provides an immense opportunity to reveal the processes of horizontal gene transfer.

8.4 Conclusions

There is a need to know the mode of interaction of pathogenic microbes with eukaryotic hosts before their application as bioinoculants. If they show a bivalent interaction, then it becomes all the more crucial to assess their risk. Such microbial species are known as crossover or cross-kingdom pathogens, viz. *Pseudomonas aeruginosa, Burkholderia cepacia, Dickeya* spp., *Enterococcus faecalis* and *Serratia marcescens*. Risk assessment should be done not only on the basis of genomic analysis (by analysing the presence of sequences responsible for pathogenesis) but also on the basis of interaction studies and proteomics, as there are chances of horizontal gene transfer and recombination, which facilitate bacteria to attain new physiological attributes. Also, bacteria associated with plant and humans contain similar 'interaction elements', but their presence is not a requisite for pathogenicity. In fact, better association has been displayed at protein level as shown in case of epidemic bacteria. In a nutshell, gaining the knowledge with respect to root colonization of human pathogens, and their means of transmission to humans is the need of the hour, which will be helpful in their further biotechnological applications.

8.5 Future Perspective

To decrease the incidences of diseases in plants and humans, rhizospheric engineering is a technique which seems to be promising in amelioration of the negative impact of these pathogens by preventing their germination, growth and attachment to the root surface. Similar to the probiotics for humans, 'probiotics for plants' has been proposed. This will be possible by strategically shaping the rhizosphere microbiome by designing a 'core microbiome' which will be effective against various soil-borne pathogens in varied agricultural fields. This microbiome should consist of microbes which display activities to combat the pathogens (Turnbaugh et al. 2009; Ursell et al. 2012) together with possessing PGP properties. This can be achieved by employing microbial genera that not only colonize the rhizosphere but also have their genomic data available publically. They should also be flexible in terms of genetic engineering attempts, viz. *Pseudomonas*, *Bacillus*, *Streptomyces*, *Paenibacillus* and *Rhizobium*.

Acknowledgements The study was supported by grant received from the Department of Biotechnology, Government of India (BT/PR5499/AGR/21/355/2012). RS wishes to acknowledge the fellowship received from CSIR, Government of India.

References

Barret M, Morrissey JP, O'Gara F (2011) Functional genomics analysis of plant growth-promoting rhizobacterial traits involved in rhizosphere competence. Biol Fertil Soils 47:729–743

Berg G (2009) Plant-microbe interactions promoting plant growth and health: perspectives for controlled use of microorganisms in agriculture. J Appl Microbiol Biotechnol 84:8–11

Berg G, Knaape C, Ballin G, Seidel D (1994) Biological control of *Verticillium dahliae* KLEB by naturally occurring rhizosphere bacteria. Arch Phytopathol Dis Protect 29:249–262

Berg G, Marten P, Ballin G (1996) *Stenotrophomonas maltophilia* in the rhizosphere of oilseed rape – occurrence, characterization and interaction with phytopathogenic fungi. Microbiol Res 151:19–27

Berg G, Roskot N, Smalla K (1999) Genotypic and phenotypic relationship in clinical and environmental isolates of *Stenotrophomonas maltophilia*. J Clin Microbiol 37:3594–3600

Berg G, Roskot N, Steidle A, Eberl L, Zock A, Smalla K (2002) Plant-dependent genotypic and phenotypic diversity of antagonistic rhizobacteria isolated from different *Verticillium* host plants. Appl Environ Microbiol 68:3328–3338

Berg G, Eberl L, Hartmann A (2005) The rhizosphere as a reservoir for opportunistic human pathogenic bacteria. Environ Microbiol 71:4203–4213

Berg G, Zachow C, Cardinale M, Mũller H (2010) Ecology and human pathogenicity of plant-associated bacteria. In: Ehlers RU (ed) Regulation of biological control agents. Springer, Berlin, pp 175–189

Bernier SP, Silo-Suh L, Woods DE, Ohman DE, Sokol PA (2003) Comparative analysis of plant and animal models for characterization of *Burkholderia cepacia* virulence. Infect Immun 71:5306–5313

Binks PR, Nicklin S, Bruce NC (1995) Degradation of RDX by *Stenotrophomonas maltophilia* PB1. Appl Environ Microbiol 61:1813–1322

Bonkowski M, Villenave C, Griffiths B (2009) Rhizosphere fauna: the functional anö structural diversity of intimate interactions of soil fauna with plant roots. Plant Soil 321:213–233

Bulgarelli D et al (2012) Revealing structure and assembly cues for Arabidopsis root-inhabiting bacterial microbiota. Nature 488:91–95

Cao H, Baldini RL, Rahme LG (2001) Common mechanisms for pathogens of plants and animals. Annu Rev Phytopathol 39:259–284

Chelius MK, Triplett EW (2000) Immunolocalization of dinitrogenase reductase produced by *Klebsiella pneumoniae* in association with *Zea mays* L. Appl Environ Microbiol 66:783–787

Cheuk W, Woo PCY, Yuen KY, Yu PH, Chan JKC (2000) Intestinal inflammatory pseudotumour with regional lymph node involvement: identification of a new bacterium as the etiological agent. J Pathol 192:289–292

Coenye T, Vandamme P (2003) Diversity and significance of *Burkholderia* species occupying diverse ecological niches. Environ Microbiol 5:719–729

Cook RJ, Tomashow LS, Weller DM, Fujimoto D, Mazzola M, Bangera G, Kim DS (1995) Molecular mechanisms of defense by rhizobacteria against root disease. Proc Natl Acad Sci U S A 92:4197–4201

Cooley MB, Miller WG, Mandrell RE (2003) Colonization of *Arabidopsis thaliana* with *Salmonella enterica* and enterohemorrhagic *Escherichia coli* O157:H7 and competition by *Enterobacter asburiae*. Appl Environ Microbiol 69:4915–4926

Coutinho TH, Venter SN (2009) Pathogen profile. *Pantoea ananatis*: an unconventional plant pathogen. Mol Plant Pathol 10:325–335

Critzer FJ, Doyle MP (2010) Microbial ecology of foodborne pathogens associated with produce. Curr Opin Biotechnol 21:125–130

Cruz AT, Andreea C, Allen CH (2007) *Pantonea agglomerans*, a plant pathogen causing human disease. J Clin Microbiol 45:1989–1992

Dalmastri C, Chiarini L, Cantale C, Bevivino A, Tabacchioni S (1999) Soil type and maize cultivar affect the genetic diversity of maize root-associated *Burkholderia cepacia* populations. Microb Ecol 38:273–284

De Souza JT, De Boer M, De Waard P, Van Beek TA, Raaijmakers JM (2003) Biochemical, genetic, and zoosporicidal properties of cyclic lipopeptide surfactants produced by *Pseudomonas fluorescens*. Appl Environ Microbiol 69:7161–7172

Delmotte N et al (2009) Community proteogenomics reveals insights into the physiology of phyllosphere bacteria. P Natl Acad Sci USA 106:16428–16433

Dennis PG, Miller AJ, Hirsch PR (2010) Are root exudates more important than other sources of rhizodeposits in structuring rhizosphere bacterial communities? FEMS Microbiol Ecol 72:313–327

Denton M, Kerr KG (1998) Microbiological and clinical aspects of infections associated with *Stenotrophomonas maltophilia*. Clin Microbiol Rev 11:7–80

Dörr J, Hurek T, Reinhold-Hurek B (1998) Type IV pili are involved in plant–microbe and fungus–microbe interactions. Mol Microbiol 30:7–17

Dowe MJ, Jackson ED, Mori JG, Bell CR (1997) *Listeria monocytogenes* survival in soil and incidence in agricultural soils. J Food Prot 60:1201–1207

Eberl L, Vandamme P (2016) Members of the genus *Burkholderia*: good and bad guys. F1000 Res 5: 1–10

Eckburg PB, Relman DA (2007) The role of microbes in Crohn's disease. Clin Infect Dis 44:256–262

Fravel DR (1988) Role of antibiosis in the biocontrol of plant diseases. Annu Rev Phytopathol 26:75–91

Germida JJ, Siciliano SD (2001) Taxonomic diversity of bacteria associated with the roots of modern, recent and ancient wheat cultivars. Biol Fertil Soils 33:410–415

Goris J, Boon N, Lebbe L, Verstraete W, De Vos P (2003) Diversity of activated sludge bacteria receiving the 3-chloroaniline degradative plasmid pC1gfp. FEMS Microbiol Ecol 46:221–230

Govan JRW, Hughes JE, Vandamme P (1996) *Burkholderia cepacia*: medical, taxonomic and ecological issues. J Med Microbiol 45:395–407

Govan JRW, Balendreau J, Vandamme P (2000) *Burkholderia cepacia* – friend and foe. ASM News 66:124–125

Graner G, Persson P, Meijer J, Alstrøm S (2003) A study on microbial diversity in different cultivars of *Brassica napus* in relation to its wilt pathogen, *Verticillium longisporum*. FEMS Microbiol Lett 29:269–276

Gupta CP, Sharma A, Dubey RC, Maheshwari DK (2001) Effect of metal ions on growth of *Pseudomonas aeruginosa* and siderophore and protein production. Indian J Exp Biol 39:1318–1321

Gyaneshwar P, James EK, Mathan N, Reddy PM, Reinhold-Hurek B, Ladha JK (2001) Endophytic colonization of rice by a diazotrophic strain of *Serratia marcescens*. J Bacteriol 183:2634–2645

Hacker J, Hentschel U, Dobrindt U (2003) Prokaryotic chromosomes and diseases. Science 301:790–793

Hadley WM et al (1987) Five month oral (diet) toxicity/infectivity study of *Bacillus thuringiensis* insecticides in sheep. Fundam Appl Toxicol 8:236–242

Hartmann A, Gantner S, Schuhegger R, Steidle A, Dürr C, Schmid M et al (2004) N-acyl homoserine lactones of rhizosphere bacteria trigger systemic resistance in tomato plants. In: Lugtenberg B, Tikhonovich I, Provorov N (eds) Biology of molecular plant–microbe interactions, vol 4. MPMI, St Paul, MN, pp 554–556

Hartmann A, Schmid M, van Tuinen D, Berg G (2009) Plant-driven selection of microbes. Plant Soil 321:235–257

Hauben L, Vauterin L, Moore ERB, Hoste M, Swings J (1999) Genomic diversity of the genus *Stenotrophomonas*. Int J Syst Bacteriol 49:1749–1760

Hebbar KP, Martel MH, Heulin T (1998) Suppression of pre- and postemergence damping-off in corn by *Burkholderia cepacia*. Europ J Plant Pathol 104:29–36

Hinsinger P, Marschner P (2006) Rhizosphere – perspectives and challenges – a tribute to Lorenz Hiltner. Plant Soil 283:vii–viii

Hinsinger P, Bengough AG, Vetterlein D, Young IM (2009) Rhizosphere: biophysics, biogeochemistry and ecological relevance. Plant Soil 321:117–152

Holden N, Pritchard L, Toth I (2009) Colonization outwith the colon: plants as an alternative environmental reservoir for human pathogenic enterobacteria. FEMS Microbiol Rev 33:689–703

Holmes A, Govan J, Goldstein R (1998) Agricultural use of *Burkholderia (Pseudomonas) cepacia*: a threat to human health? Emerg Infect Dis 4:221–227

Ikemoto S, Suzuki K, Kaneko T, Komagata K (1980) Characterization of strains of *Pseudomonas maltophilia* which do not require methionine. Int J Syst Bacteriol 30:437–447

Jelveh N, Cunha BA (1999) *Ochrobactrum anthropic* bacteremia. Heart Lung 28:145–146

Jensen GB, Hansen MB, Eilenberg J, Maillon J (2003) The hidden lifestyle of *Bacillus cereus* and relatives. Environ Microbiol 5:631–640

Jones DL, Nguyen C, Finlay RD (2009) Carbon flow in the rhizosphere: carbon trading at the soil-root interface. Plant Soil 321:5–33

Juhas M, vander Meer JR, Gaillard M, Harding RM, Hood DW, Crook DW (2009) Genomic islands: tools of bacterial horizontal gene transfer and evolution. FEMS Microbiol Rev 33:376–393

Kaestli M et al (2012) Out of the ground: aerial and exotic habitats of the melioidosis bacterium *Burkholderia pseudomallei* in grasses in Australia. Environ Microbiol 14:2058–2070

Kang YW, Carlson R, Tharpe W, Schell MA (1998) Characterization of genes involved in biosynthesis of a novel antibiotic from *Burkholderia cepacia* BC11 and their role in biological control of *Rhizoctonia solani*. Appl Environ Microbiol 64:3939–3947

Klerks MM, Franz E, van Gent-Pelzer M, Zijlstra C, van Bruggen AHC (2007) Differential interaction of *Salmonella enterica* serovars with lettuce cultivars and plant–microbe factors influencing the colonization efficiency. ISME J 1:620–631

Knief C et al (2011) Metaproteogenomic analysis of microbial communities in the phyllosphere and rhizosphere of rice. ISME J 6:1378–1390

Knudsen GR, Walter MV, Porteous LA, Prince VJ, Amstrong JL, Seidler RJ (1988) Predictive model of conjugated plasmid transfer in the rhizosphere and phyllosphere. Appl Environ Microbiol 54:343–347

Kobayashi DY, Gugliemoni M, Clarke BB (1995) Isolation of chitinolytic bacteria *Xanthomonas maltophilia* and *Serratia marcescens* as biological control agents for summer patch disease of turf grass. Soil Biol Biochem 27:1479–1487

Krechel A, Faupel A, Hallmann J, Ulrich A, Berg G (2002) Potato-associated bacteria and their antagonistic potential towards plant pathogenic fungi and the plant parasitic nematode *Meloidogyne incognita* (Kofoid and White) Chitwood. Can J Microbiol 48:772–786

Kumar A, Munder A, Aravind R, Eapen SJ, Tűmmler B, Raaijmakers JM (2013) Friend or foe: genetic and functional characterization of plant endophytic *Pseudomonas aeruginosa*. Environ Microbiol 15:764–779

Lambert B, Frederik L, Van Rooyen L, Gossele F, Papon Y, Swings J (1987) Rhizobacteria of maize and their antifungal activities. Appl Environ Microbiol 53:1866–1871

Lee EY, Jun YS, Cho KS, Ryu HW (2002) Degradation characteristics of toluene, benzene, ethylbenzene, and xylene by *Stenotrophomonas maltophilia* T3-c. J Air Waste Manag Assoc 52:400–406

Lottmann J, Berg G (2001) Phenotypic and genotypic characterization of antagonistic bacteria associated with roots of transgenic and non-transgenic potato plants. Microbiol Res 156:75–82

Lottmann J, Heuer H, Smalla K, Berg G (1999) Influence of transgenic T4-lysozyme-producing plants on beneficial plant-associated bacteria. FEMS Microb Ecol 29:365–377

Lugtenberg BJJ, Dekkers LC (1999) What makes *Pseudomonas* bacteria rhizosphere competent? Environ Microbiol 1:9–13

Lundberg DS et al (2012) Defining the core *Arabidopsis thaliana* root microbiome. Nature 488:86–90

Lynch JM (1990) Introduction: some consequences of microbial rhizosphere competence for plant and soil. In: Lynch JM (ed) The Rhizosphere. Wiley, Chichester, pp 1–10

Mark GL et al (2005) Transcriptome profiling of bacterial responses to root exudates identifies genes involved in microbe–plant interactions. P Natl Acad Sci USA 102:17454–17459

Meeting FB (1992) Soil microbial ecology: applications in agricultural and environmental management. Marcel Dekker, New York

Mehnaz S, Mirza MS, Haurat J, Bally R, Normand P, Bano A, Malik KA (2001) Isolation and 16S rRNA sequence analysis of the beneficial bacteria from the rhizosphere of rice. Can J Microbiol 47:110–117

Mendes A, Garbeva P, Raaijmakers JM (2013) The rhizosphere microbiome: significance of plant beneficial, plant pathogenic, and human pathogenic microorganisms. FEMS Microbiol Rev 37:634–663

Miller KJ, Wood JM (1996) Osmoadaption by rhizosphere bacteria. Ann Rev Microbiol 50:101–136

Moller LV, Arends JP, Harmsen HJ, Talens A, Terpstra P, Slooff MJ (1999) *Ochrobactrum intermedium* infection after liver transplantation. J Clin Microbiol 37:241–244

Morales A, Garland JL, Lim DV (1996) Survival of potentially pathogenic human-associated bacteria in the rhizosphere of hydroponically grown wheat. FEMS Microb Ecol 20:155–162

Nakayama T, Homma Y, Hashidoko Y, Mitzutani J, Tahara S (1999) Possible role of xanthobaccins produced by *Stenotrophomonas* sp. strain SB-K88 in suppression of sugar beet damping-off disease. Appl Environ Microbiol 65:4334–4339

Neumann G, Römheld V (2001) The release of root exudates as affected by the plant's physiological status. In: Pinton R, Varanini Z, Nannipieri P (eds) The rhizosphere. Marcel Dekker, New York, NY, pp 41–93

Nithya A, Babu S (2017) Prevalence of plant beneficial and human pathogenic bacteria isolated from salad vegetables in India. BMC Microbiol 17:1–16

Overbeek R, Olson R, Pusch GD, Olsen GJ, Davis JJ, Disz T, Edwards RA, Gerdes S, Parrello B, Shukla M, Vonstein V, Wattam AR, Xia F, Stevens R (2014) The SEED and the rapid annotation of microbial genomes using subsystems technology (RAST). Nucleic Acids Res 42:206–214

Parke JL, Gurian-Sherman D (2001) Diversity of the *Burkholderia cepacia* complex and implications for risk assessment of biological control strains. Annu Rev Phytopathol 39:225–258

Pierret A, Doussan C, Capowiez Y, Bastardie F, Pagès L (2007) Root functional architecture: a framework for modeling the interplay between roots and soil. Vadose Zone J 6:269–281

Raaijmakers JM, Vlami M, de Souza JT (2002) Antibiotic production by bacterial biocontrol agents. Antonie Leeuwenhoek 81:537–547

Raaijmakers JM, Paulitz TC, Steinberg C, Alabouvette C, Moënne-Loccoz Y (2009) The rhizosphere: a playground and battlefield for soilborne pathogens and beneficial microorganisms. Plant Soil 321:341–361

Rahme LG, Stevens EJ, Wolfort SF, Shoa J, Tompkins RG, Ausubel FM (1995) Common virulence factors for bacterial pathogenicity in plants and animals. Science 268:1899–1902

Reiter B, Pfeifer U, Schwab H, Sessitsch A (2002) Response of endophytic bacterial communities in potato plants to infection with *Erwinia carotovora* subsp. *atroseptica*. Appl Environ Microbiol 68:2261–2268

Riesenfeld CS, Goodman RM, Handelsman J (2004) Uncultured soil bacteria are a reservoir of new antibiotic resistance genes. Environ Microbiol 6:981–989

Saunders JR, Allison H, James CE, McCarthy AJ, Sharp R (2001) Phage-mediated transfer of virulence genes. J Chem Technol Biotechnol 76:662–666

Schwieger F, Tebbe CC (2000) Effect of field inoculation with *Sinorhizobium meliloti* L33 on the composition of bacterial communities in rhizospheres of a target plant (*Medicago sativa*) and a non-target plant (*Chenopodium album*)–linking of 16S rRNA gene-based single-strand conformation polymorphism community profiles to the diversity of cultivated bacteria. Appl Environ Microbiol 66:3556–3565

Sessitsch A, Reiter B, Berg G (2004) Endophytic bacterial communities of field-grown potato plants and their plant growth-promoting abilities. Can J Microbiol 50:239–249

Sharma R, Paliwal JS, Chopra P, Dogra D, Pooniya V, Bisaria VS, Swarnalakshmi K, Sharma S (2017) Survival, efficacy and risk assessment of bacterial inoculants in *Cajanus cajan*. Agric Ecosyst Environ 240:244–252

Sørensen J (1997) The rhizosphere as a habitat for soil microorganisms. In: Van Elsas JD, Trevors JT, EMH W (eds) Modern soil microbiology. Marcel Dekker, New York, NY, pp 21–45

Steinkamp G, Wiedemann B, Rietschel E, Krahl A, Giehlen J, Barmeier H, Ratjen F (2005) Prospective evaluation of emerging bacteria in cystis fibrosis. J Cyst Fibros 4:41–48

Strawn LK et al (2013) Landscape and meteorological factors affecting prevalence of three food-borne pathogens in fruit and vegetable farms. Appl Environ Microbiol 79:588–600

Suckstorff I, Berg G (2003) Evidence for dose-dependent effects on plant growth by *Stenotrophomonas* strains from different origins. J Appl Microbiol 95:656–663

Tabacchioni S, Bevivino A, Dalmastri C, Chiarini L (2002) *Burkholderia cepacia* complex in the rhizosphere: a minireview. Ann Microbiol 52:103–117

Tan MW, Rahme LG, Sternberg JA, Tompkins RG, Ausubel FM (1999) *Pseudomonas aeruginosa* killing of *Caenorhabditis elegans* used to identify *P. aeruginosa* virulence factors. Proc Natl Acad Sci U S A 96:2408–2413

Teplitski M, Barak JD, Schneider KR (2009) Human enteric pathogens in produce: un-answered ecological questions with direct implications for food safety. Curr Opin Biotechnol 20:166–171

Thomashow LS, Bonsall RF, Weller DM (1997) Antibiotic production by soil and rhizosphere microbes in situ. In: Hurst CJ, Knudson GR, MJ MI, Setzenbach LD, Walter MV (eds) Manual of environmental microbiology. American Society for Microbiology Press, Washington, DC, pp 493–499

Toth IK, Pritchard L, Birch PR (2006) Comparative genomics reveals what makes an enterobacterial plant pathogen. Annu Rev Phytopathol 44:305–336

Tripathi AK, Verma SC, Ron EZ (2002) Molecular characterization of a salt-tolerant bacterial community in the rice rhizosphere. Res Microbiol 153:579–584

Troxler J, Azelvandre P, Zala M, Defago G, Haas D (1997) Conjugative transfer of chromosomal genes between fluorescents pseudomonads in the rhizosphere of wheat. Appl Environ Microbiol 63:213–219

Turnbaugh PJ et al (2009) A core gut microbiome in obese and lean twins. Nature 457:480–484

Tyler HL, Triplett EW (2008) Plants as a habitat for beneficial and/or human pathogenic bacteria. Annu Rev Phytopathol 46:53–73

Ursell LK, Metcalf JL, Parfrey LW, Knight R (2012) Defining the human microbiome. Nutr Rev 70: S38–S44

van Baarlen P, van Belkum A, Summerbell RC, Crous PW, Thomma B (2007) Molecular mechanisms of pathogenicity: how do pathogenic microorganisms develop cross-kingdom host jumps? FEMS Microbiol Rev 31:239–277

van Loon LC, Bakker PA, Pieterse CM (1998) Systemic resistance induced by rhizosphere bacteria. Annu Rev Phytopathol 36:453–483

Wang HB et al (2011) Characterization of metaproteomics in crop rhizospheric soil. J Proteome Res 10:932–940

Wei B, Huang T, Dalwadi H, Sutton CL, Bruckner D, Braun J (2002) *Pseudomonas fluorescens* encodes the Crohn's disease-associated I2 sequenceand T-cell superantigen. Infect Immun 70:6567–6575

Whipps J (2001) Microbial interactions and biocontrol in the rhizosphere. J Exp Bot 52:487–511

Wu L, Wang H, Zhang Z, Lin R, Lin W (2011) Comparative metaproteomic analysis on consecutively *Rehmannia glutinosa*-monocultured rhizosphere soil. PLoS One 6:e20611

Yabuuchi E, Kosako Y, Oyaizu H, Yano I, Hotta H, Hashimoto Y, Ezaki T, Arakawa M (1992) Proposal of *Burkholderia* gen. nov. and transfer of seven species of the genus *Pseudomonas* homology group II to the new genus, with the type species *Burkholderia cepacia* (Palleroni and Holmes 1981) comb. nov. Microbiol Immunol 36:1251–1275

Chapter 9
Effect of Organic Farming on Structural and Functional Diversity of Soil Microbiome: Benefits and Risks

Vijay Laxmi Shrivas, Upma Singh, L. Weisskopf, P. Hariprasad, and Shilpi Sharma

Abstract The significance of microbial diversity in soil has been demonstrated since decades. Assorted variety of microorganisms in soil is fundamental to support soil health, as an extensive diversity of microorganisms is associated with essential soil functions. Several abiotic and biotic factors have been reported to affect the structural and functional diversity of soil microbes. This chapter centers around information related to the application of different biological amendments, and their influential role in shaping soil microbial community diversity, and health. Specifically, it focuses on the application of different organic fertilizers, and their influence on the diversity of microbial communities in arable soil, and upon the risks associated with such amendments, since they can be a source of heavy metals, antibiotics, antibiotic-resistant bacteria, and antibiotic-resistant genes carried by mobile genetic elements.

V. L. Shrivas
Department of Biochemical Engineering and Biotechnology, Indian Institute of Technology Delhi, Hauz Khas, New Delhi, India

Centre for Rural Development and Technology, Indian Institute of Technology Delhi, Hauz Khas, New Delhi, India

U. Singh · S. Sharma (✉)
Department of Biochemical Engineering and Biotechnology, Indian Institute of Technology Delhi, Hauz Khas, New Delhi, India
e-mail: shilpi@dbeb.iitd.ac.in

L. Weisskopf
Department of Biology, University of Fribourg, Fribourg, Switzerland

P. Hariprasad
Centre for Rural Development and Technology, Indian Institute of Technology Delhi, Hauz Khas, New Delhi, India

A. Varma et al. (eds.), *Plant Biotic Interactions*,
https://doi.org/10.1007/978-3-030-26657-8_9

9.1 Introduction

Rapid increase in global population has led to an enhancement in the consumption of agricultural products, which has in turn led to the exploration of newer methods and techniques for food production. In order to obtain better quality products, various agricultural management practices are in use. There are numerous factors that determine farming practices such as cultivars, soil types, irrigation pattern, and climatic changes. Fertilizer application is considered as one of the most essential processes for enhancing nutrient availability to the plants, and for the alteration of soil properties and its associated microflora (Marschner et al. 2003), which in turn might affect physical, chemical and biological properties of soil.

The practice of new technologies and modern agriculture during the period between 1950 and 1960 was termed as "Green revolution," which replaced the use of all natural and organic fertilizers and pesticides. Tremendous increase in amendment with chemical fertilizers in turn compromises with the soil health, and its fertility (Nicholson and Hirsch 1998). Moreover, such prolific use of chemicals has negatively influenced the ecosystem and biodiversity. Therefore, there is a great need for the development of sustainable and eco-friendly approach that excludes or minimizes the application of agrochemicals and pesticides in conventional farming.

Sustainable agriculture aims at using the soil and ecosystem resources for food production without jeopardizing the capacity of future generations to do the same. It implies a systemic approach of farm management and food production, considering the welfare of the complete ecosystem, including its human and animal components. A "healthy" soil is a crucial component of sustainable agriculture. Preserving it requires reduction in the use of synthetic fertilizers, and reallocation towards the employment of natural resources for improving crop productivity (Reganold and Wachter 2011). Ecosystem restoration is one of the major aspects that can be achieved through nutrient recycling and improved soil quality by preventing soil erosion and/or by higher vegetation diversity, which promotes natural control of pests, and prevents land degradation (Reganold and Wachter 2011). A number of organic amendments are being used such as manure, compost, biogas slurry, and crop residues that enhance organic carbon, water holding capacity, and cumulative stability of soil. Thus, organic farming promises to maintain the physicochemical and biological parameters of soil, which increases its productivity (Pérez and Sommaruga 2006; Johnston et al. 2009; García-Orenes et al. 2010; Brienen et al. 2015; Prosdocimi et al. 2016). An increased biological activity has direct as well as indirect effect on plant health (Lazcano et al. 2013). Microbes are involved in different ecological functions, such as nutrient cycling (Bian et al. 2014; Mendes et al. 2015), organic matter decomposition, bioremediation, bioaugmentation, and plant growth promotion (Mazzola et al. 2004), which supports the proper functioning of agricultural systems, its productivity, and sustainability (Bonkowski and Roy 2005). Hence, a balanced composition of microbes is essential to maintain soil health. Their composition in soil varies from region to region on the basis of biotic and abiotic factors. Some studies have reported that organic farming, along with other sustainable farming practices, such as reduced tillage or zero tillage, results in the improvement of soil fertility (Hobbs et al. 2008; TerAvest et al. 2016). Studies have also supported that such a practice enhances crop productivity by exerting a positive impact on active soil microflora (Mäder et al. 2002). Hartman et al. (2018) proposed that understanding the dynamics of soil microbes with respect to various agricultural practices could help in developing management strategies for sustainable farming. However, the effect of different agricultural management practices on microbial diversity is still not well understood. Therefore, in order to achieve proper understanding of microbial dynamics in the soil ecosystem, there is a need for in-depth assessment of impact of different agricultural management practices on structure and function of microbial communities (Tardy et al. 2015).

Nowadays, numerous molecular techniques are being used to evaluate the diversity of microbes in natural environments, which serve as a critical indicator for changes in environmental conditions, and quality assessment of agriculture systems. According to Hugenholtz (2002), more than 99% of microbial fraction that exists in the environment, which can be observed microscopically, is not cultivable, and only up to and estimated 1% of the total microbial load can be cultured under laboratory conditions (Rondon et al. 1999; Valentine et al. 2010). In order to identify the diversity of these non-cultivable microorganisms under different agro-climatic zones, various cultivation-independent techniques are being utilized, but the exact role of these microorganisms in the environment is yet to be elucidated (Lyngwi et al. 2013). Nevertheless, molecular methods and phylogenetic analysis help in developing a proper understanding of their distribution and function in the environment (Bailón-Salas et al. 2017).

The present chapter discusses the various aspects related to microbial diversity under organic management farming practice.

9.2 Organic Farming

Conventional farming is one of the crop-cultivating frameworks, which incorporates the utilization of fertilizers, pesticides, and herbicides of chemical origin, concentrated animal feeding operation, massive irrigation systems, etc. Organic farming is generally differentiated from conventional agriculture by management practices that encourage the cycling of assets, include advanced environmental adaptation, and foster biodiversity. Rather than utilizing engineered composts, chemical fertilizers, and pesticides, organic farming depends on crop rotation, plant and animal excrements as manures, natural weeding, and/or via biological pest control (Williams 2002). Hence, organic farming limits the negative impacts of intensive agricultural practices by controlling weeds, pests, and diseases through natural methodologies, and by enhancing biodiversity.

A survey done by the Research Institute of Organic Agriculture (FiBL), in Switzerland on the worldwide organic agriculture (2018), states that around 57.8 million hectares (1.2%) of agricultural fields are currently under organic farming management (Willer and Lernoud 2018). In addition to this, the report includes the relative surface of land used under organic farming in different countries, and the number of organic crop producers. Among the 170 countries investigated, India ranks first in terms of number of organic producers, and ninth in terms of area covered under organic agriculture (Willer and Lernoud 2018).

There are a number of organic amendments that can help in improving soil quality and its health, through modification of physicochemical conditions, which includes water-holding capacity of soil, cationic exchange ability, or change in organic C:N ratio. In a study by You et al. (2018), an organic amendment like sugarcane mulch was found responsible for the elevation of carbon and nitrogen contents, microbial respiration, enzymatic activities, and microbial diversity. These changes in the soil properties seemed to be independent of plant species. Similarly, biochar is also considered as one of the most eco-friendly and sustainable regimes for improving soil quality (Liu et al. 2018). Biochar is charcoal (carbon rich) obtained from plant matter. It is porous in nature with high surface area, thus helps in enhancing nutrients, and maintaining water retention in the soil (Laird 2008). In an experiment by Liu et al. (2018), the impact of biochar on denitrifying microbial communities and soil properties was investigated, and it was found that biochar application helped in managing reclaimed soil in the subsidence area of coalmines.

Several studies have monitored the effect of inoculating one or more plant growth-promoting rhizobacteria (PGPR) or plant growth-promoting bacteria (PGPB) on the soil microbiome in organic farming (Bashan et al. 2014; Lupatini et al. 2017). The application of these consortia along with organic amendments enable the soil to provide favorable environment for the functional indigenous microbial community. By exhibiting an affirmative effect on the metabolic actions of soil microbes, these consortia strengthened the plant's health. The indirect effect of these consortia was toward controlling various diseases in plants by the enrichment of beneficial microbes in the soil, which potentially eliminates the pathogenic microbes from the soil. For instance, *Bacillus* and *Trichoderma* in the soil were shown to help in eradicating the *Fusarium* wilt disease from the plant (Xiong et al. 2017).

9.3 Impact of Organic Farming on Microbial Community Dynamics

Impact of different agricultural management practices on the microbial community diversity is still a matter of concern due to limited information (Zhang et al. 2016). Studies suggest that community composition and functions of microbes can be altered by different agricultural practices, or even with the type of soil structure (Bossio et al. 1998). Henneron et al. (2015) described that organic farming can lead to an upsurge in the abundance and biomass of soil microorganisms from 30 to 70%, and can help in improving the bacterial pathways in food web operational in soil. For instance, the interaction between the microbes and *microbivorous* nematodes take part in a micro-food network, which can play a vital role in nutrient cycling and in the regulation of microbial population. According to Wardle et al. (1995), the elevated grazing of *microbivorous* nematodes creat higher burden on bacteria as compared to fungi. Various studies have proven the sensitivity of fungal diversity with respect to agricultural management practices. A study conducted by Schneider et al. (2015) in dairy farms in Ontario reported variations in the communities of arbuscular mycorrhizal fungi (AMF) under organic and conventional practices. AM fungi are known for their plant growth-promoting activities by assisting the plant in phosphorus uptake. Results indicated that organically managed fields have enhanced AMF community than conventional farms, as judged by plant development and P-use efficiency. Another study also found a similar trend in terms of Shannon diversity index, where the diversity of AM fungi was considerably higher under organic orchards' farms (Beenhouwer et al. 2015).

In an experiment by Ashworth et al. (2017), the application of different types of nutrients (inorganic vs. organic) and crop rotations showed a resilient influence on the structure of microbial community, which had direct impact on various ecological processes beneficial to agricultural farms. Maximum diversity was observed in fields amended with poultry litter (organic), compared to inorganic amendments. The major bacterial phylum that dominated under this treatment was *Bacteroidetes*, followed by *Proteobacteria* and *Acidobacteria.* Zhang et al. (2016) investigated the changes occurring in soil microbial community composition, along with the microbial biomass of bacteria, fungi, AMF, and actinomycete by using phospholipid fatty acid (PLFA) analysis. Results signified a strong connection among microbial community structure and environmental factors such as soil moisture, temperature, pH, and soil organic carbon and nitrogen (both labile C:N and recalcitrant C:N).

Henneron et al. (2015) stated that microbial community under conventional tillage practices was less diversified and active in comparison to that observed under organic systems. This phenomenon can be a result of inefficient use of microbial energy, which is required for coping with the disturbance in their natural habitat by tillage (Wardle 1995). Similarly, alteration in bacterial community was strongly observed under tillage, whereas an alteration in the fungal communities was a result of the different management types as well as tillage (Wang et al. 2012).

A change in microbial community configuration was noticed in response to cellulose and hemicelluloses as prime carbon source (Dumontet et al. 2017). The study was conducted using tomato as a model plant, and biochar and other organic fertilizers (cattle manure, dairy waste, vineyard compost, and poultry or slaughter-house waste) were used along with control soil. Analysis of soil properties and microbial community composition was done using denaturing gradient gel electrophoresis (DGGE) and BIOLOG® assay, which revealed that biochar amendment promoted the growth and diversity of cellulose-degrading bacteria. In another study, an experiment was designed using paddy as a model crop, and the role of different livestock manure (cattle and swine) on the productivity and soil quality was assessed (Das et al. 2017). Enhanced diversity of *α-Proteobacteria, β-Proteobacteria,* and *Firmicutes,* and species richness were noticed under composted cattle manure amendment by sequencing of 16S rRNA amplicon. Examination of 16S rRNA (bacterial) and 18S rRNA (fungal) DGGE profiles of 19-year-old flower farms was performed under conventional, ecological, and intermediate management practices (Santamaría et al. 2017). The results concluded that bacteria were more sensitive to the changes in the soil quality and, thus, would serve as better indicators for alterations in soil quality compared to fungi. Wang et al. (2016) stated that organic farming had more stable and uniform bacterial community structure with increased nutrient availability and soil enzyme activities. Bacteriain organically managed farms belonged mostly to *Rhizobiales, Myxococcales, Thiotrichaceae, Micromonosporaceae,* and *Desulfurellaceae,* while conventional farming practices enhanced the abundance of acid- and alkali-resistant microbes (*Acidobacteriaceae* and *Sporolactobacillaceae*). Velmourougane (2016) evaluated the long-term effects of organic vs. conventional farming on Arabica coffee. Analysis of physicochemical and biological activity of soil revealed higher organic carbon, respiration rate, and fluorescein diacetate activity under organically managed fields. Organic amendments led to an enhancement of up to 34% in the abundance of total microbial community in comparison to the conventional practices. Evaluation of organic and conventionally managed apple farms revealed the consequence of adopting these agricultural management practices on the composition of fungal communities associated with fruits (Abdelfattah et al. 2016). Phylum *Ascomycota* was found to be dominant in comparison to other fungal phyla, and accounted for about 91.6% of the total fungal community. Several unique taxa were detected under organically managed apple farms. Study also suggested that management practices might exert a selection on community composition.

At present, studies have employed high-throughput molecular tools for detection of microbial diversity under natural environment, such as GeoChip, which is a microarray-based technique consisting of around 28,000 probes that helps in detecting genes involved in most of the biogeochemical cycles such as carbon, nitrogen, sulfur, and phosphorus cycles. This is one of the techniques widely used to study functional microbial diversity in different agricultural management practices. A study conducted by Xue et al. (2013) using GeoChip as a prime technique revealed that the expression of functional genes, which are involved in N/P/C cycles, were higher under organic farming, than conventional farming. In contrast, the assimilatory pathways involved in N reduction and other metabolic processes, such as methane oxidation and lignin degradation, were found to be unaffected by the type of the farming practices. Ding et al. (2018) also used GeoChip to investigate the shifts in functional microbial diversity found in paddy fields, which were under different fertilization regimes. The results suggested a change in the community structure in treatments amended with the fertilizer, such as rice straw combined with nitrogen, phosphorus, and potassium (NPK). Members of certain families, like *Bradyrhizobiaceae* and *Rhodospirillaceae*, were found to be responsible for the increase in crop productivity. Abundance of functional genes were also found to be in correlation with consequent soil enzymatic activities and crop yield, suggesting that fertilization might be responsible for community shift due to an enhancement in soil nutrient turnover, which in turn affected the crop yield. Overall, the study indicated a change in bacterial population along with an improvement of functional diversity. The shift was directly linked to the combination of rice straw amended with balanced fertilizer dosage. Implementation of a lucrative fertilization regime incorporated with rice straw and low chemical fertilizer was suggested for ecological nutrient management.

Thus, it can be concluded that the type of farming practice has a direct impact on structural and functional diversity of microbial community, with organic farming exerting an overall positive role on microbial diversity.

9.4 Risk Assessment of Organic Farming

Organic farming promotes the usage of manure or compost in place of chemical fertilizers (Bengtsson et al. 2005). These organic amendments include various agricultural wastes such as harvest waste and wastes from farms, poultry houses, and animals. Animal waste coming from various animal-feeding operations enhances the risk of release of antibiotics into the environment, as these wastes are directly applied as fertilizers in the agricultural fields (Hu et al. 2010). Such manure has also been found to be a reservoir of various heavy metals, antibiotics, antibiotic-resistant bacteria, genes, and other mobile genetic elements (Chen et al. 2017). So antibiotic stress has become one of the major emerging issues in agricultural systems nowadays, not only in India but also worldwide.

Manure enhances soil microbial activities by increasing the levels of organic matter, soil porosity, moisture, and nutrients, which promotes the growth of a plant and grain yield. The consistent recycling of organic litter in soil is a proficient method for sustaining optimal levels of soil organic matter. Thus, an addition of organic manure in soil was shown to refurbish degraded cropland (Zhen et al. 2014). Organic manure mostly includes animal wastes, urban organic wastes, crop residues, biogas-spent slurry, and green manures. Sewage sludge and industrial wastes are also used in agriculture applications. However, such manure has been found to be reservoirs of heavy metals, antibiotics, antibiotic-resistant genes and bacteria, and mobile genetic elements (Chen et al. 2017), which raises the question of safety issues linked to the application of such organic manure.

Antibiotics are heavily used in farm animals as compared to humans. Animals partially metabolize the antibiotics and a major fraction is excreted in feces. Overuse of antibiotics leads to increased occurrence of antibiotic-resistant bacteria (ARB), which is one of the major threats to global health. Antibiotic resistance increases because of adaptation by some fundamental mechanisms, like alteration in membrane permeability toward antibiotics, changes in bacterial proteins that have antibacterial targets, and the enzymatic degradation of antimicrobial drugs. This antimicrobial resistance can be conserved in either bacterial chromosome or plasmid target sites (Dever and Dermody 1991). Manure applications support the entry of ARB together with nutrients and organic matter into arable soils and water bodies, which poses a risk not only for indigenous microorganisms but also for human health. Table 9.1 compiles studies related to detection of ARB and ARGs in agricultural systems.

Table 9.1 Effect of antibiotics on the occurrence of ARB and ARGs in agricultural systems

S. No.	Classes of antibiotics	Antibiotics	Manure types	Methods	Results	References
1.	Aminoglycosides	Gentamicin, Amikacin	Animal manure	Metagenomic library construction	• Manure amendments accounted for approx. 70% increase in ARGs in soil bacteria	Su et al. (2014)
		Streptomycin	Swine manure, laying hens manure	qPCR, pyrosequencing	• *StrA* and *sul2* genes were significantly higher in cattle and pig excreta microbiota than laying hen microbiota • Cattle and pig manure represented an important reservoir and carrier of antibiotic-resistant bacteria	Faldynova et al. (2013)
		Aminoglycosides	Dairy manure	qPCR, Sanger sequencing	• Abundance of ARG lower in composted manure than untreated, mechanically dewatered and anaerobically digested dairy manures • ARGs were detected more frequently in soil with raw or digested manure compared to unmanured soil	Tien et al. (2017)
2.	Cephalosporins	Cephalothin	Cow manure	qPCR, pyrosequencing	• Higher number of β-lactam-resistant bacteria was detected in manure vs. inorganically fertilized soil • β-lactamase harboring populations of soil bacteria, including *Psychrobacter pulmonis* and *Janthinobacterium* sp. increased in manured soil	Udikovic-Kolic et al. (2014)
		Ceftazidime, Cefoxitin	Cattle manure, sheep manure, pig manure	RT-PCR	• Antimicrobial resistance phenotype in animal-origin *E. coli* was mediated by wide array of resistance genes (blaTEM, blaSHV, blaOXA, strAB, aadA1, sul1, sul2, tet(A), tet(B),	Enne et al. (2008)

					dfrA12, and dfrA17), mostly in pigs • Bacterial isolates from pig manure showed common resistance to ceftazidime (0.1%) which is relatively rare	
3.	Fluoroquinolones	Norfloxacin, Ciprofloxacin, Fleroxacin, Enrofloxacin	Cow manure, pig manure, chicken manure	Ultrasonic extraction, HPLC	• High antibiotic concentrations were detected in manures samples from China – Ciprofloxacin: ~33.98 and 29.59 mg/kg in cow and pig manure, respectively – Enrofloxacin: 33.26 and 46.70 mg/kg in cow and pig manure, respectively • Maximum concentrations in chicken manure: – Enrofloxacin: 1420.76 mg/kg – Norfloxacin: 225.45 mg/kg – Fleroxacin: 99.43 mg/kg – Ciprofloxacin: 45.59 mg/kg	Zhao et al. (2010)
		Difloxacin (dif)	Pig slurry	Denaturing gradient gel electrophoresis (DGGE), PLFA fatty acid profiling	• Slurry-derived fecal bacteria survived in mesocosm soil 14 days after manure application • Showed the effects of differently medicated pig's slurries on bulk and rhizosphere microbial communities, with a significant shift to Gram-positive fungi and bacteria	Reichel et al. (2013)
4.	Penicillins	Amoxicillin	Piggery manure	qPCR, restriction profiling of plasmid, Dot-blot and Southern blot, DGGE	• Piggery manure at field scale was a source of broad-host range plasmids • Total CFU counts and amoxicillin-resistant CFU counts increased after	Binh et al. (2008)

(continued)

Table 9.1 (continued)

S. No.	Classes of antibiotics	Antibiotics	Manure types	Methods	Results	References
					manure application • Abundance of *bla-TEM* gene encoding for amoxicillin resistance was high in manure-treated soil samples	
		Ampicillin, Penicillin, Amoxicillin-clavulanic acid	Dairy and swine manure		• No coherent increase in the abundance of ARB enumerated from vegetable grown in manured soil • Detected numerous antibiotic-resistant elements in DNA from vegetables grown in unmanured soil	Marti et al. (2013)
5.	Sulfonamides	Sulfamethoxazole, Sulfadiazine (sdz), Sulfamethazine	Liquid slurry	LC-MS-MS	• Sulfamethazine concentration was lower than tetracycline by two orders of magnitude in the plow layer. SDZ application might have primarily affected nitrite oxidation by reducing nitrite-oxidizing bacteria	Hamscher et al. (2005)
		Sulfonamide	Pig slurry	qPCR	• SDZ application may primarily affect nitrite oxidation by NOB	Ollivier et al. (2013)
		Sulfonamide	Pig slurry	Denaturing gradient gel electrophoresis (DGGE), PLFA fatty acid profiling	• Showed the effects of differently medicated pig slurries on bulk and rhizosphere microbial communities, with a substantial shift to Gram-positive bacteria and fungi	Reichel et al. (2013)
6.	Streptogramin	Streptogramin B (MLSB)	Pig manure	High-throughput quantitative PCR (HT-qPCR)	• Resident bacterial composition shifted markedly, and significantly correlated with profiles of ARGs	Chen et al. (2017)

					• Abundance of ARG in manured soil decreased with time but was higher than control	
7.	Tetracyclines	Chlortetracycline (CTC), Tetracycline	Liquid manure	LC-MS	• In soil, tetracycline concentration was higher than 150 μg kg^{-1} soil	Hamscher et al. (2005)
		Tetracycline	Pig manure	High-throughput quantitative PCR (HT-qPCR)	• ARGs' abundance and diversity increased in soil after manure application	Chen et al. (2017)
		Minocycline	Animal manure	Metagenomic library construction, Sanger sequencing	• Manure amendments accounted for approx. 70% increase in ARGs in soil bacteria	Su et al. (2014)
		Chlortetracycline (CTC)	Pig manure	HPLC, LC-ESI-MS/MS analysis	• CTC-resistant bacteria were present in significant level in manure-amended soil, but declined to same levels as detected in beginning	Halling-Sørensen et al. (2005)
		Oxytetracycline (OTC)	Cow manure	16S rRNA qPCR, LC-ESI-MS/MS	• 16S rRNA gene analysis showed persistence of resistant *Arthrobacter* spp., *Pseudomonas* spp., and *Rhodococcus* spp.	Yeom et al. (2017)

Centre for Science and Environment (CSE), India, stated that unsafe dumping of poultry litters and waste in agriculture farms is a severe cause of ARB dispersal in environment. They collected litter and soil samples from 12 randomly selected farms, mainly located in Uttar Pradesh, Haryana, Rajasthan, and Punjab states. From these samples, 217 bacterial isolates were obtained of which 187 isolates belonged to three bacterial genera, i.e., *Klebsiella pneumoniae*, *Escherichia coli,* and *Staphylococcus lentus.* These isolates were further evaluated for resistance against 16 different commonly used antibiotics (doxycycline hydrochloride, nitrofurantoin, amoxyclav, levofloxacin, cefuroxime, ceftriaxone, ciprofloxacin, chloramphenicol, amikacin, gentamicin, co-trimoxazole, meropenem, linezolid, azithromycin and clindamycin), among which ten are critical for human health. The bacterial isolates also displayed multidrug resistance against selected antibiotics: 100% *E. coli*, 78% *S. lentus,* and 92% *K. pneumoniae* were multidrug resistant. Out of these, 40% *E. coli* and 30% *K. pneumoniae* had multidrug resistance for ten antibiotics from the class of fluoroquinolones, penicillins, cephalosporins, and carbapenems. Due to the high prevalence of multiple drug resistance, these bacteria can cause untreatable infections (Bhushan et al. 2017).

The occurrence of antimicrobial resistance (AMR) in bacterial pathogens has harmed soil health as well as public health. Exposure of these bacterial pathogens to antimicrobials enhances the resistance in indigenous microbes and also impacts various microbial species by the dispersal and transfer of antimicrobial resistance genes. The primary route of the emergence of antimicrobial resistance gene (ARG) in microbes is horizontal gene transfer (HGT) of plasmids, integrons, and transposons. Horizontal gene transfer by means of plasmids is known to be the major cause of drug-resistant proliferation. Pornsukarom and Thakur (2017) detected plasmids carrying AMR gene in multiple *Salmonella* serotypes across varied swine lands in North Carolina. The main reason for AMR dissemination was the venting of manure generated by swine production systems. Manure and soil samples were picked on 0, 7, 14, and 21 days after manure application from six farms in North Carolina and plasmid analysis performed with 14 different isolates. They presented evidence on horizontal transmission of resistance through plasmids by identifying 90 kb and 100 kb IncI1 plasmids in *Salmonella* Rissen and *Salmonella* Johannesburg, which carried tet (A) and blaCMY-2 (genes encoding resistance against tetracycline and cephalosporin) genes, respectively. *IncF* plasmid was also widely dispersed in different serotypes and farms. Simultaneously, a microcosm experiment was performed by Chen et al. (2017) to investigate the underlying mechanisms of antibiotic-resistant gene dispersal and its impact on manure-borne bacterial pathogens. The experiment was conducted for a period of 2 months with four treatments comprising of control soil, soil with pig manure, irradiated soil with pig manure, and soil with irradiated pig manure. Two hundred forty unique ARGs from almost all the main classes of ARGs were identified through high-throughput quantitative PCR (HT-qPCR), and significant increment was found in diversity and abundance of ARGs in the soil with manure application. In addition, the shift in soil bacterial composition was remarkably correlated with ARGs' profiles. An elevated number of ARGs were observed due to the inclusion of manure-derived antibiotic-resistant bacteria in soil, and by horizontal gene transfer of mobile genetic elements in the native soil microbes from ARB. Although the quantity of ARGs decreased in manured soil over time, it was still high when compared to control soil. Some antibiotics like tetracycline, oxytetracycline, and sulfathiazole antibiotics are commonly found in veterinary manures, and are non-degradable in the aquatic environment. Use of these water bodies in agricultural fields may cause severe environmental effects. Yeom et al. (2017) assessed the residual concentration of these antibiotics in cow manure amended soil and its relationship with ARB. Soil samples were collected from farms at depths of 0, 7, and 15 cm. 16S rRNA gene analysis confirmed that *Arthrobacter* spp., *Pseudomonas* spp., and *Rhodococcus* spp. had persistent resistance to the antibiotics tested. Both *Arthrobacter* spp. and *Rhodococcus* spp. possessed resistance to tetracycline and oxytetracycline, whereas *Pseudomonas* spp. gained resistance against sulfathiazole. Antibiotic biodegradation test revealed the ARB's ability to grow in the residual antibiotic soil during 13 days of incubation.

Animal manures are not only loaded with drug-resistant bacteria but also have raised concentrations of heavy metals such as Cu, Zn, As, Cd, etc., which accumulate in soils because of repeated applications over long periods of time. This in turn affects soil fertility and food quality, and promotes migration of these heavy metals by runoff and leaching. Therefore, heavy metal contamination in animal manures poses potential risks to the environment. Northeast China widely used heavy metal-contaminated animal manures in farms. Thus, to determine heavy metal (Cu, Cr, Zn, Cd, Pd, and As) concentration in feeds and animals, Zhang et al. (2012) investigated 104 livestock feeds plus 118 animal manure samples collected from different farms with different herd sizes. They found that pig manure had higher heavy metal pollution than poultry, and cattle manure, thereby posing a higher risk of heavy metal contamination to farmlands and water bodies. Not only Cd concentration but also elevated Cu concentration in animal manure is a primary concern in China. Fengsong et al. (2011) determined Cu concentration from intensive farms of three Northeast Provinces and sampled 224 animal manures and feeds from farms. Results showed a wide range of Cu concentration from 1.5 to 1521.2 mg kg^{-1} dm^{-1} animal manures. Average value of Cu metal ranked more than 10 times higher in pig manure than chicken and cattle manure. Long-term agricultural application of Cu-containing animal manure can upsurge the risk in soil and surface water. Animal manure not only contains enhanced heavy metal levels, but also increases phosphorous and nitrogen level in soils. He et al. (2009) investigated the over-application of poultry litter in cotton and corn farms maintained for a period of 10 years. Cotton crop needs low nitrogen uptake, so the poultry litter application rates for cotton were low. Thus, after the cotton crop, P, Zn, Cu, or Mg concentrations did not increase. On the other hand, corn production always needs greater nitrogen demand; therefore, the poultry litter application was raised up to four times. Therefore, the content of not only total phosphorus in soil surface was enhanced to about 560 mg kg^{-1} dry soil, but also Cu and Zn contents were reported to rise from 7 to 22 mg kg^{-1} and from 17 to 32 mg kg^{-1} of dry soil, respectively. Level of phosphorus content was more than double the initial rate, which rose to the threshold point beyond extractable phosphorus. In addition, nutrient accessibility was significantly altered because the accumulation of phosphorus exceeded the soil buffer capacity.

9.5 Conclusion

Soil ecosystem is a huge reservoir of microorganisms. With the recent advancement in soil community analysis, molecular strategies have added a completely new dimension to the already existing information with respect to soil microbial diversity. After taking into account the complete scenario of different agricultural management practices, it is noticeable that organic farming enhances microbial richness, which is responsible for the enhancement of soil quality under different conditions. However, organic amendments, which constitute an important pillar of organic farming, can also cause substantial environmental and human health hazards related to the input, accumulation, and spread of heavy metals and antibiotic-resistant bacteria. The potential effect of antibiotics on the ecology of microbes is of notable concern, both from the perspective of upgrading the ecological reservoir of antibiotic-resistant genes, and through the reduced growth of microorganisms that are essential for various ecological functions. These aspects need to be taken into account when striving to develop sustainable cropping practices and strategies to lower these risks, e.g., proper composting of the amendment, and detailed analysis of its composition may decrease the level of antibiotics in manure and the consequent dispersion of environmental ARGs into the food chain.

Acknowledgment The study was supported by grants received from the Department of Biotechnology, Government of India (BT/PR27680/BCE/8/1434/2018), and Science and Engineering Research Board, Government of India (YSS/2015/001437). US wishes to acknowledge the fellowship received from the University Grant Commission, Government of India.

References

Abdelfattah A, Wisniewski M, Droby S, Schena L (2016) Spatial and compositional variation in the fungal communities of organic and conventionally grown apple fruit at the consumer point-of-purchase. Hort J 3:16047

Ashworth AJ, DeBruyn JM, Allen FL, Radosevich M, Owens PR (2017) Microbial community structure is affected by cropping sequences and poultry litter under long-term no-tillage. Soil Biol Biochem 114:210–219

Bailón-Salas AM, Medrano-Roldán H, Valle-Cervantes S, Ordaz-Díaz LA, Urtiz-Estrada N, Rojas-Contreras JA (2017) Review of molecular techniques for the identification of bacterial communities in biological effluent treatment facilities at pulp and paper mills. Bio Res 12:4384–4409

Bashan Y, de-Bashan LE, Prabhu SR, Hernandez JP (2014) Advances in plant growth-promoting bacterial inoculant technology: formulations and practical perspectives (1998–2013). Plant Soil 378:1–33

Beenhouwer MD, Van Geel M, Ceulemans T, Muleta D, Lievens B, Honnay O (2015) Changing soil characteristics alter the arbuscular mycorrhizal fungi communities of Arabica coffee (*Coffea arabica*) in Ethiopia across a management intensity gradient. Soil Biol Biochem 91:133–139

Bengtsson J, Ahnström J, Weibull AC (2005) The effects of organic agriculture on biodiversity and abundance: a meta-analysis. J Appl Ecol 42:261–269

Bhushan C, Khurana A, Sinha R, Nagaraju M (2017) Antibiotic resistance in poultry environment: spread of resistance from poultry farm to agricultural field. CSE:1–36

Bian R, Joseph S, Cui L, Pan G, Li L, Liu X, Marjo C (2014) A three-year experiment confirms continuous immobilization of cadmium and lead in contaminated paddy field with biochar amendment. J Hazard Mater 272:121–128
Binh CTT, Heuer H, Kaupenjohann M, Smalla K (2008) Piggery manure used for soil fertilization is a reservoir for transferable antibiotic resistance plasmids. FEMS Microbiol Ecol 66:25–37
Bonkowski M, Roy J (2005) Soil microbial diversity and soil functioning affect competition among grasses in experimental microcosms. Oecologia 143:232–240
Bossio DA, Scow KM, Gunapala N, Graham KJ (1998) Determinants of soil microbial communities: effects of agricultural management season and soil type on phospholipid fatty acid profiles. Microb Ecol 36:1–12
Brienen RJ, Phillips OL, Feldpausch TR, Gloor E, Baker TRL, Loyd J, Martinez RV (2015) Long-term decline of the Amazon carbon sink. Nature 519:344
Chen QL, An XL, Li H, Zhu YG, Su JQ, Cui L (2017) Do manure-borne or indigenous soil microorganisms influence the spread of antibiotic resistance genes in manured soil ? Soil Biol Biochem 114:229–237
Das S, Jeong ST, Das S, Kim PJ (2017) Composted cattle manure increases microbial activity and soil fertility more than composted swine manure in a submerged rice paddy. Front Microbol 8:1702
Dever LA, Dermody TS (1991) Mechanisms of bacterial resistance to antibiotics. Arch Intern Med 151:886–895
Ding LJ, Su JQ, Sun GX, Wu JS, Wei WX (2018) Increased microbial functional diversity under long-term organic and integrated fertilization in a paddy soil. Appl Microbiol Biotechnol 102:1969–1982
Dumontet S, Cavoski I, Ricciuti P, Mondelli D, Jarrar M, Pasquale V, Crecchio C (2017) Metabolic and genetic patterns of soil microbial communities in response to different amendments under organic farming system. Geoderma 296:79–85
Enne VI, Cassar C, Sprigings K, Woodward MJ, Bennett PM (2008) A high prevalence of antimicrobial resistant *Escherichia coli* isolated from pigs and a low prevalence of antimicrobial resistant *E. coli* from cattle and sheep in Great Britain at slaughter. FEMS Microbiol Lett 278:193–199
Faldynova M, Videnska P, Havlickova H, Sisak F, Juricova H, Babak V, Steinhauser L, Rychlik I (2013) Prevalence of antibiotic resistance genes in faecal samples from cattle pigs and poultry. Vet Med - Czech 58:298–304
Fengsong Z, Yanxia LI, Ming Y, Wei LI, Weijin YAN (2011) Copper residue in animal manures and the potential pollution risk in northeast China. Mol Ecol Resour 2:91–96
García-Orenes F, Guerrero C, Roldán A, Mataix-Solera J, Cerdà A, Campoy M, Caravaca F (2010) Soil microbial biomass and activity under different agricultural management systems in a semiarid Mediterranean agroecosystem. Soil Tillage Res 109:110–115
Halling-Sørensen B, Jacobsen AM, Jensen J, Sengeløv G, Vaclavik E, Ingerslev F (2005) Dissipation and effects of chlortetracycline and tylosin in two agricultural soils: a field-scale study in southern Denmark. Environ Toxicol Chem 24:802–810
Hamscher G, Pawelzick HT, Höper H, Nau H (2005) Different behavior of tetracyclines and sulfonamides in sandy soils after repeated fertilization with liquid manure. Environ Toxicol Chem 24:861–868
Hartman K, van der Heijden MGA, Wittwer RA, Banerjee S, Walser JC, Schlaeppi K (2018) Cropping practices manipulate abundance patterns of root and soil microbiome members paving the way to smart farming. Microbiome 6:1–14
He Z, Endale DM, Schomberg HH, Jenkins MB (2009) Total phosphorus zinc copper and manganese concentrations in cecil soil through 10 years of poultry litter application. Soil Sci 174:687–695
Henneron L, Bernard L, Hedde M, Pelosi C, Villenave C, Chenu C, Blanchart E (2015) Fourteen years of evidence for positive effects of conservation agriculture and organic farming on soil life. Agron Sustain Dev 35:169–181
Hobbs PR, Sayre K, Gupta R (2008) The role of conservation agriculture in sustainable agriculture. Philos Trans R Soc Lond Ser B Biol Sci 363:543–555

Hu X, Zhou Q, Luo Y (2010) Occurrence and source analysis of typical veterinary antibiotics in manure, soil, vegetables and groundwater from organic vegetable bases, northern China. Enviro Pol 158:2992–2998

Hugenholtz P (2002) Exploring prokaryotic diversity in the genomic era. Genome Biol 3:0003.1–0003.8

Johnston AE, Poulton PR, Coleman K (2009) Soil organic matter: its importance in sustainable agriculture and carbon dioxide fluxes. Adv Agron 101:1–57

Laird DA (2008) The charcoal vision: a win–win–win scenario for simultaneously producing bioenergy permanently sequestering carbon while improving soil and water quality. Agron J 100:178–181

Lazcano C, Gómez-Brandón M, Revilla P, Domínguez J (2013) Short-term effects of organic and inorganic fertilizers on soil microbial community structure and function. Biol Fert Soils 49:723–733

Liu Y, Zhu J, Ye C, Zhu P, Ba Q, Pang J, Shu L (2018) Effects of biochar application on the abundance and community composition of denitrifying bacteria in a reclaimed soil from coal mining subsidence area. Sci Total Environ 625:1218–1224

Lupatini M, Korthals GW, de Hollander M, Janssens TK, Kuramae EE (2017) Soil microbiome is more heterogeneous in organic than in conventional farming system. Front Microbiol 7:2064

Lyngwi NA, Koijam K, Sharma D, Joshi SR (2013) Cultivable bacterial diversity along the altitudinal zonation and vegetation range of tropical Eastern Himalaya. Rev Biol Trop 61:467–490

Mäder P, Fliessbach A, Dubois D, Gunst L, Fried P, Niggli U (2002) Soil fertility and biodiversity in organic farming. Science 296:1694–1697

Marschner P, Kandeler E, Marschner B (2003) Structure and function of the soil microbial community in a long-term fertilizer experiment. Soil Biol Biochem 35:453–461

Marti R, Scott A, Tien YC, Murray R, Sabourin L, Zhang Y, Topp E (2013) The impact of manure fertilization on the abundance of antibiotic-resistant bacteria and frequency of detection of antibiotic resistance genes in soil, and on vegetables at harvest. Appl Environ Microbiol 79:5701–5709

Mazzola M, Funnell DL, Raaijmakers JM (2004) Wheat cultivar-specific selection of 2 4-diacetylphloroglucinol-producing fluorescent Pseudomonas species from resident soil populations. Microb Ecol 48:338–348

Mendes LW, Tsai SM, Navarrete AA, De Hollander M, van Veen JA, Kuramae EE (2015) Soil-borne microbiome: linking diversity to function. Microb Ecol 70:255–265

Nicholson PS, Hirsch PR (1998) The effects of pesticides on the diversity of culturable soil bacteria. J Appl Microbiol 84:551–558

Ollivier J, Schacht D, Kindler R, Groeneweg J, Engel M, Wilke BM, Kleineidam K, Schloter M (2013) Effects of repeated application of sulfadiazine-contaminated pig manure on the abundance and diversity of ammonia and nitrite oxidizers in the root-rhizosphere complex of pasture plants under field conditions. Front Microbiol 4:1–14

Pérez MT, Sommaruga R (2006) Differential effect of algal-and soil-derived dissolved organic matter on alpine lake bacterial community composition and activity. Limnol Oceanogr 51:2527–2537

Pornsukarom S, Thakur S (2017) Horizontal dissemination of antimicrobial resistance determinants in multiple Salmonella serotypes following isolation from the commercial swine operation environment after manure application. J Appl Environ Microbiol 83:1–14

Prosdocimi M, Jordán A, Tarolli P, Keesstra S, Novara A, Cerdà A (2016) The immediate effectiveness of barley straw mulch in reducing soil erodibility and surface runoff generation in Mediterranean vineyards. Sci Total Environ 547:323–330

Reganold JP, Wachter JM (2011) Agron 3610 | farming system and sustainable agriculture. Organic agriculture in the twenty-first century. Nat Plants 2:15221

Reichel R, Rosendahl I, Peeters ETHM, Focks A, Groeneweg J, Bierl R, Schlichting A, Amelung W, Thiele-Bruhn S (2013) Effects of slurry from sulfadiazine- (SDZ) and difloxacin-

(DIF) medicated pigs on the structural diversity of microorganisms in bulk and rhizosphere soil. Soil Biol Biochem 62:82–91

Rondon MR, Goodman RM, Handelsman J (1999) The Earth's bounty: assessing and accessing soil microbial diversity. Trends Biotechnol 17:403–409

Santamaría J, Parrado CA, López L (2017) Soil microbial community structure and diversity in cut flower cultures under conventional and ecological management. Rev Bras Ciênc Solo 42

Schneider KD, Lynch DH, Dunfield K, Khosla K, Jansa J, Voroney RP (2015) Farm system management affects community structure of arbuscular mycorrhizal fungi. Appl Soil Ecol 96:192–200

Su JQ, Wei B, Xu CY, Qiao M, Zhu YG (2014) Functional metagenomic characterization of antibiotic resistance genes in agricultural soils from China. Environ Int 65:9–15

Tardy V, Spor A, Mathieu O, Leveque J, Terrat S, Plassart P, Régnier T, Bardgett RD, van der Putten W, Roggero PP, Seddaiu G, Bagella S, Lemanceau P, Ranjard L, Maron PA (2015) Shifts in microbial diversity through land use intensity as drivers of carbon mineralization in soil. Soil Biol Biochem 90:204–2013

TerAvest D, Carpenter-Boggs L, Thierfelder C, Reganold JP (2016) Crop production and soil water management in conservation agriculture no-till and conventional tillage systems in Malawi. Agric Ecosyst Environ 212:285–296

Tien YC, Li B, Zhang T, Scott A, Murray R, Sabourin L, Marti R, Topp E (2017) Impact of dairy manure pre-application treatment on manure composition soil dynamics of antibiotic resistance genes and abundance of antibiotic-resistance genes on vegetables at harvest. Sci Total Environ 581–582:32–39

Udikovic-Kolic N, Wichmann F, Broderick NA, Handelsman J (2014) Bloom of resident antibiotic-resistant bacteria in soil following manure fertilization. Proc Natl Acad Sci USA 111:15202–15207

Valentine DL, Kessler JD, Redmond MC, Mendes SD, Heintz MB, Farwell C, Chan EW (2010) Propane respiration jump-starts microbial response to a deep oil spill. Science 330:208–211

Velmourougane K (2016) Impact of organic and conventional systems of coffee farming on soil properties and culturable microbial diversity. Scientifica 2016:3604026–3604026

Wang JJ, Li XY, Zhu AN, Zhang XK, Zhang HW, Liang WJ (2012) Effects of tillage and residue management on soil microbial communities in North China. Plant Soil Environ 58:28–33

Wang W, Wang H, Feng Y, Wang L, Xiao X, Xi Y, Luo X, Sun R, Ye X, Huang Y, Zhang Z (2016) Consistent responses of the microbial community structure to organic farming along the middle and lower reaches of the Yangtze river. Sci Rep 6:35046

Wardle DA (1995) Impacts of disturbance on detritus food webs in agro-ecosystems of contrasting tillage and weed management practices. Adv Ecol Res 26:105–185

Wardle DA, Yeates GW, Watson RN, Nicholson KS (1995) The detritus food-web and the diversity of soil fauna as indicators of disturbance regimes in agro-ecosystems. Plant Soil 170:35–43

Willer H, Lernoud J (eds) (2018) The world of organic agriculture. Statistics and emerging trends 2018. Research Institute of Organic Agriculture (FiBL), Frick, and IFOAM – Organics International, Bonn

Williams CM (2002) Nutritional quality of organic food: shades of grey or shades of green? Proc Nutr Soc 61:19–24

Xiong W, Guo S, Jousset A, Zhao Q, Wu H, Li R, Shen Q (2017) Bio-fertilizer application induces soil suppressiveness against Fusarium wilt disease by reshaping the soil microbiome. Soil Biol Biochem 114:238–247

Xue K, Wu L, Deng Y, He Z, van Nostrand J, Robertson PG, Zhou J (2013) Functional gene differences in soil microbial communities from conventional low-input and organic farmlands. Appl Environ Microbiol 79:1284–1292

Yeom JR, Yoon SU, Kim CG (2017) Quantification of residual antibiotics in cow manure being spread over agricultural land and assessment of their behavioral effects on antibiotic resistant bacteria. Chemosphere 182:771–780

You F, Dalal R, Huang L (2018) Biochar and biomass organic amendments shaped different dominance of lithoautotrophs and organoheterotrophs in microbial communities colonizing neutral copper (Cu)-molybdenum (Mo)-gold (Au) tailings. Geoderma 309:100–110

Zhang F, Li Y, Yang M, Li W (2012) Content of heavy metals in animal feeds and manures from farms of different scales in Northeast China. Int J Environ Res Public Health 9:2658–2668

Zhang Q, Wu J, Yang F, Lei Y, Zhang Q, Cheng X (2016) Alterations in soil microbial community composition and biomass following agricultural land use change. Sci Rep 6:36587

Zhao L, Dong YH, Wang H (2010) Residues of veterinary antibiotics in manures from feedlot livestock in eight provinces of China. Sci Total Environ 408:1069–1075

Zhen Z, Liu H, Wang N, Guo L, Meng J, Ding N, Wu G, Jiang G (2014) Effects of manure compost application on soil microbial community diversity and soil microenvironments in a temperate cropland in China. PLoS One 9:e108555

Chapter 10
Plants for Biocontrol and Biological Control of Plant Pathogens

Prachi Saxena, Jyoti Srivastava, Shrishti Pandey, Shreya Srivastava, Neha Maurya, Niharika Chand Kaushik, Shubham Mishra, Garima Asthana, Prachi Bhargava, Rajesh Kumar, and Siddharth Vats

Abstract For maintaining the quality and quantity of food produced around the world, the plant diseases are needed to be controlled. To prevent and control plant diseases, different methods may be employed including chemical fertilizers and pesticides which have significantly improved the crop productivity and quality over the past many years. However, there are strict regulations on the use of chemical because of its hazardous effect. But some pest management researchers have developed alternatives which are referred to as biological controls for controlling pests and diseases by the application of biocontrol plants and also by the use of plant–microbe interactions for controlling plant pathogens. This chapter focuses on the various plants which are used for biocontrol and how they can be helpful in sustainable development. Similarly, how to use microbes and other organisms for the control of plant diseases has also been studied in detail.

10.1 Introduction

The negative effects of overuse of chemicals in agriculture like pesticides, insecticides, herbicides and fertilizers affecting health of its consumers, biomagnification of the chemicals and the economic burden caused by their cost and many other side effects have raised a serious alarm (Tandon and Vats 2016; Kaur et al. 2010; Vats and Bhargava 2017). Food is important and to prevent it from being spoiled during cultivation stages requires chemical-based treatment. But this chemical-based treatment can be avoided (Sharma et al. 2018). The term "biological control" also known as "biocontrol" has been used in the fields of both entomology and plant pathology. An organism that prevents the growth of pathogen is referred to as the biological

P. Saxena · J. Srivastava · S. Pandey · S. Srivastava · N. Maurya · N. C. Kaushik · S. Mishra · G. Asthana · P. Bhargava · S. Vats (✉)
Institute of Biosciences and Technology, Shri Ramswaroop Memorial University, Lucknow, Uttar Pradesh, India

R. Kumar
University Institute of Engineering and Technology, Kurukshetra University, Kurukshetra, Haryana, India

A. Varma et al. (eds.), *Plant Biotic Interactions*,
https://doi.org/10.1007/978-3-030-26657-8_10

control agents (BCAs) in both entomology and plant pathology. Since the term biocontrol can refer to different forms of ideas, before applying it to the review of any particular work, it is important to specify the broad array of the term. Biological control is defined as the living organisms other than disease-resistant host plants which are amended in a soil to suppress the activities of plant pathogens (Pal and Gardener 2006).

10.2 Plants as Biocontrol

Allelopathy is one of the mechanisms where biochemical released by one plant influences the growth of the other plants. This phenomenon has been observed on cases where crop rotation is practised. The next crop does not grow properly due to presence of chemical released by the first crop. Based on the same strategy, some plants are used to control the growth of the weeds, pests and even microbes. The plants used are termed as biocontrol plants. A biocontrol plant is deliberately joined to a cropping system with the purpose of improving the crop productivity (Brennan 2016). They also play a major role with the aim of increasing the regulation of biological control systems, eventually leading to increased crop efficiency. There are many plant species which are cost-effective in enhancing biocontrol, and there is also a rapid increase of biocontrol arthropods. These plants, biocontrol carrying plants, are multitalented but are not specialist. There are various factors which also influence their release of chemicals. Generally, plants which are used for biocontrol are species which are not eliminated easily by bio-aggressors and have characteristics of invasive species. Mechanism by which biocontrol plant supports and inhibits the growth of pest is dependent on various factors. Microorganisms and plants move with each other for the natural process and trade of supplements. Plants which are used for biocontrol are mentioned in Table 10.1.

Plant-based biocontrol is a common biological phenomenon among plants for their growth, survival, competition facing, reproduction and development (Fig. 10.1). The chemicals released by one plant to inhibit or promote the growth of other organisms are termed as allelochemicals. Plant-based biocontrol is one part of allelopathy. These allelopathic plants are used for crop management, weed control, reestablishment of the crops and agricultural management. For sustainable agriculture, the use of allelopathic plants is important and takes advantage of their stimulatory and inhibitory activities but avoids their toxicity. Allelochemicals secreted by plant biocontrol find significant use of insecticides, herbicides, growth regulators and antimicrobial crop protection against pathogens. Allelopathic chemicals have effects on micro- and ultra-structure of plants, elongation, maturity and divisions of cells, permeation ability of the cell membrane, cellular respiration, system of growth regulation, metabolism and metabolic pathway control system, oxidative and antioxidant system, enzyme inhibition and activation, enzyme synthesis, synthesis of biomolecules, photosynthesis, reproduction activity of the plants and plant cells, uptake of nutrients, structural characteristics, microbes interaction, interaction with the other plants and how the allelochemical actions. Out of various factors, plants' interaction

Table 10.1 Biocontrol plants and the species affected

S. No.	Plant biocontrol	Allelochemicals released by plants	Affected pest
1	*Eucalyptus*	Phenolics	Shrubs, herbs, grasses, weeds
2	*Balsam poplar*	Terpenoids and phenolics	Green aldar
3	*Sassafras*	Phenolics	Silver maple, boxelder, weeds
4	*Juniper*	Phenolics	Grasses, weeds
5	*Sycamore*	Coumarins	Yellow birch, weeds, grasses, herbs
6	*Southern red oak*	Terpenoids, phenolics	Sweetgum
7	Sugar maple	Phenolics	Spruce, weeds, grasses, yellow birch
8	Hackberry	Coumarins	Grasses, weeds, herbs
9	Black walnut	Quinone	Apple, birch, pines, birch, hackberry, azalea, basswood
10	*Oaks*	Coumarins	Herbs, grasses
11	*Secale cereal*	Phenolic acids, beta hydroxyl butyric acid, benzoxazolinone	Weeds, grasses. Shrubs
12	*Triticum aestivum*	Phytotoxic phenolic acids, simple acids	Weeds
13	*Helianthus Annuus*	Terpenes, phenolics, heliannuol	Shrubs, weeds
14	*Sorghum bicolour*	Dhurrin, phenolic, sorgoleone	Weeds
15	Sugarcane	Phenolic, malic acid, tartaric acid, oxalic acid, vanillic acid, ascetic acid	Shrubs, weeds
16	Barley	Phenolics,	*Nicotiana tabacum, Stellaria media, Solanum ptycanthum*
17	Rye	Phenolics	*Lactuca sativa, wild oat, Bilderdykia convolvulus*
18	Rice	Phenolics	*Purple ammannia, Lactuca sativa, wild oat, weeds*

Table adapted from Makoi et al.

has been the one reason why there is a distribution of species in which some are abundant while some are scarcely distributed. Invasive species are successful because of the allelopathy (Fig. 10.2).

10.2.1 Allelochemicals Involved in the Plant Biocontrol

Allelochemicals are phytochemicals which are nonnutritive, are secondary metabolites in nature of the plant metabolic pathways or are produced by the transformation/decomposition caused by the microbes. Main families of allelochemicals are provided in Table 10.2.

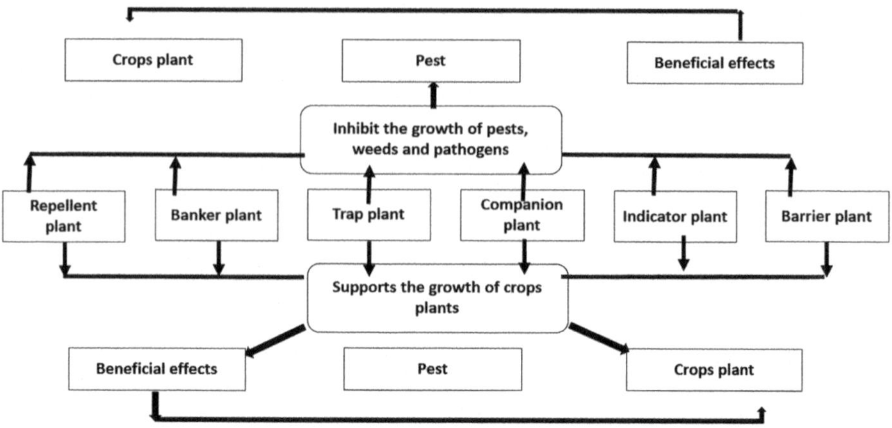

Fig. 10.1 A general role of biocontrol plants in supporting the growth of plant crops and inhibiting the growth of pest

Fundamentally based bio-manures wherever plants deposits improved with organisms shapes a reliant association with plants. They increment the arrangement of the little and macronutrients to the plants and yields by colonizing the premise zone or rhizosphere (Gupta et al. 2018; Bhargava et al. 2017; Goel et al. 2017). Association of plants and microorganisms have decreased the reliance on substance basically based on manures and advanced the origination of natural cultivating. Predominant of plants unwellness like root hitch unwellness in soybean by use of bio-fungicide and BINA manures Application of metallic component solubilizes, synthetic component fixers, substance component solubilizes, the mix of molds, organisms and microorganism in consortia in rising the soundness of the plants and expanding supplements comfort and worthiness to the plants. Azotobacter, mycorrhiza, phosphate solubilizing microorganism, *Azospirillum* and *Rhizobium* territory unit the organisms used in eco-friendly farming practices (Bhargava et al. 2017). Biopesticides from organisms and plants is furthermore a bio-asset discovering application in property horticulture and biocontrol. Plants like family *Azadirachta* and *chrysanthemum*, while organisms like *eubacterium thuringiensis*, *Trichoderma* and *Nuclear polyhedrosis* infection territory unit utilized for biocontrol. Microorganisms essentially based biopesticides (biocontrol operators) neutralize phytopathogens by assembling biomolecules like hydrolytic compounds, anti-infection agents, administration the action of phytopathogens by the generation of siderophores and HCN (Bhattacharyya and Jha 2012; Saha et al. 2016; Sharma et al. 2014; Sharma and Johri 2003; Vats and Miglani 2011).

10.3 Plant–Microbe Interaction for Protection of Crops

Several factors have been studied which govern the ability of biocontrol agents to establish themselves on roots. Rhizosphere competency is one such factor (Harman and Kubicek 2014). The "rhizosphere" term was first coined by Lorenz Hiltner, the

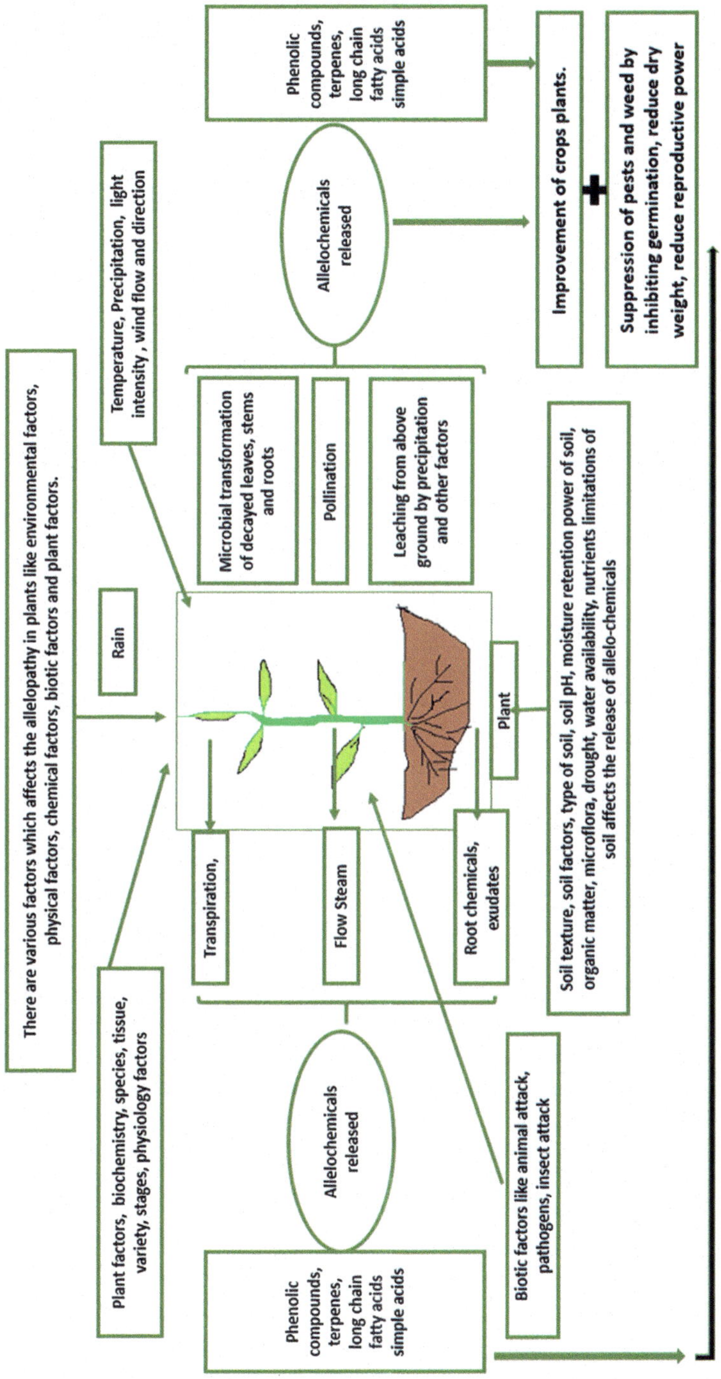

Fig. 10.2 Mechanism of action of biocontrol plants

Table 10.2 Different types of phytochemicals (allelochemicals) used by biocontrol

S. No.	Allelochemicals: the phytochemicals behind plant-based biocontrol
1	Water-soluble organic acids, straight chain alcohols, aliphatic aldehydes and ketones
2	Simple unsaturated lactones
3	Long-chain fatty acids and polyacetylenes
4	Benzoquinone, anthraquinone and complex quinones
5	Sulphide and glucosinolates
6	Alkaloids and cyanohydrins
7	Terpenoids and steroids
8	Coumarin
9	Tannins
10	Flavonoids
11	Amino acids and peptides
12	Purines and nucleosides
13	Simple phenols, benzoic acid and its derivatives
14	Cinnamic acid and its derivatives

Adapted from Cheng and Cheng (2015), Vats (2017), Negi and Vats (2014)

German Agronomist and plant physiologist in 1904 to plant–root interface. The term rhizosphere is derived from the Greek word “rhiza”, meaning root. The rhizosphere was defined by Lorentz Hiltner as “the soil compartment influenced by the roots of growing plants”. These soil regions are directly influenced by soil microorganisms. Another factor which governs the biocontrol agent’s activity is by involving food base with the biocontrol agents to enhance its activity (Bailey and Den 2014). The food base like green manures, composts and stable manures can support the activity of biological control activity (Monaco et al. 2004). Mutualism is an association where both the species derive benefit. In such association, there occurs obligatory interaction which is lifelong and beneficial for survival of each species such as lichens, a symbiotic association of fungus and alga. These types of association can enhance host defences. Proto-cooperation, unlike mutualism, organism do not depend on each other for survival in proto-cooperation. These are considered as non-obligatory mutualism. Commensalism is a kind of symbiotic interaction between two living organisms where only one is benefitted and neither is harmed. Most plant-associated microorganisms are considered as commensals because their presence neither harms nor benefits. The above three interactions, i.e. mutualism, commensalism and proto-cooperation, are considered as the positive interaction/ beneficial interaction, where either one or both the species are benefitted. These benefits may be continuous, obligatory or facultative. Neutralism shows a type of interaction where one species has no effect on the other, i.e. neither of the population affects the other. Antibiosis refers to the complete or partial inhibition of one organism by another. Antibiosis shows the antagonistic effect which is negative interaction. The various types of interaction are given in Table 10.3.

Table 10.3 Mechanism of plant biocontrol (Cheng and Cheng 2015; Thomashow 2013; Harman and Kubicek 2014)

Antagonism	This is basically referring to the type of agent which helps in reducing the numbers or it may also help in decreasing the availability of that resources to others
Antibiosis	It is the condition in which one or more than one antibiotic substances or metabolic products are secreted by an organism which have harmful effect. Here, the action of making antibiotics is common in bacteria of soil
Nutrient competition	It is a type of competition which occurred between the microorganisms. Major quality of this category is that it always occurred for narrow origin. This is the most ordinary way by which growth of organisms may be limited as compared to others. For example, carbon, nitrogen, oxygen, etc.
Parasitism	It refers to the interconnection between the host and pathogens
Destructive myco-parasitism	It discovers when antagonist captures the pathogens, for example, cellulose, glucanase, chitinase, etc. This all happens by secreting enzymes by others. Another can be the parasitism of one fungus by another which can only occur by the direct association, by undefined enzymes which came from cell wall. They may form some of antibiotics, i.e. *Trichoderma harzianum* which is commonly known for the treatment of seed
Predation	It refers to the relationship that is found between organisms in which one organism is captured and provides food on the other

10.4 Mode of Functioning of Biocontrol Agents

Biocontrol using adversary has extremely reduced the use of pesticides as a promising alternative. However, it is essential to know about the mode of action of antagonists (microbial) so that it can further help in knowing the procedure and results of the known competitor and in providing the knowledge to use the desirable and effective strain of antagonists to achieve better result (Babychan 2017).

- Conflict for space and nutrition: Competition for nutrients and space among the antagonists is the most important mode of functioning which inhibits the activity of pathogens causing decaying and damage to the harvested plant produces. Microbial adversary grows faster than disease-causing pathogens; hence, rapid growth of antagonists easily colonizes the fruit wound and enhances biological control. Conflict between antagonists for several important micronutrients like iron is essential for the control of biological diseases (Goel et al. 2017). Plants act in response to various environmental encouraging factors like phototropism, geotropism, temperature, etc. These factors prevent the activity of pathogens to the harvested produces (Audenaert et al. 2002).
- Antibiosis: Antibiotic production is the second most essential factor through which microbial adversary inhibits the pathogens in harvested plant produces. Several microbes release some compounds having antibiotic property. Many plant diseases and disease-causing pathogens are inhibited by the activity of these antibiotics produced by microbial community. Antibiotic named iturin produced by specific bacterial antagonists like *Bacillus subtilis* and *Pseudomonas cepacia* Burkh is used to destroy the pathogens. Antibiosis is a major technique

which reduces the postharvest losses of fruits and vegetables (Spadaro and Droby 2016).

- Parasitism: It is another mode of action of biocontrol agents in which microbial antagonists directly interface with the disease-causing pathogens. Antagonist *Pichia guilliermondii* strongly adheres to the mycelium of *Botrytis cinerea* through a lectin link which directly inhibits the activity of this pathogen to the plant (Wisniewski et al. 1991). Several compounds are also produced by microbial antagonists which cause the breakdown of cell wall of phytopathogenic fungi to destroy them; these compounds are known as lytic enzymes. Microbial antagonists' strongest adhesion with the disease-causing pathogens with the property of cell wall degradation compounds (enzymes) increases the potency of biocontrol agents to decrease the diseases caused to fruits and vegetables (Coda et al. 2013).
- Induced resistance: It is defined as the state of modified defensive ability which is developed by plants when they are given appropriate stimulus. Effectiveness of several yeasts used as biocontrol agents is recognized before the inoculation of pathogen. And it is observed that resistance is induced in the fruit skin by the yeast cell against pathogens. As in avocado fruit, resistance is induced by the antifungal compound production against disease-causing pathogen (Prusky et al. 1994; Sharma et al. 2009).

Competition is a type of interaction which occurs within (intraspecific) or between (interspecific) species that results in the decreased growth and fecundity of the interacting species. Parasitism is a symbiotic association of two phylogenetically unrelated organisms. Competition, antibiosis, parasitism and predation show negative interaction where one or both species are harmed. The introduction of synthetic inorganic fertilizers and pesticides in disease-resistant varieties resulted in byproduct such as manures which became solid wastes (Vats and Bhargava 2017; Vats and Kumar 2015; Vats et al. 2014; Maurya et al. 2014). Inappropriate disposal practices resulted in water pollution and spread of animal pathogens into the environment (Vats and Negi 2013;Vats 2017; Vats et al. 2012, 2013a, b, 2017; Maurya et al. 2013). Those cropping systems where adequate management programs based on genetic host resistance or cultural practices were not available resulted in declining of soil structure, mineralization of soil organic matter and numerous diseases caused by soilborne plant pathogens, e.g. *Phytophthora* root rot of avocado (Baker and Cooker 1974). To avoid many types of problems, today composting proved to have advantage over other organic solid waste disposal practices and is one of the environment-friendly sound waste treatment processes (Vats and Mishra 2016).

10.5 Some Useful Characteristics of Biocontrol Plants

As we know, plants grow really fast and they are economical. All they need is some of the agriculture requirements like temperature, sunlight, oxygen, etc. (Kim et al. 2008). The main classification which are closely connected to the implementation of biocontrol plants is that they can provide nutrition for long period of time. Furthermore, the requirements of the crop must be close so that which can be protected by temperature, light, nutrients, etc. (Brennan 2016).

What Makes a Successful Biocontrol?
Any biocontrol would be put into the successful category on the bases of its production and formation.

Production of Highly Effective Biocontrol

1. Must have the ability of approach and persevere
2. Must have the ability to colonize
3. Must have the ability to rapidly increase in numbers
4. Always play as a non-pathogenic role to host plant and as well as to the environment

Formation and Manufacturing of Various Agents Must Be

1. Easier and cost-effective.
2. With magnificent shelf-life.
3. With production details mentioned in biomass.
4. Some important aspects for successful agent: inexpensive and ability of generating in extreme quantities and viability should be maintained properly.

Before a successful agent becomes commercially available, it has to file an application which contains about the mutual benefits and losses. So, application must permit full expression of the agent:

(a) Most importantly, that all of the agents will produce and rise at large scale for achieving their levels must be ensured.
(b) It must have ability to survive in changing environmental condition.
(c) Potential for reproduction must be fast.
(d) It should be a good contender.
(e) Rate of survival should be at higher level in soil.
(f) Pathogen control scale must be high as always.

10.6 Biological Control of Plant Diseases

The term "biological control" and its abbreviated synonym "biocontrol" that can be consume in various fields of biology, most importantly in plant pathology and entomology. In entomology, it relates to the predatory insects, entomo-pathogenic nematodes or microbial pathogens (Makoi and Ndakidemi 2007). The term plant pathology relates to the microbial antagonists which help to overcome many diseases, and it also involves host particular pathogens which control weed populations. In terms of both fields, the organisms which conquer the pest or pathogen are referred to as the biological control agents (BCA) (Harman and Kubicek 2014). Moreover, the phrase biological control also played a major role in the natural products fermented from different origin. These types of alignments are simple mixtures of natural ingredients with the complex mixtures that may affect the host of the pathogens. Biological control determines the introduction of living organisms, and also the resistant host plants, to extinguish the activities and increasing number of pathogens of plants.

10.7 Interrelationship Between Biocontrol for Plants

As we know, there is a huge interaction between plants and pathogens with a wide variety of organisms. Such interaction may give harmful results to plant growth. In contemplation of biocontrol mechanism, it played a major role in interaction of organisms in many differ ways that can be direct and indirect forms. In 1953 Odum (1994) who is widely considered the father of modern ecology said that by the outcomes of each, we can understand the interrelationship between inhabitants. These types of interrelationships can be in the forms of mutualism, proto-cooperation, commensalism, neutralism, competition, amensalism, parasitism and predation. And this was evolved for macrology, means that all of the interrelations that have been discovered in the world at the microscopic and macroscopic scale. And many of the diseases occurred in plants includes the plants and as well as microbes, and these interrelationships can be occurred at multiple forms of scale (Harman and Kubicek 2014). If we consider plant's outlook, biocontrol can be formed by the variations occurring in specified and unspecified interrelationships. By the terminology of Odum, we can easily understand and differentiate the various elements of ecosystem which are playing major role to biocontrol. There are various types of interaction that takes place in the biological control of plant pathogens.

10.8 Application of *Bacillus thuringiensis* in Agriculture

In the 1950s *Bacillus thuringiensis* had replaced the business of chemical pesticides because it was a valuable alternative for chemical pesticides as it is harmless to environment and shows high activity against insects. During this period of time, many *Bt*-based products were launched in the market, and they were commercially successful. On the other hand, after the discovery of *Bt israelensis*, it is used to control mosquitoes and flies to stop the disease spreading. In many countries like Africa, the USA, USSR, etc., it is used in vector control and *Bt tenebrionis* is used against *Colorado*, i.e. against potato beetle (Entwistle et al. 1993). *Bt* was only the biopesticide which was highly specific towards its host. In the 1990s *Bt* was the most commercially successful pesticide as it has covered the 90% of the global market of biopesticide. In 2012 this figure decreased to 55% because of development of resistant genes in insects. *Bt* products is mostly made up of dried crystal mixture which is produced by fermenting the different types of media together with formulation ingredients. According to the information from US market, there are 13 active ingredients used in 123 registered products (Walker et al. 2003). *Bt* strain improvement was done to get better insecticidal activity, and this was done by using recombinant DNA technology in which plasmids are transferred by the process of conjugation (Sanchis 2011). Although *Bt* was a good biopesticide with specificity in its action and benefits for environment, it has never covered the large part of insecticidal market. This is because of limited activity of *Bt*-based products, i.e. *Bt* do not target major insect orders like white flies and aphids. Along with it *Bt*-based products do not act on the insects which attack the plant roots, i.e. *Bt*-based products are mostly used to protect the plant from the pests present outside and above the ground. This problem leads to the formation of transgenic crop with insecticidal crystal protein forming gene in it (de Maagd 2015). A company of Belgium Plant Genetic System was the first one to develop a transgenic tobacco plant which has insect tolerance activity because of the presence of Cry gene taken from *Bt.* But this attempt was not so successful because of limited expression of insecticidal crystal proteins (Vaeck et al. 1987). In the same year, a group of scientists inserted Cry genes of *Bt* into tomato plant and cotton plant. They transferred the Cry gene of *Bt* using Ti plasmid of *Agrobacterium tumefaciens*, and the plasmid was inserted into plant cell by the process of transformation. Before, this method was used only for dicotyledonous plants (e.g. tobacco), but by the development of new techniques like electroporation, it became possible to transfer DNA in monocotyledonous plants (e.g. maize) (Koziel et al. 1993). A variety of transgenic crops were developed using Cry gene of *Bt*, and the most popular of the were *Bt* cotton and *Bt* maize which were commercially launched in 1996 (de Maagd 2015).

10.8.1 **Bacillus thuringiensis** ***as Biocontrol Agent***

Bacillus thuringiensis, commonly known as *Bt*, is a unique naturally occurring bacterium which has a long history of research. It is the oldest and most widely used biopesticide which is used as the pest control agent. The special quality of *Bt* which makes it unique is the presence of insecticidal and parasporal crystal-forming genes on its plasmid. These crystal proteins are toxic in nature and lead to the death of insects by forming pores in intestine of the host pests. *Bt* is a spore-forming aerobic bacterium that has a highly specific insecticidal activity. They have been used as a biological agent in fields of forestry, agriculture and medicine for eliminating the disease-causing vectors and crop-destroying pests. Many *Bt*-based products and *Bt* crops are commercially available in the market and they are proven to be commercially successful as a biopesticide.

Bacillus thuringiensis (*Bt*) is a Gram-positive bacterium which forms endospores. This aerobic bacterium was first isolated by a Japanese microbiologist Shigetane Ishiwata in 1901. Ishiwata named it "Sottokin Bacillus" which means "sudden death Bacillus" (Ishiwata 1901). In 1915 Ernst Berliner, a German biologist, named it *Bacillus thuringiensis* and discovered the presence of crystal in *Bt* (Berliner 1915). As *Bt* was isolated from infected *Bombyx mori*, i.e. silkworm, it was considered a threat to silk industry for a long time. But in the 1950s, insect pathologists focused their studies on insecticidal activity of *Bt*. Sporeine was the first commercially used product which was made from *Bt* in 1938 in France. Soon the USA, USSR and some European countries like France and Germany started to produce the products made from *Bt* on large scale (Sanchis 2011). Till the 1970s it was thought that *Bt* only targets Lepidoptera (i.e. larvae of moths and the butterflies). But this belief was broken by Goldberg and Margalit (1977) by the isolation of new subspecies of *Bt*, i.e. *israelensis*, in 1977. *Bt israelensis* targets the Dipteran insects (i.e. larvae of mosquitoes and the black-flies (Goldberg and Margalit 1977). In the 1980s *Bt israelensis* was extensively used in medical field for controlling vector especially against mosquitoes. In 1983 in Germany one more subspecies was isolated, i.e. *morrissoni* with var. *tenebrionis*. It targets coleopteran insect's larvae (i.e. beetles). This discovery greatly enhanced the commercial use and the study of insecticidal activity of *Bt*. In 2003, activity of crystal protein was studied against larvae of nematode, which demonstrated that the crystals also target the phylum Nematoda (Wei et al. 2003). Now, thousands of strains of *Bt* have been discovered from different parts of the world which targets different organisms including Protozoa, Hymenoptera (i.e. wasp and bees), Hemiptera and Mallophaga. The discovery of genetically modified organism leads to the formation of plants which are insect resistant by inserting the genes of *Bt*. The first pest-resistant transgenic crop was made in 1996 (de Maagd et al. 1999).

10.8.2 Phylogeny and Life Cycle of **Bacillus thuringiensis**

Bt is a type of Gram-positive bacterium which taxonomically comes under family *Bacillaceae* and group of *Bacillus cereus*. It is a free-living and parasporal crystal-producing bacterium whose life cycle consists of two phases, i.e. spore formation phase and vegetative phase. In adverse conditions *Bt* forms endospores which are tough, non-reproductive and dormant structures. Spores formed have a protective and hard capsule around the *bacilli*. In favourable conditions these endospores again start their growth and vegetative division (Ibrahim et al. 2010). From the genetic point of view, *Bt* is very similar to the *Bacillus cereus* and *Bacillus anthracis*. *B. cereus* is a pathogenic soil bacterium which is a common cause for food poisoning, and *B. anthracis* causes anthrax which is a fatal and lethal disease. The difference between *Bt* and *B. cereus* is the presence of insecticidal toxin-forming genes in its plasmid of *Bt*, and the difference between *Bt* and *B. anthracis* is the absence of mammalian active toxin-forming genes in its plasmid (Helgason et al. 2000). Only *Bt* forms crystals during spore formation phase because of which it became a facultative pathogen for insects (Glare and O'Callaghan 2000). There are many other species of bacteria with crystal proteins-producing genes which have insecticidal activity similar to *Bt*. For example, *Lysinibacillus sphaericus* has insecticidal activity against larvae of mosquito, various species of *Paenibacillus* target the larvae of honeybee and Japanese beetle, *Brevibacillus laterosporus* targets Diptera and *Clostridium bifermentans* targets mosquitoes (de Maagd et al. 2003).

10.8.3 **Bacillus thuringiensis***: Toxin Diversity, Specificity and Classification*

Bacillus thuringiensis is a species with a vast protection of various toxins which is produced from parasporal and insecticidal crystal protein-producing genes (de Maagd et al. 2003). Tom Angus showed that in *Bt* insecticidal crystal proteins (commonly called as Cry proteins) are formed during the sporulation phase (Angus 1954). It was recognized that different strains of *Bacillus thuringiensis* show their insecticidal activity on different classes of insects; this showed the specificity of activity of insecticidal crystal proteins (Bravo et al. 2017). The first nomenclature and classification of crystal proteins were given by Hofte and Whiteley in 1989. They divided the crystal proteins in groups on the basis of their molecular relationship and insecticidal activity (i.e. CryI, CryII, CryIII, Cry IV, Cry V, etc.) (Hofte and Whiteley 1989). This classification leads to the better understanding of insecticidal activity of various crystal proteins, but as the new strains were found, this classification failed. A new classification on the basis of identity of amino acid and evolutionary divergence was given for crystal proteins, according to which crystal proteins were divided into 300 subgroups (Ruan et al. 2015). Now, Cry genes are classified on basis of homology sequence in four levels of hierarchy, because of which today we see about 60 major classes of Cry proteins (Sanchis 2011).

10.8.4 Mode of Action of Insecticidal Crystal Protein of **Bacillus thuringiensis**

The specificity of different strains of *Bt* to show their activity against different insects led to various researches and studies on specificity and mode of action of *Bt*. Knowles discovered the mode of action of *Bt* and its mechanism to cause toxicity (Knowles 1994). The Cry proteins cause poison in the stomach, i.e. it is responsible for the lysis of epithelial cells of the stomach. After ingestion, crystal gets dissolved in the intestinal tract and then crystal is solubilized according to pH of the intestinal tract. For the insecticidal activity of the crystal proteins, the pH should be highly alkaline, and then only the crystal will solubilize (Hofmann et al. 1988). Therefore, the pH of the intestinal tract decides the toxicity of *Bt* on host insect. After dissolving, the inactive toxin molecule (protoxin molecules) gets activated by the cleavage of proteolytic enzymes (proteases). By the action of proteases, the molecular weight of crystal protein decreases from around 130–140 kDa to 55–65 kDa, because of which Cry protein can easily pass through epithelium of midgut (Richards and Richards 1979). Activated crystal proteins bind to their receptor present on the brush border of intestinal microvilli in host insect (Van Rie et al. 1990). Because of binding, the toxin inserts into the cell membrane of host insect leading to the pore formation and the lysis of epithelial cell of the intestine due to selective permeability of cation (English and Slatin 1990). On physiological level, the lysis of intestinal epithelial cell results in paralyzed digestive system of host insect and insect stops eating. This may lead to the death of insect within 3 days of ingestion of protein crystals (Sanchis 2011).

10.8.5 Ecological Issues and Development of Resistance with the Bt

The first and foremost concern was that Cry toxins produced in transgenic plants are in active state, whereas they are produced in the form of inactive molecule in case of bacteria, because of which the action of crystal protein on specific host decreases and it may lead to the activity of toxin against nontarget organisms (Hilbeck 2002; Stotzky 2000). Another risk is that the crops produced from these transgenic plants may be infertile and cannot give rise to new plant (Sanchis 2011). Few reports are there which discuss the harmful effects of *Bt* sprays. Its similarities with *B. cereus*, a soil pathogen responsible for causing food diseases, have raised the question on the safety of using *Bt* crops. Some *Bt* strains produce beta exotoxins which target mammals and fishes. Still the researches are done on the safety of transgenic crops made from *Bt* (de Maagd 2015). The authorities of the USA and Europe like Environmental Protection Act (EPA) and European Food Safety Agency (EFSA) have approved the use of transgenic crops having Cry gene of *Bt* as safe (EFSA Panel on Genetically Modified Organisms 2010). Continuous use of *Bt* sprays and *Bt* crops leads to the development of resistance in the pest insect. Due to long-term use of *Bt*-based products, the insects develop the resistance against Cry toxins of *Bt*. *Bt* crops create a pressure of selection on insects which lead to the evolution of resistance in insects. In 1985, the first report on resistance *Bt* was published. According to it a very small level of resistance was seen in Indian meal moth (*Plodia interpunctella*) found in storage bin of grain treated by *Bt* (Lacey et al. 2015). Still the rate of emergence of resistance is quite slow; this may be because of the strategies of resistance management. Still researches are going to find the side effects of *Bt* crops and *Bt*-based products, and it's a topic of great discussion, whether *Bt*-based products are safe or not.

10.9 Entomopathogenic Nematodes: Another Way of Controlling Insect Pests

Entomopathogenic nematodes, also known as EPNs, are nematodes which parasitize insects. These have been introduced from 23 families of nematodes (Koppenhöfer 2007). Out of these 23, the 2 most important families are Steinernematidae and Heterorhabditidae. These two families got the high attention because they are successfully producing the desired product for controlling the insects and pests

(Kaya and Gaugler 1993; Koppenhöfer 2007). Enormous research and findings from the last three decades have clarified the success and failures of EPNs when they are being used as controlling agents for insect pests related to crops and other ornamental and turfs (Shapiro-Ilan et al. 2006). There are some members of the species which can recover and carry on their normal life cycle in the environment. These species have both direct and indirect impacts on nematodes of parasitic plants and their pathogens, and besides this, they also improve the quality of soil. They can withstand range of pesticides either chemical or biological.

10.9.1 Advantages and Disadvantages

There are various advantages and disadvantages of EPNs. Advantages like a wide range of insect host, their fast killing, active seeking and trapping of host can be produced both in vitro and in vivo, and they can be used without any registration or with little registration. Disadvantages are production requires high cost, their shelf life is limited, they also require storage in cold or refrigerated place and also they are affected by various environmental factors like temperature, moisture, sensitivity to harmful rays like UV, etc. which limit their activity (infectivity).

10.9.2 Production of EPNs

These EPNs are produced in living organism, i.e. in vivo, within a host which is an insect that acts as media for the growth and making of nematode-bacteria. The whole process is in constant need of insects which are fit and have good health as a source. This demand of constant source, their sensitivity to various biological variations and high cost required for production restrict the method of production discussed above. So another method of production, i.e. in vitro methods, has been acquired. *Steinernema* spp., one of the families of EPNs, have been cultured in liquid media which is monoxenic. This production requires fermentation technology. This way of production has been the high-profit-making methods of all the other known methods. Large tanks of high volume like up to 80,000 litres are being used for the production of nematodes. This result in lowering of cost of production can be easily launched in markets demanding for nematodes with low cash value of crop.

10.9.3 Bacterial Symbionts of EPNs

The collaborative relationship, i.e. symbiotic relationship between earthly animal and microbe, is explained by mutualistic partnership among the bacteria which are associated with the EPNs (entomopathogenic nematodes). These bacteria are

Xenorhabdus and the *Photorhabdus* spp. with their respective nematode hosts: *Steinernema* and the other *Heterorhabditis* spp. These nematodes along with their partners form an insecticidal complex. This insecticidal complex is the functional component which can be used against various insect hosts (Heungens et al. 2002; Shi and Bode 2018). Phase variation is the phenomenon which is the property of symbionts of the genera *Xenorhabdus* and *Photorhabdus*. IJ nematodes and the infected insects give the primary phase, whereas secondary phase is obtained from the developmental process of nematodes at the late stage. The cells which are being obtained from the secondary stage produce secondary metabolites in low amount and with low nutritional value. The bacteria are cultured in standard conditions in their primary phase, but there can be a sudden switch to secondary stage once they got exposed to stressful conditions. This shifting to another stage has an adverse impact on the development of nematode (Ehlers 2005). So phase variation must be avoided by taking effective measures.

The curiosity of knowing more about this new complex formed from nematode and bacteria known as entomopathogenic complex came into context because these have some specific potential that can be applied for the biological check of various insects and other pests from arthropods (Tanada and Kaya 2012). Nowadays different laboratories from all over the world are experimenting on this complex of nematode–bacteria so that they can be effectively used in various industries and also in the field of medicines and agriculture. This symbiotic relationship has become an outstanding model for fundamental research in various subjects and fields like environmental science, biochemistry, evolution and genetics (Burnell and Stock 2000; Goodrich-Blair and Clarke 2007).

10.9.4 Life Cycle of EPNs Symbionts

The infective juvenile, i.e. IJ, nematode which is at the third stage and is the free-living structure carries the bacterium which is symbiotic in its gut. Steinernematid nematode lives in the soil, and there it nurses the symbiotic bacteria which are present in their intestine. It bears it until it seizes a receptive insect host. The IJ releases the symbionts in the body cavity (hemocoel) at one stage of their life cycle, and interestingly this is the stage at which the nematode and the associated bacteria live separately. Still they are somehow in close proximity with each other. The bacteria which have been released help in destroying the insect host by producing various extracellular toxins, and they increase its population in the resulting dead body of insect host. This particular bacterium is the most important source for the growth of nematode and also helps in its development as it is supposed to act as a direct food source and also provide nutrients by the degradation of the insect cadaver. The dead body of the insect host which is killed by *Steinernema*-*Xenorhabdus* turns into yellowish brown colour and the host which is killed by *Heterorhabditis*-*Photorhabdus* gives red colour. Larvae of *Steinernema* genus develop into adults which are amphimictic and those of *Heterorhabditis* are hermaphrodites. As the number of nematodes increases, the amount of nutrients taken by them for their growth becomes limited and hence they again re-associate with the bacteria. After the re-association, the progeny of nematodes get differentiated into colonized and nonfeeding form of IJ. Generally, this occurs within 7–14 days after the infection has occurred. But this usually depends on various factors like temperature, size of the host body, etc., and after the completion of this cycle, the infective juveniles again appear in the soil in search for a new hunt (host). When isolating entomopathogenic nematode, the species show some significant variation which can be surprising sometimes. These variations occur in behaviour, the range of host, their ability of causing infection, reproduction and ability to tolerate environmental factors. This type of variation has revived the interest in describing their (nematodes) genetic diversity because it was supposed that there may be various other species or strains present which may carry different traits related to biological or ecological traits which may come out to be more beneficial other than those biological controlling agents which are being used in recent times against the pests which are agriculturally important (El-Borai et al. 2007; Oestergaard et al. 2006; Shapiro-Ilan et al. 2006; Vinciguerra and Clausi 2006). In recent times, with the help of advancements in the techniques have made this system of nematode–bacterium tractable to study about various aspects of structure and development and other physiological and chemical aspects which was not possible earlier (Murfin et al. 2012). All the above-mentioned aspects and the genetic diversity of nematode–bacterium system along with the expansion of various molecular tools and the enhancement of laboratories for being able to reconstruct the life cycle of the organisms which is a natural biological phenomenon have made a perfect model system of symbiotic partnership. This system is helpful in knowing the beneficial interactions of host and microbe in details.

10.9.5 Host Range

It has been observed that in experiments performed in laboratories, the EPNs are able to invade insects in large amount along with some other arthropods (Poinar 1979), while when they are exposed in natural surroundings, the range of insect host becomes narrower and is restricted to insects only. So the range of host covered by EPNs is still not clear. Naturally infected insects are rarely found. The infected larvae of Coleoptera, Diptera, Hymenoptera and Lepidoptera conquer (Peters 1996). In 1996 Peter observed that there are some species which have a broader range of hosts (generalists, e.g. *S. carpocapsae*, *S. feltiae*) and some have limited range of hosts (specialists, e.g. *S. glaseri*, *S. scapterisci*). The elaterid larva which is a wireworm protects itself from invasion by the morphological barriers. Likewise many other insects also protect them from invasion; some shows aggressive behaviour (scarab larvae), while other very often excrete to clear out the gut from the nematodes (fly larvae). Besides these ways there are some who move to a preventive place for their protection. All the above-mentioned defence strategies are strong enough to build an insect which will be resistant to infections caused by EPNs.

10.9.6 Foraging Strategies of EPNs

Foraging strategies mean strategies in searching and obtaining food, so from the biocontrol belief it can be quite interesting to know about the strategies of EPNs because there are various types of insect pests for which appropriate nematodes are present. The range of strategies varies from actively foraging to sitting and waiting (Lewis et al. 1992). Cruisers are the ones which pass across the soil in search of a stationary host. On the other hand, ambushers are the ones which persist around the surface of the soil, and they attack the insects which move around it. There are various species in majority who prefer an intermediate strategy. There are some ambushers (e.g. *Steinernema carpocapsae*) who can acquire the quality of a cruiser in different habitats and can successfully control the sitting or inactive pests.

10.9.7 The Methods of Application and Formulation of EPN

The controlling agents obtained from EPNs are very expensive, and hence the techniques for their application should be enhanced to attain a cost-effective benchmark. For their successful introduction in the market, they demand for a stable formulation. This formulation is a difficult job as large markets demand of having a limited duration of shelf-life, i.e. 6 months, when kept under room temperature. The living animal which are carried by nematodes requires specific factors for their survival like moisture, temperature and oxygen. EPNs are cultured in classical aq. solution, but in recent times, other methods have also been introduced for their formulation like nematode wool, vermiculite, gels, clay, sponge, etc. The products obtained from EPN have shelf-life which depends on the species of nematode and their formulation. Some nematodes can remain infective for 4–6 months under their storage conditions, whereas some which are being formulated in gels can remain infective for 9 months. EPNs formulated in powders which are water soluble can remain infective for 1 year. They are mixed with water and can be used lie sprayers or other irrigation systems. The persistence of EPN is enhanced by the dead bodies of insect hosts (Shapiro-Ilan et al. 2006). The most favourable perspective is the introduction of genetic engineering for the improvement of strains of EPNs. Two most important methods include hybridization and selective breeding which help in improvement of specific traits which are essential for their survival. These also improve certain traits like tolerance to environmental factors, shelf-life, pathogenicity, nematicide resistance and desiccation. From the safety point of view, it has been observed by Ehlers (2005) from the last few decades that EPNs are not harmful to humans and animals. In various fields and laboratory test, it was found that arthropods which are not targeted are not influenced by EPNs. However more exploration is needed according to Grewal (2012) on some bacteria like *H. bacteriophora* and *H. megidis* which have been commercially produced. This arises come useful concern for the safety of humans towards EPNs. The main barrier is the regulation of EPNs in their commercialization as biopesticides. Although there is no registration process for the EPNs in some countries like the USA, while some countries demand for the registration for the foreign exotic species. The recent directive of EU 2009/128/EC was on sustainable utilization of pesticides which said to expand the use of methods which involve no chemicals. This may help in increasing the demand for EPN in the near future. A variety of pests in soil and foliar and cryptic habitats can be controlled by EPNs which are the natives of the soil and this property which has given attention to various insects for their control. Weevils are the chief beetles that the EPNs target other fungus gnats and maggots are also controlled by EPNs. Cryptic habitats are the most favourable habitats for the survival of the EPNs as they receive minimum amount of UV rays and desiccation. Wood-boring moth (*Euzophera semifuneralis*) and the codling moth (*Cydia pomonella*) are targeted pests found here. On the other hand, foliar habitats are the most unsuitable habitat for their survival due to the stressful conditions associated, and hence they are less efficient. Larvae of diamondback moth (*Tuta absoluta*) and dipteran leafminers are targeted pests.

10.10 Microbial Control of Phytopathogenic Nematodes: Nematodiasis and Biocontrol

The organisms which are multicellular and are abundant on the earth are nematodes. Many are plant parasitic and cause serious losses to crop worldwide and are known to be most harmful pest to agriculture (Koenning et al. 2004; Murfin et al. 2012). Nematodes mainly live in the soil and they usually attack the plant part which are underground, and hence the management of nematodes becomes difficult than bacteria and viruses (Tu et al. 2003). Two major groups of the nematodes which are plant parasitic are root-knot nematodes and cyst knot nematodes (i.e. *Meloidogyne* spp. *Globodera* spp., respectively) (Molinari 2011). *Meloidogyne* spp. infect around 3000 species of flowering plant which belong to 114 different families that result in 10–20% crop loss per year (Wang 1997); they cause root knots and giant cell formation by preferentially invading the roots of plants. It leads to fracture in epidermis and cortical tissue and results in delayed maturity, toppling, reduced yields and quality of crop produce, high costs of production and therefore loss of income due to slowing of respiration, transpiration and photosynthesis because of giant cell formation. Furthermore infection by this species may weaken the immune system and protective enzyme system in plants, and the plants which are damaged are more prone to infection caused by other pathogen (i.e.bacteria and fungi).

Chemical nematicides are usually fast, effective, and easily applicable, but some developed countries have stopped their use because of environmental safety and public health (Schneider et al. 2003); therefore it has become important to search for a novel and environment-friendly alternative to manage plant–parasitic nematodes.

Nematodes which are in soil are subjected to infection caused by fungi and bacteria. This generates the possibility of using microorganisms present in soil to control the nematodes which are plant parasitic (Akhtar and Malik 2000). Biological control by exploiting the interaction within the nematode and antagonistic microorganism and with their host is an alternative approach against chemical control.

10.10.1 Microbial Populations Antagonizing Phytopathogenic Nematodes

There are a large number of species on earth which belong to different phylogenetic groups which are antagonist to nematodes such as bacteria, fungi, viruses, insects and some invertebrates. These microorganisms which are nematicidal have developed physical-chemical and also the biochemical mechanism by which they can kill nematodes (Akhtar and Malik 2000).

10.10.2 Nematophagous Fungi

Nematophagous fungi are also known as *carnivorous fungi and are enemies of nematodes and they have highly* sophisticated mode of infection; they are specialized in trapping and digesting the nematodes. Around 700 species of nematophagous fungi from several phyla are known, namely, *Basidiomycota, Zygomycota, Ascomycota and Chytridiomycota*. Moreover, even the organisms which belong to the *Oomycota* had their nematophagous activity described (Li et al. 2007). These comprise mainly three groups of fungi: endoparasitic, nematode-trapping/predatorial and opportunistic or ovicidal fungi. Endoparasitic fungi are the obligate parasite of nematodes and for infection it uses spores (zoospores, conidia) following mycelia growth from the spore inside the nematodes spore may adhere to the nematode cuticle. Till date, about 120 species of nematode-endoparasitic have been found in the genera *Myzocytium* Schenk and *Drechmeria* (Huang et al. 2015). Predators produce modified hyphae which are known as traps; by using this and with the help of mechanical and enzymatic process, they bind to the larvae and digest it. There are many different traps, for example, three-dimensional adhesive networks, adhesive nodules, non-differentiated adhesive hyphae and constricting and non-constricting rings. The major part of nematode-trapping fungi belongs to asexual taxa, and they are also known as hyphomycetes. Three hundred and forty-seven species of nematode-trapping fungi have been reported, and these are mainly found in *Basidiomycota*, *Ascomycota* and *Zygomycota*. The genera which represent this group include *Cystopage* Drechsler, *Dactylella* Grove, *Arthrobotrys* Corda, *Dactylellina* M. Morelet, *Drechslerella* Subram, *Hohenbuehelia* Schulzer, *Nematoctonus* Drec, *Hyphoderma, Orbilia* Fr., *Zoophagus* Sommerst, *Stylopage* Drechsler and *Tridentaria* Preuss (Huang et al.

2015). The ovicidal group are fungi also known as cyst- and egg-parasitizing fungi as this group can colonize reproductive structures of nematodes and also affect reproductive capabilities of nematodes. It uses traps in the predation process. However, their target groups include cysts, eggs and nematode females. They make use of spores (zoospores, conidia) for infection structures, which may get attached on the cuticle of nematodes and get ingested (Braga et al. 2013).

In recent studies, producers of special attack devices (structures mechanically damaging the nematodes cuticle) and toxin-producing fungi are two new groups which are reported (Liu et al. 2013).

10.10.3 Nematophagous Bacteria

They occur worldwide and have been reported from at least 51 countries. Members of the genus have been reported to infect 323 nematode species belonging to 116 genera, including both plant–parasitic nematodes and free-living nematodes (Tian et al. 2007). Nematophagous bacteria are gaining more consideration day by day as they are novel and safe and provide potential tool, and they also have many advantages over nematophagous fungi as they multiply fast and are easily cultivated and also give mass production. Following bacterial strain from genera, *Agrobacterium*, *Arthrobacter*, *Actinomycetes*, *Alcaligenes*, *Azotobacter*, *Bacillus*, *Aureobacterium*, *Beijerinckia*, *Clavibacter*, *Clostridium*, *Comamonas*, *Chromobacterium*, *Curtobacterium*, *Enterobacter*, *Flavobacterium*, *Desulfovibrio*, *Gluconobacter*, *Pseudomonas*, *Klebsiella*, *Methylobacterium*, *Hydrogenophaga*, *Stenotrophomonas*, *Pasteuria*, *Phyllobacterium*, *sphingobacterium*, *Rhizobium* and *Variovorax* (Liu et al. 2013). On the basis of nematicidal action, nematophagous bacteria are classified into three different groups. They are parasporal crystal-forming bacteria, obligate parasitic bacteria and opportunistic parasitic bacteria. Genus *Pasteuria* contains members which are mycelial, obligate, endospore-forming bacterial parasites of nematodes which are plant parasitic (Sayre and Wergin 1977). Four species of *Pasteuria* that have been reported are *Pasteuria ramose, P. thornei, P. nishizawae* and *P. penetrans*. Among these *Pasteuria ramosa* species have been described from the water fleas (Ebert et al. 1998). The rest three species which infect nematode are:

- *P. penetrans*: firstly they parasitize giant cell or root knot nematodes, for example, *Meloidogyne* spp.
- *P. thornei*: it parasitizes the nematodes that are root lesion, for example, *Pratylenchus* spp.
- *P. nishizawae*: it is found on cyst nematodes of genera *Globodera* and *Heterodera* (Atibalentja et al. 2000).

Many bacterial species which come under this genus had showed their huge potential as the biocontrol agent antagonist to nematodes which are plant parasitic.

10.10.4 Mechanisms of Infection

Pasteuria penetrans infects the nematodes which are root-knot, for example, *Meloidogyne* spp. Spores formed by *Pasteuria* get attached to cuticles of second phase of juveniles, and when juvenile enters the roots of plant, it starts germinating and feeding. The germ tubes penetrate the cuticles, and then vegetative microcolonies are formed. These colonies proliferate in the body of the female nematode (developing) and degenerate it releasing the mature endospores in the soil (Mankau et al. 1976; Sayre and Wergin 1977).

10.10.5 Virulence Factors as Well as Molecular Mechanisms Used by Nematophagous Fungi

The process by which nematophagous fungi infect the nematodes includes different steps. First step is attraction and recognition, second step adhesion, then penetration and last step digestion. Attraction and recognition which is the first step of infection process occurs by the cell-to-cell communication, which requires a range of physiological, morphological or biochemical interactions between nematodes and fungi. When nematodes come in contact or pass through nematophagous fungi may get attached on the nematode by adhesive materials present on the surface and spores. After attachment with the nematodes, cuticles break down from which the hypha of fungus penetrates inside nematodes. As the current evidence suggests, extracellular hydrolytic enzymes (chitinases, serine proteases and collagenases) and mechanical forces are involved for the penetration. From a long time. It is known that nematodes get attracted by living mycelia and culture filtrates of many nematophagous fungi. The attraction between pinewood nematode (PWN) that the fungus continuously produces few volatile organic compounds (VOCs) that are involved in this process. After recognition, the nematophagous fungi and their host nematodes get attached by adhesive proteins. Adhesive proteins are the main components of extracellular fibrillar polymers in the nematophagous fungi, and they get accumulated on the surface of spores, and they play important role for the fungi to get adhered to cuticle of nematodes (Su et al. 2017). Apart from adhesion, these proteins plays roles in, trap development, nutrient storage nematode and defence against competitors, attraction and recognition, and defence against competitors (de Ulzurrun and Hsueh 2018). Studies have shown that these extracellular enzymes can break down the physiological integrity of eggshell and nematode cuticle which facilitates the fungal penetration.

10.10.6 Nematophagous Bacteria: Cry Protein-Mediated Infection

Bacillus thuringiensis (*Bt*) is a bacterium that forms spores, and it produces crystals that are proteinaceous and protoxic (known as cry protein or crystal protein) during the sporulation process. *Bt* is considered as the most suitable biopesticide as it shows toxin activity on beetles, nematodes and caterpillars, but it does not affect the vertebrates (Darban et al. 2017). Cry proteins are nowadays expressed in transgenic plant that are involved in protection of crop from nematode infestation. Till date, six Cry proteins have been reported they are (Cry5, Cry6, Cry12, Cry13, Cry14, and Cry21) that have nematotoxic activities phytopathogenic nematodes and nematodes which are free-living (Kotze et al. 2005) these are grouped as Cry5, Cry12, Cry13, Cry14 and Cry21 in the Cry5 family), (Cry6 in the Cry6 family). Once the nematode larvae ingest the Cry5 toxin, the crystals that are present in it get dissolved inside its gut and toxin molecules get bind to a receptor present in the epithelial cell. This binding gives rise to pore and vacuole formation, pitting and finally degradation of the intestine (Crickmore 2005).

10.10.7 Commercial Biocontrol Nematicides

Over the last 20 years, a number of researches have been done to know the applications of microorganisms that can be used as agents for biocontrol against nematodes. Many bacteria have been recognized as plant-parasitic pathogens nematodes, and they have shown suppressible effects on nematode populations. However, only few biocontrol products are commercially launched from the microorganisms with nematicidal effects, which can be used in the field of agriculture system (Whipps and Davies 2000). The processes of development of agents used for biocontrol are unpredictable and very much variable to be implemented on large scale (Tian et al. 2007). It does not matter how effective a nematicide is against nematode in laboratory, but in practice its effects may differ. In order to study practical effects of microorganisms on nematodes as biocontrol agent, we need to have knowledge and proper understanding of interactions of nematodes with biocontrol strains and plants and its effect on environment.

10.10.8 Postharvest Control

Harvested plant produces (like fruits, vegetables, pomes and grains) bear microbiota which affect the quality, texture and shelf-life and made them safe for human consumption. But these plant produces get affected by several postharvest diseases which affect its quality and quantity, and consumption of unhealthy food can cause health issues in humans and its livestock (Ojha et al. 2013; Pal and Gardener 2006; Sharma et al. 2009; Singh and Vats 2019). Spoilage to these fruits and vegetables occurs due to fungal and bacterial rot and production of mycotoxins. These plant produces need some proper approaches, methods and procedures which decrease these types of current challenges. Lack of awareness and functioning of equipment in the chain of postharvest gives the result in the form of great postharvest losses. These problems of postharvest losses need to be overcome, and this can be possible only by understanding the factors which cause spoilage of plant produces. Currently, by the use of methods of disinfection, sanitation and chemical pesticides, these problems of postharvest losses are controlled. Several different strains of microorganisms (like viral, bacterial, fungal and yeast) are used as biocontrol agents to control the postharvest diseases of plant produces. The action mechanism of these microorganisms is very diverse and their actions are simultaneous. Mechanism involves several competition and conflict among microorganisms for nutrition, space, antibiosity, enzymes for the lysis of cell wall, decrease in pH, parasitism and induction of simulative responses (defence) in harvested fruits and vegetables. In place of chemical pesticide uses, several strains of biocontrol agents are used in several products to reduce postharvest losses. Complex microbial communities are provided with a physio-chemical environment by the plant, e.g. bacteria, fungi and viruses. Establishment of specific microbial communities is determined by the different properties between aerial and soil-plant parts. Condition of storage, farming practices, species of plants, type of soil, season of the year and cultivar also change the composition of microbiota (Setati et al. 2012). According to metagenomics studies, it is revealed that in fresh fruits and vegetables, plant microbiota diversity is very high (e.g. 100–1000 operational taxonomic units, OTUs) with some dominant groups (20–50 operational taxonomic units) (Gupta et al. 2018).

Harvested produce has the greater influence of microbiota as the plants, which are the major source of fresh products. Raw products of plants are highly rich in nutrients, and these products at the field level and during postharvest processing become contaminated. Various biological functions (e.g. respiration) and physiological activities (e.g. ripening of fruits) are maintained by harvested products and during postharvest process its microbiota is influenced. The quality and quantity are mainly affected by these microorganisms (Magan and Aldred 2007). Agricultural production losses involve the much higher costs on the farm economy causes the occurrence of postharvest (estimated worldwide of about 10–40%). It is most severe in developing countries rather than developed nations (Abdullahi 2015). These production losses contributed by several factors include crops initial quality, humidity, temperature, mechanical injury atmosphere of the storage and crop handling type. The main goals of postharvest technology to harvested produce are to maintain quality (texture, flavour, nutritional value), food safety protection and reduction of losses among harvest and produce consumption (Sudheer and Indira 2007). Rotting of fresh fruits is caused by various strains of fungi which affect the different kinds of stone, pokeberries, cucurbits tropical fruits as well as citrus fruits like lemon, oranges, grapes, etc. Some of the fungal pathogens found in different fruits are *Penicillium* species in apple and citrus fruits, *Colletotrichum* in berries and *Rhizopus* and *Monilinia* in stone. More common vegetables are caused bacterial rot. Fungi and bacteria also affect the dried grains and nuts which result in starch and carbohydrate degradation due to which quality, nutritional properties and weight of produce are also affected. All fruits and vegetables are having two major concerns, i.e. safety and stability. Pathogenic and toxic chemical presence in fruits and vegetables principally determines its safety. Various activities of enzymes, degradation by microbes and chemical reactions cause perishability of plant produces due to which product may undergo several undesirable changes like changes in texture and nutrient properties. Perishability of fruits and vegetables directly depends upon its composition, postharvest processing, product packaging and its storing environment (Abdullahi 2015). By increasing shelf-life of fruits and storage, low O_2 and high CO_2 condition postharvest losses can be prevented.

10.10.9 Preharvest Implications on Postharvest Quality

Preharvest factors largely affect the postharvest quality of plant produce (fruits and vegetables) like productivity area, type of soil, irrigation facility, rootstock, insect pests and selection of vegetable variety. Preharvest factors are categorized into primary and secondary factors. Primary factors comprise of weather conditions, nutrition and regulators of plant growth, and the secondary factors involve quality of soil, rootstock, manipulation of crop load and irrigation. Consideration of preharvest factors is most important to influence or maximize the quality of fruits and vegetables (Thaker et al. 2017). Soil type, irrigation and rootstock are the factors which directly affect the supply of water and nutrients to the plant which is responsible for the quality and composition of the plant produce (texture, aroma, nutrient content and taste). These quality attributes are also affected by some practices of field management like control of pest, fertility of soil and management of water (Wang 1997). He also reported that some of the mineral elements like nitrogen and calcium have direct effect on production of ethylene, storage life, respiration, ripening and rotting of diverse plant produces (fruits and vegetables) depending on preharvest factors. The two most important factors like position of fruit on tree and the temperature of fruits also lead to postharvest disorders (such as bitter pit). Fruit and vegetable exposure to low temperature is also one of the major postharvest disorders which leads to hardening of fruits (Zhang et al. 2003).

For the maintenance of good quality and high quantity of plant produce (fruits and vegetables) postharvest disorders should be controlled. Decaying of fruits and vegetables causes a level of postharvest losses (Fig. 10.4). In developed countries, about 20–25% of plant products are decayed by pathogenic effects (Babychan 2017). But in developing countries, it is about 50%. And the management of these losses is done by fungicides which increase the quality and quantity of plant produces in the past years. But the use of agrochemicals (fungicides, weedicides, pesticides, etc.) causes environmental destruction and various health issues. To inhibit the actions and populacy of disease-causing plant pathogens by the use of living organism (disease resistant to host) is referred to as biocontrol (Fig. 10.3). Utilization of some microbial adversary (such as fungi, yeast and bacteria) is quite favourable (Barea 2015). Strains isolated from wild plant microbiota and postharvest plant produces are used to study the biocontrol of rotting of produces by fungus (Babychan 2017). Many bacterial strains (*Pseudomonas syringae, P. fluorescens, P. graminis*) and yeast strains (*Candida famata, C. oleophila, C. sake*) are used as biocontrol agents to prevent fungal rot (Magan and Aldred 2007).

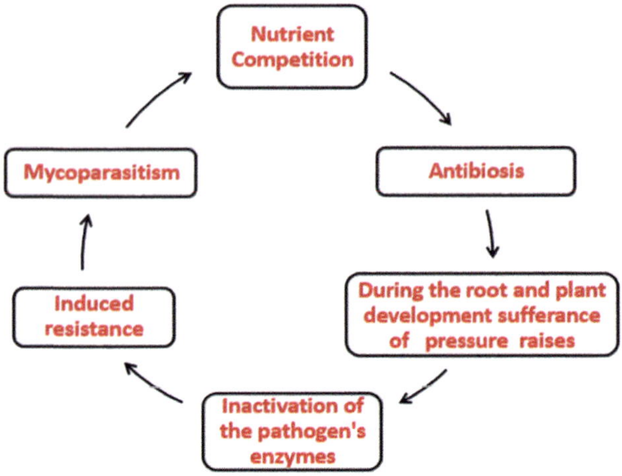

Fig. 10.3 Postharvest biochemical change in crop

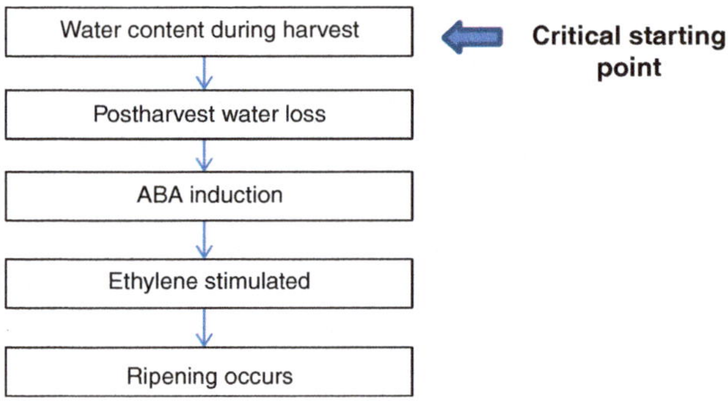

Fig. 10.4 Biocontrol plants and weed interaction

10.10.10 Production, Formulation and Utilization of Biopesticides

Biopesticides are the pesticides produced by the microbes for the elimination of disease-causing pathogens. Cells are grown on large scale, are preserved and get formulated for the storage of the postharvest biopesticides for the commercial development (Montesinos 2003; Sharma et al. 2014). On industrial scale for the production of microbial pesticides, solid or liquid phase fermentation methods are used. But it totally depends upon the essential qualities and features of microorganisms used (Vats and Kumar 2015; Jain et al. 2011; Kaur et al. 2010). By liquid fermentation process, several bacterial and yeast strains are grown in bioreactors and some of fungal strains growth occurred through solid-state fermentation. Harvest of desired cells is carried out by the centrifugation process from liquid cultures or from solid-state culture; suspension of concentrated cells or spores is synthesized and get cleaned which contains some metabolites in supernatant such as lytic enzymes and antimicrobials. Bacteriophage production is a two-stage process; the first stage consists of preparation of bacterial target (e.g. *Escherichia coli*) which is used as host cell for the multiplication of the lytic bacteriophage. Materialistic biopesticide formulation generally contains suspension of liquid phase in frozen form or dried products sustained under refrigeration. Condition for optimum storage, handling and dispersal is permitted by dehydration, but especially lyophilization, spray drying and fluidized bed drying are very costly processes. Sometimes viability of cells or spores is lost by the spray drying process due to thermal treatment. Composition of final formulations consists of ingredients which are active (components of cultures, spores or cells), inert materials which are helpful in supporting cells and adjuvants. Stability of products is maintained for many months and years. Expiry of the cells of biological control agents after the occurrence of dehydration and delivery because of the sudden changes in the condition of optimal laboratory culture to the process of stressing dehydration and the fruit surfaces which are growth-limiting or growth inhibitors (Zhang et al. 2003).

10.10.11 Limitations and Future Trend

The management of losses of postharvest uses less effective procedures and strategies whose impact is lower than the products of synthetic antimicrobial such as fungicides. Therefore, for the optimization of control levels, strategies with different barriers are essential. And these barriers are soft chemicals which directly affect the spoilage microbes like bicarbonates, silicates, surface disinfectants such as O3, chlorine, electrolyzed water, and physically used methods (hot water, microwaves, UV light). A part of the list of these technologies are formed by biological controls. The use of the biological control agents is incompatible for many of these systems because the colonization, growth and metabolic activities are inhibited by them. Others maintain the compatibility as simultaneous treatment (defence inducers). A drawback of some biological control agents of diseases caused to postharvest produces is directly associated with the safety of antagonistic microbes. In laboratories most of the BCAs perform their activity much better, but they fail to give best result in the field, and this is due to some physiological and ecological limitations which affect the efficiency of biocontrol agents at certain level. Various methods of genetic engineering (like mutation or PEG using protoplasm) are used to overcome these types of problems in the field.

10.11 Conclusions

In this chapter we have briefly overviewed the current insights onto understanding the interactions among microorganisms and their plant hosts, as well as the areas which are less studied and needed to be explored for future research. Technological advances that enable sensitive detection of metabolites including root exudate components produced in situ will continue to be critical to unravelling the complex molecular and organismal interrelationships in the rhizosphere habitat. Better knowledge of the microbe–plant dialogue is essential given the need for increased agricultural productivity to provide food and biofuel feedstocks in the face of climate change, the increasing world population and the loss of productive lands. Increasing population demands for increase in agricultural production, pests destroy around 30% of the commercially important crops and it's a great challenge to protect the crops from pests. In such a situation, *Bt* has been proven to be a boon as it is harmless for environment and has a good insecticidal activity. In the era of biopesticides, *Bt* crops and its products cover a large ratio of the global market. Study of *Bt* has given us valuable knowledge on structure of toxin and its function. After hundreds of years of its discovery, still researchers find it fascinating to study more about *Bt* strains.

References

Abdullahi N (2015) Postharvest management of fruits and vegetables. Tech Sci Afr J 11:93–100

Akhtar M, Malik A (2000) Roles of organic soil amendments and soil organisms in the biological control of plant-parasitic nematodes: a review. Bioresour Technol 74(1):35–47

Angus TA (1954) A bacterial toxin paralyzing silkworm larvae. Nature 173:545–546

Atibalentja N, Noel GR, Domier LL (2000) Phylogenetic position of the north American isolates of Pasteuria that parasitizes the soybean cyst nematodes, Heteroderaglycines, as inferred from 16S rDNA sequence analysis. Int J Syst Evol Microbiol 50:605–613

Audenaert K, Pattery T, Cornelis P, Hofte M (2002) Introduction of systemic resistance to Botrytis cinerea in tomato by *Pseudomonas aeruginosa* 7NSK2: role of salicylic acid, Pyochelin, and Pyocyanin. Mol Plant-Microbe Interact 2:47–56

Babychan M (2017) Biocontrol agents in management of post-harvest diseases. Life Sci Int Res J 4 (1):51–53

Bailey BA, Den RL (2014) Gliocladium on plant growth and resistance to pathogens. In: Harman GE, Kubicek CP (eds) Trichoderma and gliocladium, Volume 2: Enzymes, biological control and commercial applications. CRC Press, Boca Raton, FL, p 185

Baker KF, Cooker RJ (1974) In: Kelman A, Sequiera L (eds) Biological control of plant pathogens. Freeman, San Francisco, CA, 433 pp

Barea JM (2015) Future challenges and perspectives for applying microbial biotechnology in sustainable agriculture based on a better understanding of plant-microbiome interactions. J Soil Sci Plant Nutr 15(2):261–282

Berliner E (1915) Uber die Schlaffsucht der Mehlmottenraupe (EphestiaKuhniella, Zell.) und ihrenErreger *Bacillus thuringiensis*, n. sp. Z Angew Entomol 2:29–56

Bhargava P, Singh AK, Goel R (2017) Microbes: bioresource in agriculture and environmental sustainability. In: Singh DP, Singh HB, Prabha R (eds) Plant-microbe interactions in agro-ecological perspectives. Springer, Singapore, pp 361–376

Bhattacharyya PN, Jha DK (2012) Plant growth-promoting rhizobacteria (PGPR): emergence in agriculture. World J Microbiol Biotechnol 28(4):1327–1350

Braga FR, Araujo JM, Araújo JVD, Soares FEDF, Tavela ADO, Frassy LN et al (2013) In vitro predatory activity of conidia of fungal isolates of the *Duddingtonia flagrans* on *Angiostrongylus vasorum* first-stage larvae. Rev Soc Bras Med Trop 46(1):108–110

Bravo A, Pacheco S, Gómez I, Garcia-Gómez B, Onofre J, Soberón M (2017) Insecticidal proteins from *Bacillus thuringiensis* and their mechanism of action. In: Fiuza LM, Polanczyk RA,

Crickmore N (eds) *Bacillus thuringiensis* and *Lysinibacillus sphaericus*. Springer, Cham, pp 53–66

Brennan EB (2016) Agronomy of strip intercropping broccoli with alyssum for biological control of aphids. Biol Control 97:109–119

Burnell AM, Stock SP (2000) Heterorhabditis, Steinernema and their bacterial symbionts—lethal pathogens of insects. Nematology 2(1):31–42

Cheng F, Cheng Z (2015) Research progress on the use of plant allelopathy in agriculture and the physiological and ecological mechanisms of allelopathy. Front Plant Sci 6:1020

Coda R, Rizzello CG, Di Cagno R, Trani A, Cardinali G, Gobbetti M (2013) Antifungal activity of *Meyerozyma guilliermondii*: identification of active compounds synthesized during dough fermentation and their effect on long-term storage of wheat bread. Food Microbiol 33(2):243–251

Crickmore N (2005) Using worms to better understand how *Bacillus thuringiensis* kills insects. Trends Microbiol 13:347–350

Darban DA, Gowen SR, Pembroke B (2017) Soil bioassay with time interval and spore density factors affecting the infection of *Meloidogyne javanica* by *Pasteuria penetrans*. Pak J Nematol 35(2):209–213

de Maagd RA (2015) *Bacillus thuringiensis*-based products for insect pest control. In: Lugtenberg B (ed) Principles of plant-microbe interactions. Springer, Cham, pp 185–192

de Maagd RA, Bosch D, Stiekema WJ (1999) *Bacillus thuringiensis* toxin mediated insect resistance in plants. Trends Plant Sci 4:9–13

de Maagd RA, Bravo A, Berry C, Crickmore N, Schnepf HE (2003) Structure, diversity, and evolution of protein toxins from spore-forming entomopathogenic bacteria. Annu Rev Genet 37(1):409–433

de Ulzurrun GVD, Hsueh YP (2018) Predator-prey interactions of nematode-trapping fungi and nematodes: both sides of the coin. Appl Microbiol Biotechnol 102(9):3939–3949

Ebert D, Zschokke-Rohringer CD, Carius HJ (1998) Within–and between–population variation for resistance of *Daphnia magna* to the bacterial endoparasite *Pasteuria ramosa*. Proc R Soc Lond B Biol Sci 265(1410):2127–2134

EFSA Panel on Genetically Modified Organisms (2010) Guidance on the environmental risk assessment of genetically modified plants. EFSA J 8:1879–1990

Ehlers RU (2005) Forum on safety and regulation. In: Grewal PS, Ehlers RU, Shapiro-Ilan DI (eds) Nematodes as biocontrol agents. CABI, Oxford, pp 107–115

El-Borai FE, Zellers JD, Duncan LW (2007) Suppression of *Diaprepes abbreviatus* in potted citrus by combinations of entomopathogenic nematodes with different lifespans. Nematropica 37:33–41

English L, Slatin SL (1990) Mode of action of delta-endotoxins from *Bacillus thuringiensis*: a comparison with other bacterial toxins. Insect Biochem Mol Biol 22:1–7

Entwistle P, Bailey MC, Cory JC, Higgs SC (1993) *Bacillus thuringiensis*: an environmental biopesticides: theory and practice. Wiley, Chichester

Glare TR, O'callaghan M (2000) *Bacillus thuringiensis*: biology, ecology and safety. Wiley, New York

Goel R, Bhargava P, Gupta N, Vats S (2017) Health issues and heavy metals. Austin J Environ Toxicol 3(1):1018

Goldberg LJ, Margalit J (1977) A bacterial spore demonstrating rapid larvicidal activity against *Anopheles sergentii*, *Uranotaenia unguiculata*, *Culex univittatus*, *Aedes aegypti* and *Culex pipiens*. Mosq News 37:355–358

Goodrich-Blair H, Clarke DJ (2007) Mutualism and pathogenesis in *Xenorhabdus* and *Photorhabdus*: two roads to the same destination. Mol Microbiol 64:260–268

Grewal PS (2012) Entomopathogenic nematodes as tools in integrated pest management. In: Abrol DP, Shankar U (eds) Integrated pest management: principles and practice. CABI, Wallingford, pp 162–236

Gupta N, Vats S, Bhargava P (2018) Sustainable agriculture: role of metagenomics and metabolomics in exploring the soil microbiota. In: Choudhary DK, Kumar M, Prasad R, Kumar V (eds) In Silico approach for sustainable agriculture. Springer, Singapore, pp 183–199

Harman GE, Kubicek CP (2014) Trichoderma and Gliocladium, Vol 2, Enzymes, biological control and commercial applications. CRC, Boca Raton, FL

Helgason E, Økstad OA, Caugant DA, Johansen HA, Fouet A, Mock M et al (2000) *Bacillus anthracis*, *Bacillus cereus*, and *Bacillus thuringiensis*—one species on the basis of genetic evidence. Appl Environ Microbiol 66(6):2627–2630

Heungens K, Cowles CE, Goodrich-Blair H (2002) Identification of *Xenorhabdus nematophila* genes required for mutualistic colonization of *Steinernema carpocapsae* nematodes. Mol Microbiol 45:1337–1353

Hilbeck A (2002) Transgenic host plant resistance and non-target effects. In: Letourneau DK, Burrows BE (eds) Genetically engineered organisms: assessing environmental and human health effects. CRC, Boca Raton, FL, pp 167–185

Hofmann C, Vanderbruggen H, Hofte H, Van Rie J, Jansens S, Van Mellaert H (1988) Specificity of *Bacillus thuringiensis* delta-endotoxins is correlated with the presence of high-affinity binding sites in the brush border membrane of target insect midguts. Proc Natl Acad Sci U S A 85:7844–7848

Hofte H, Whiteley HR (1989) Insecticidal crystal proteins of *Bacillus thuringiensis*. Microbiol Rev 53:242–255

Huang X, Zhang K, Yu Z, Li G (2015) Microbial control of phytopathogenic nematodes. In: Lugtenberg B (ed) Principles of plant-microbe interactions. Springer, Cham, pp 155–164

Ibrahim MA, Griko N, Junker M, Bulla LA (2010) *Bacillus thuringiensis*: a genomics and proteomics perspective. Bioeng Bugs 1(1):31–50

Ishiwata S (1901) On a kind of severe flacherie (sotto disease). Dainihon Sanshi Kaiho 114:1–5

Jain P, Miglani K, Vats S (2011) Aptamers-potential applications in diagnostics and therapeutics. Everyman's Sci XLV(6):361

Kaur A, Vats S, Rekhi S, Bhardwaj A, Goel J, Tanwar RS, Gaur KK (2010) Physico-chemical analysis of the industrial effluents and their impact on the soil microflora. Procedia Environ Sci 2:595–599

Kaya HK, Gaugler R (1993) Entomopathogenic nematodes. Annu Rev Entomol 38:181–206

Kim YC, Jung H, Kim KY, Park SK (2008) An effective biocontrol bioformulation against Phytophthora blight of pepper using growth mixtures of combined chitinolytic bacteria under different field conditions. Eur J Plant Pathol 120(4):373–382

Knowles BH (1994) Mechanism of action of *Bacillus thuringiensis* δ-endotoxins. Adv Insect Physiol 24:275–308

Koenning SR, Wrather JA, Kirkpatrick TL, Walker NR, Starr JL, Mueller JD (2004) Plant-parasitic nematodes attacking cotton in the United States: old and emerging production challenges. Plant Dis 88(2):100–113

Koppenhöfer AM (2007) Nematodes. In: Lacey LA, Kaya HK (eds) Field manual of techniques in invertebrate pathology. Springer, Dordrecht, pp 249–264

Kotze AC, O'grady J, Gough JM, Pearson R, Bagnall NH, Kemp DH, Akhurst RJ (2005) Toxicity of *Bacillus thuringiensis* to parasitic and free-living life-stages of nematode parasites of livestock. Int J Parasitol 35(9):1013–1022

Koziel GM, Beland GL, Bowman C, Carozzi NB, Crenshaw R, Crossland L, Dawson J, Desai N, Hill M, Kadwell S, Launis K, Maddox D, McPherson K, Heghji M, Merlin E, Rhodes R, Warren G, Wright M, Evola S (1993) Field performance of elite transgenic maize plants expressing an insecticidal protein derived from *Bacillus thuringiensis*. Nat Biotechnol 11:194–200

Lacey LA, Grzywacz D, Shapiro-Ilan DI, Frutos R, Brownbridge M, Goettel MS (2015) Insect pathogens as biological control agents: back to the future. J Invertebr Pathol 132:1–41

Lewis EE, Gaugler R, Harrison R (1992) Entomopathogenic nematode host finding—response to host contact cues by cruise and ambush foragers. Parasitology 105:309–315

Li GH, Zhang KQ, Xu JP et al (2007) Nematicidal substances from fungi. Recent Pat Biotechnol 1:212–233

Liu Z, Budiharjo A, Wang P et al (2013) The highly modified microcin peptide plantazolicin is associated with nematicidal activity of *Bacillus amyloliquefaciens* FZB42. Appl Microbiol Biotechnol 97:10081–10090

Magan N, Aldred D (2007) Post-harvest control strategies: minimizing mycotoxins in the food chain. Int J Food Microbiol 119:131–139

Makoi JH, Ndakidemi PA (2007) Biological, ecological and agronomic significance of plant phenolic compounds in rhizosphere of the symbiotic legumes. Afr J Biotechnol 6(12):1358–1368

Mankau R, Imbriani JL, Bell AH (1976) SEM observations on nematode cuticle penetration by *Bacillus penetrans*. J Nematol 8:179–181

Maurya DP, Vats S, Rai S, Negi S (2013) Optimization of enzymatic saccharification of microwave pretreated sugarcane tops through response surface methodology for biofuel. Indian J Exp Biol 51(11):992–996

Maurya DP, Singh D, Vats S (2014) Cellulase production and utilization. In: Aurora J (ed) Chemical technology. Lambert Academic Publishing, p 80

Molinari S (2011) Natural genetic and induced plant resistance, as a control strategy to plant parasitic nematodes alternative to pesticides. Plant Cell Rep 30:311–323

Monaco C, Sisterna M, Perelló A, Dal Bello G (2004) Preliminary studies on biological control of the blackpoint complex of wheat in Argentina. World J Microbiol Biotechnol 20(3):285–290

Montesinos E (2003) Development, registration and commercialization of microbial pesticides for plant protection. Int Microbiol 6(4):245–252

Murfin KE, Dillman AR, Foster JM, Bulgheresi S, Slatko BE, Sternberg PW, Goodrich-Blair H (2012) Nematode-bacterium symbioses—cooperation and conflict revealed in the "Omics" age. Biol Bull 223(1):85–102

Negi S, Vats S (2014) Pine forest litter based bio-refinery for biofuels and value-added phytochemicals. In: Singh RS, Pandey A, Larroche C (eds) Advances in industrial biotechnology. IK International Publishing House, New Delhi, pp 98–116

Odum HT (1994) Ecological and general systems: an introduction to systems ecology. University Press of Colorado, Niwot, CO

Oestergaard J, Belau C, Strauch O, Ester A, van -Rozen K, Ehlers RU (2006) Biological control of *Tipulapaludosa* (Diptera: Nematocera) using entomopathogenic nematodes (*Steinernema*spp.) and *Bacillus thuringiensis* subsp. *israelensis*. Biol Control 39:525–531

Ojha AK, Forster S, Kumar S, Vats S, Negi S, Fischer I (2013) Synthesis of well–dispersed silver nanorods of different aspect ratios and their antimicrobial properties against gram positive and negative bacterial strains. J Nanobiotechnol 11(1):42

Pal KK, Gardener BM (2006) Biological control of plant pathogens. Plant Health Instr 2:1117–1142

Peters A (1996) The natural host range of *Steinernema* and *Heterorhabditis* spp. and their impact on insect populations. Biocontrol Sci Tech 6:389–402

Poinar GO Jr (1979) Nematodes for biological control of insects. CRC, Boca Raton, FL, p 249

Prusky D, Freeman S, Rodriguez RJ, Keen NT (1994) A nonpathogenic mutant strain of *Colletotrichum magna* induces resistance to *C. gloeosporioides* in avocado fruits. Mol Plant Microbe Interact 7(3):326–333

Richards AG, Richards PA (1979) The cuticular protuberances of insects. Int J Insect Morphol Embryol 8(3–4):143–157

Ruan L, Crickmore N, Peng D, Sun M (2015) Are nematodes a missing link in the confounded ecology of the entomopathogen *Bacillus thuringiensis*? Trends Microbiol 23(6):341–346

Saha M, Sarkar S, Sarkar B, Sharma BK, Bhattacharjee S, Tribedi P (2016) Microbial siderophores and their potential applications: a review. Environ Sci Pollut Res 23(5):3984–3999

Sanchis V (2011) From microbial sprays to insect-resistant transgenic plants: history of the biopesticide *Bacillus thuringiensis*. A review. Agron Sustain Dev 31(1):217–231

Sayre RM, Wergin WP (1977) Bacterial parasite of a plant nematode: morphology and ultrastructure. J Bacteriol 129:1091–1101

Schneider SM, Rosskopf EN, Leesch JG, Chellemi DO, Bull CT, Mazzola M (2003) Research on alternatives to methyl bromide: *pre-plant and post-harvest*. Pest ManagSci 59:814–826

Setati ME, Jacobson D, Andong UC et al (2012) The vineyard yeast microbiome, a mixed model microbial map. PLoS One 7(12):e52609

Shapiro-Ilan DI, Gouge DH, Piggott SJ, Fife JP (2006) Application technology and environmental considerations foruse of entomopathogenic nematodes in biological control. Biol Control 38:124–133

Sharma A, Johri BN (2003) Growth promoting influence of siderophore-producing Pseudomonas strains GRP3A and PRS^ sub 9^ in maize (*Zea mays* L.) under iron limiting conditions. Microbiol Res 158(3):243

Sharma RR, Singh D, Singh R (2009) Biological control of postharvest diseases of fruits and vegetables by microbial antagonists: a review. Biol Control 50(3):205–221

Sharma KM, Kumar R, Vats S, Gupta A (2014) Production, partial purification and characterization of alkaline protease from *Bacillus aryabhattai* K3. Int J Adv Pharm Biol Chem 3(2):290–298

Sharma D, Javed S, Arshilekha PS, Babbar P, Shukla D, Srivastava P, Vats S (2018) Food additives and their effects: a mini review. Int J Curr Res 10(06):69999–70002

Shi YM, Bode HB (2018) Chemical language and warfare of bacterial natural products in bacteria–nematode–insect interactions. Nat Prod Rep 35(4):309–335

Singh M, Vats S (2019) Mathematically designed bioprocess for release of value added products with pharmaceutical applications from wastes generated from spices industries. Int J Pharm Sci Res 10(1):130–138

Spadaro D, Droby S (2016) Development of biocontrol products for postharvest diseases of fruit: the importance of elucidating the mechanisms of action of yeast antagonists. Trends Food Sci Technol 47:39–49

Stotzky G (2000) Persistence and biological activity in soil of insecticidal proteins from *Bacillus thuringiensis* and of bacterial DNA bound on clays and humic acids. J Environ Qual 29:691–705

Su H, Zhao Y, Zhou J, Feng H, Jiang D, Zhang KQ, Yang J (2017) Trapping devices of nematode-trapping fungi: formation, evolution, and genomic perspectives. Biol Rev 92(1):357–368

Sudheer KP, Indira V (2007) Post harvest technology of horticultural crops, vol 7. New India Publishing, New Delhi

Tanada Y, Kaya HK (2012) Insect pathology. Academic, New York

Tandon S, Vats S (2016) Microbial biosynthesis of cadmium sulfide (Cds) nanoparticles and their characterization. Eur J Pharm Med Res 3(9):545–550

Thaker M, Hanjabam MD, Gudipati V, Kannuchamy N (2017) Protective effect of fish gelatin-based natural antimicrobial coatings on quality of Indian Salmon fillets during refrigerated storage. J Food Process Eng 40(1):e12270

Thomashow LS (2013) Phenazines in the environment: microbes, habitats, and ecological relevance. In: Chincholkar S, Thomashow L (eds) Microbial phenazines. Springer, Berlin, pp 199–216

Tian B, Yang J, Zhang KQ (2007) Bacteria used in the biological control of plant-parasitic nematodes: populations, mechanisms of action, and future prospects. FEMS Microbiol Ecol 61(2):197–213

Tu C, Koenning SR, Hu S (2003) Root-parasitic nematodes enhance soil microbial activities and nitrogen mineralization. Microb Ecol 46(1):134–144

Vaeck M, Reynaerts A, Höfte H, Jansens S, De Beukeleer M, Dean C, Zabeau M, Van Montagu M, Leemans J (1987) Transgenic plants protected from insect attack. Nature 327:33–37

Van Rie J, Jansens S, Hofte H, Degheele D, Van Mellaert H (1990) Receptors on the brush border membrane of the insect midgut as determinants of the specificity of *Bacillus thuringiensis* delta endotoxins. Appl Environ Microbiol 56:1378–1385

Vats S (2017) Methods for extractions of value-added nutraceuticals from lignocellulosic wastes and their health application. In: Grumezescu AM, Holban A-M (eds) Ingredients extraction by physicochemical methods in food. Academic, London, pp 1–64

Vats S, Bhargava P (2017) Alternate energy: fuel for "Modi's India" and "smart cities". Int J Curr Res 9(04):49090–49097

Vats S, Kumar R (2015) Amylolytic- extremoenzymes: saviour of environments. Eur J Biomed Pharm Sci 2(5):694–702

Vats S, Miglani K (2011) Synergistic antimicrobial effect of cow urine and *Azadirachta indica* on infectious microbes. Int J Pharm Sci Res 2(7):1781

Vats S, Mishra A (2016) Soil agro-ecological management by vermicompost a potential organic nutrient source for the state of Uttar Pradesh. Eur J Pharm Med Res 3(9):604–609

Vats S, Negi S (2013) Use of artificial neural network (ANN) for the development of bioprocess using *Pinus roxburghii* fallen foliages for the release of polyphenols and reducing sugars. Bioresour Technol 140:392–398

Vats S, Kumar R, Negi S (2012) Natural food that meet antibiotics resistance challenge: in vitro synergistic antimicrobial activity of *Azadirachta indica*, *Terminalia chebula*, *Piper nigrum* and photoactivated cow urine. Asian J Pharm Biol Res 2(2):122–126

Vats S, Maurya DP, Agarwal A, Shamoonand M, Negi S (2013a) Development of a microbial consortium for the production of blend of enzymes for the hydrolysis of agricultural wastes into sugars. J Sci Ind Res 72:585–790

Vats S, Maurya DP, Jain A, Mall V, Negi S (2013b) Mathematical model-based optimization of physico-enzymatic hydrolysis of *Pinus roxburghii* needles for the production of reducing sugars. Indian J Exp Biol 51:944–953

Vats S, Kumar R, Maurya DP (2014) Alkaline amylase from multi resistant microbes and its applications. In: Alexei E (ed.), Microbiology. Lambert Academic Publishing, p 100

Vats S, Singh M, Siraj S, Singh H, Tandon S (2017) Role of nanotechnology in theranostics and personalized medicines. J Health Res Rev 4(1):1

Vinciguerra MT, Clausi M (2006) Biological control of chestnut insect pests by means of entomopathogenic nematodes. Adv Hortic Sci 20:40–44

Walker K, Mendelsohn M, Matten S et al (2003) The role of microbial Bt products in U.S. crop protection. J New Seeds 5:31–51

Wang YC (1997) Effect of preharvest factors on postharvest quality: introduction to the colloquium. HortScience 32(5):807

Wei JZ, Hale K, Carta L, Platzer E, Wong C, Fang SC, Aroian RV (2003) *Bacillus thuringiensis* crystal proteins that target nematodes. Proc Natl Acad Sci U S A 100(5):2760–2765

Whipps JM, Davies KG (2000) Success in biological control of plant pathogens and nematodes by microorganisms. In: Gurr G, Wratten SD (eds) Biological control: measures of success. Springer, Dordrecht, pp 231–269

Wisniewski M, Biles C, Droby S, McLaughlin R, Wilson C, Chalutz E (1991) Mode of action of the postharvest biocontrol yeast, *Pichia guilliermondii*. I. Characterization of attachment to *Botrytis cinerea*. Physiol Mol Plant Pathol 39(4):245–258

Zhang Y, Chen K, Zhang S, Ferguson I (2003) The role of salicylic acid in postharvest ripening of kiwi fruit. Postharvest Biol Technol 28(1):67–74

Chapter 11
Entomopathogenic Nematodes in the Biological Control of Insect Pests with Reference to Insect Immunity

Istkhar, Ashok Kumar Chaubey, and Amar Prakash Garg

Abstract Insects are the successfully living arthropods on earth and challenging humans for food and shelter from several decades. As agricultural pests, insects cause heavy losses to crops and reduce the quality and quantity of human food. Failure of chemical pesticides and development of resistance of insects have become the major problem for researchers and are challenging mankind to search out for better solutions. Biological control management has been found as an emerging key of success, and in this context, entomopathogenic nematodes are in concern as potential biological control agents. These nematodes parasitize and enter the body of insect host and kill them quickly without the development of any resistance and failure. Several species of the entomopathogenic nematode genera *Steinernema* and *Heterorhabditis* have been identified around the world, and few of them have been commercialized for the biomanagement of several insect pests. The killing capabilities of these nematodes were enhanced severalfold by their symbiont bacteria of the genera *Xenorhabdus* and *Photorhabdus* which reside in the alimentary canal of these rhabditid nematodes. The symbiont bacteria produce severe toxins which ultimately results in the death of insect hosts. For the successful establishment of a biological species as biocontrol agent of insects, it is necessary to study the immune responses of host and defensive mechanisms of the parasites. Insects comprise innate immune system where cellular and humoral components combat with the invading pathogens. When entomopathogenic nematodes enter in the haemocoel of insect's body, both the immune components are triggered and try to evade pathogenicity of invading nematode–bacterium complexes. The mechanisms of nematode immunomodulation display the variation of modulatory approaches developed by different types of parasitic nematodes to cripple the host immune responses. Both the nematode and

Istkhar (✉)
Department of Bioscience and Biotechnology, Banasthali Vidyapith, Banasthali, Rajasthan, India

A. K. Chaubey
Nematology Laboratory, Department of Zoology, Chaudhary Charan Singh University, Meerut, India

A. P. Garg
School of Biological Engineering and Life Sciences, Shobhit Deemed University, Meerut, India

A. Varma et al. (eds.), *Plant Biotic Interactions*,
https://doi.org/10.1007/978-3-030-26657-8_11

bacteria produce chemical factors which protect them from the insect's immune system attack. The cellular and humoral responses are avoided and broken by toxins produced by nematode–bacterium complexes included different proteins producing immunodeficiency in insects. These nematodes were found to be safe to the environments and soil flora and fauna and also have exemption from one area to another by several countries.

11.1 Introduction

We are living in such a biological world where several species exist and survive with better adaptations. Insects are among the earliest and a class of distinction by virtue of their diversity, worldwide distribution, ecological importance and impact on other life forms. It can be safely concluded that insects form an extremely successful group of organisms and have exploited almost all habitats on earth with the exception of the polar regions and the oceans. Plant producing food worldwide are damaged by more than 10,000 species of insects, 30,000 species of weeds, 100,000 diseases (caused by fungi, viruses, bacteria and other microorganisms) and 1000 species of nematodes (Hall 1995; Dhaliwal et al. 2007). To reduce global crop losses, it has been estimated that around US $ 40 billion are used annually worldwide for the application of 3 million metric tonnes of pesticides, plus the use of various biological and other non-chemical controls (Pimentel 2007, 2009). Out of total 70,000 estimated pests destroying 35–40% crops, insects are contributing around 14% (Pimentel 2009).

Because of their short life span and heavy progeny production rate, insects are considered r-selective. Story does not end here as they are the most adaptable and successful animals that exist in a myriad of environment where the potential of contamination by microbes and bedbugs is high. They also have developed a demonstrable ability of resistance which has been observed in more than 645 insect and mite species over one or more chemical pesticides (Rajmohan 1998). Apart from the potential effective resistance against chemical, the defensive mechanisms of the insect also play an important role with possession of genes for high level of oxidases, esterases, glutathione-s-transferases and 'insensitive' acetylcholinesterases and nerve insensitivity to pyrethroids. In a similar manner, insects are also susceptible to microbes. Researches on host specificity indicate that several insect species are naturally resistant to those microorganisms. These studies pose some of the important questions and make us to direct on some of the important things like the following: (1) whether insect can and will evolve resistance to the innovative tactics which are initially overwhelming them; (2) life history layouts and variability in genetic makeup of some species of insect make prone to evolve resistance; and (3) there is no guarantee that developed cheaper, safe and effective biopesticides like Bt (*Bacillus thuringiensis*) cannot be misused so as to rapidly induce resistance which has been observed in diamondback moth of cabbage (*Plutella xylostella*), Indian meal moth (*Plodia interpunctella*), American bollworm of cotton (*Heliothis virescens*), Oriental bollworm of cotton (*H. armigera*), beet army worm (*Spodoptera *exigua*) and tobacco caterpillar (*Spodoptera litura*). All have shown different degrees of resistance to Bt and insect viruses (Tabashnik et al. 1994; Briese 1981).

Similar to other animals, insects susceptible to various entomopathogens try to evade them through morphological, behavioural, developmental (like maturation immunity), physiological, nutritional, biochemical and molecular genetic mechanisms. Indeed, insects that are susceptible to a pathogen can show resistance to various entomopathogens and try to resist infection. In order to appreciate the insect defence and its role in microbial control of insect pests by the conventional agricultural entomologists and applied insect pathologists, some preliminary information on the different aspects of immunity, viz. passive and active defence mechanisms against foreign invaders in comparison with vertebrate immunity, has been presented (Vilmos and Kurucz 1998; Ratcliffe 1985, 1993).

Nematodes are ancient and biologically diverse moulting animals having body size ranging from very minute (0.2 mm) to greatly large (6 m), occupying in most habitat including inside and outside the plants and animals (Blaxter and Denver 2012). In terms of individual, nematodes form the most abundant group of marine and terrestrial animals. The association of nematodes with other organisms especially arthropods is plentifully harmful to beneficial such as phoretic, necromenic, facultative and obligate (Sudhaus 2008). It is to be believed that the parasitism was developed from nonparasitic to parasitic (Poinar Jr 1983; Sudhaus 1993). In terms to understanding of evolution, fossils provide the best way of the past history of a group where only a single fossil has the potential to provide the age for the origin of numerous evolutionary roots. For entomopathogenic nematodes (EPNs), discovery of fossil is difficult because they mostly infect terrestrial insect's larvae and rarely turn up in the fossil as they are destroyed by the invaders (Poinar Jr. 2003). While actual fossils of EPN are lacking, some closely related parasitic rhabditid nematodes closely resembling the genus *Rhabditophanes* have been recovered from the haemocoel of nematoceran flies in 22–26-million-year-old Mexican amber (Poinar Jr. 2003). However, EPNs are invertebrate pathogens, and their relation to vertebrate pathogens was also found (Blaxter and Denver 2012). *Heterorhabditis* to *Strongylomorpha* (Rhabditina, a major group of gut and airway parasites of vertebrates) and steinernematids to strongyloids were found to be closely related during phylogenic study.

11.2 Life Cycle of Entomopathogenic Nematodes and Their Symbiotic Bacteria

The nematode families Steinernematidae and Heterorhabditidae share several common characteristics such as mutualistic association with enteric bacteria (*Gammaproteobacteria*, *Enterobacteriaceae*) within their intestine and the need to parasitize insect host. The EPNs have three preadaptations which allowed them to evolve the lifestyles exhibited by the genera *Steinernema* and *Heterorhabditis*

(Sudhaus 1993). First, they have evolved species that have a variety of associations with insects. Second, they produce the dauer juvenile stage which confers the capacity to enter an insect and persist in the absence of food. Third, they are bacterial feeders and hence are preadapted to enter a mutualistic relationship with entomopathogenic bacteria inside the insect haemocoel. The infective juveniles (IJs) of EPNs are compatible with other biological and chemical pesticides, fertilizers and soil amendments (Krishnayya and Grewal 2002). EPNs are globally distributed and have the ability to live in drastic conditions and tolerance against harsh environmental conditions including anoxybiosis, thermobiosis and desiccation (Grewal 2000). The free living stage of these nematodes is 3rd stage juveniles (IJs) carrying bacterial symbionts in their alimentary canal once encounter the suitable insect host, it penetrates host through mouth, anus, spiracal or by tearing the skin making the way to haemocoel (Kaya and Gaugler 1993). The presence of buccal cuticular tooth in heterorhabditid nematodes helps them to penetrate the cuticle of insect (Bedding et al. 1983). EPNs also have several other ways to enter in the body of insect (Castillo et al. 2011). Once the IJ reach the haemocoel, they release bacteria in the haemocoel which start to colonize therein by releasing toxins which in turn kill insect host within 24–48 h. When the food conditions decline, the new generations of IJS, coming from the last adult generation, abandon the dead body of insect to seek a new live host (Poinar Jr. 1990).

Xenorhabdus and *Photorhabdus* are insect pathogenic bacteria (EPB) vectored by *Steinernema* and *Heterorhabditis* genera of EPNs, respectively. The lifecycles of both *Photorhabdus* and *Xenorhabdus* are similar, revolving around the free-living infective juvenile form of their specific nematode which acts as a vector for transferring the bacteria from host to host. Following penetration of the nematodes into the insect, bacteria are regurgitated directly into the haemocoel of insect host. If the infection is successful, nematodes resume development and start feeding. One of the three generations occurs inside the host, and over the course of infection, quality of the resource diminishes (Kaya and Gaugler 1993; Adams and Nguyen 2002). Subsequently, the bacteria grow unrestricted by the insect immune system (Daborn et al. 2001) releasing toxins to kill the insect host (Ffrench-Constant et al. 2003) and also serving as a food source for their nematode symbionts (Forst and Clarke 2002). As conditions decline, IJs are produced that leave the insect to seek new hosts (Ciche and Ensign 2003). The IJ is the only free-living stage and actively seeks hosts using a species-specific strategy along a continuum from ambush to cruise foraging (Campbell and Gaugler 1997; Campbell and Kaya 2002). Among the newly produced nematodes, the IJs reassociate with the mutualistic bacteria before leaving the insect cadaver (Martens et al. 2003). The nematodes carry their bacterial symbionts monoxenically in a special vesicle of the IJs known as 'vesicle of Bird and Akhurst' (Bird and Akhurst 1983) in Steinernematidae and throughout the whole intestine of Heterorhabditidae (Endo and Nickle 1991). It has been proved that the free-living stage of the EPNs is the most infective and operative stage where the free-living IJs can search and move into the soil to find insects to infect (Spence et al. 2008). As per the present scenario, the tools and techniques of molecular biology and genomics can be used to improve the efficiency and ability of EPNs by understanding their ecology, behaviour and how they respond to climate change through screening of related genes (Segal and Glazer 2000). The life cycle pattern of entomopathogenic nematode–bacterium (EPN-EPB) complex and their residing sites within insect host are depicted in Fig. 11.1.

11.3 Taxonomic Status of Entomopathogenic Nematodes

Taxonomy is an important aspect in biological studies as it revealed how an organism is related to the other organism, no matter how different they may be. Taxonomy also provides basic understanding about the components of biodiversity which is necessary for effective decision-making about conservation and sustainable use. Although the present scenario shows that, taxonomy is highly emphasizing on molecular aspects but still, classical taxonomy, based on morphological characters is extremely important for identification of the species as morphology of an organism to a great extent, is a clear manifestation of its genetic makeup or constitution, therefore, it has ever remained a key component and back for taxonomic research. The molecular technique used was adequate to identify nematode isolates, enabling knowledge of its biodiversity and contributing to the detection of new isolates that may be used in biological control programs of insect pests.

Nematodes are highly diverse and ubiquitous creature on planet earth. More than 28,000 species have been described, and the global species richness has been estimated between 500,000 and 1 million species (Hugot et al. 2001; Gaugler and Bilgrami 2004). Nematodes represent an especially abundant and species-rich phylum, with many free-living and parasitic species. Nematodes are found to occupy almost all the conceivable habitats but are often overlooked because most of them are microscopic in size. For instance, a square yard of woodland or agricultural habitat may contain several million nematodes. They occupy almost every habitat be it terrestrial, fresh water or marine. Many species are highly specialized parasites of vertebrates, including humans, or of insects and other invertebrates. The most studied nematode *Caenorhabditis elegans* (Rhabditidae) is currently considered a eukaryote model organism for research in biology and biomedical fields. However, many other nematode species are emerging at present as powerful model organisms for studying diverse disciplines including ecology, physiology, developmental biology and evolution. On the other hand, nematodes are also utilized as biocontrol agents of insect pests. Natural enemies of arthropods fall into three major categories: predators, parasitoids and pathogens. The nematodes fall into the pathogens category because of causing sickness to insect which ultimately results in the death of the host insect.

Among the vast variety of parasitic nematodes, some have evolved an association with insect pathogenic bacteria. Together the bacteria and nematode are a lethal duo. These nematodes are called 'entomopathogenic nematodes' (EPNs). They are lethal obligatory insect parasitoids, inhabiting the soil. The term entomopathogenic comes from the Greek word *entomon*, meaning insect, and *pathogenic*, meaning causing disease. Although many other parasitic nematodes cause diseases in plants, livestock

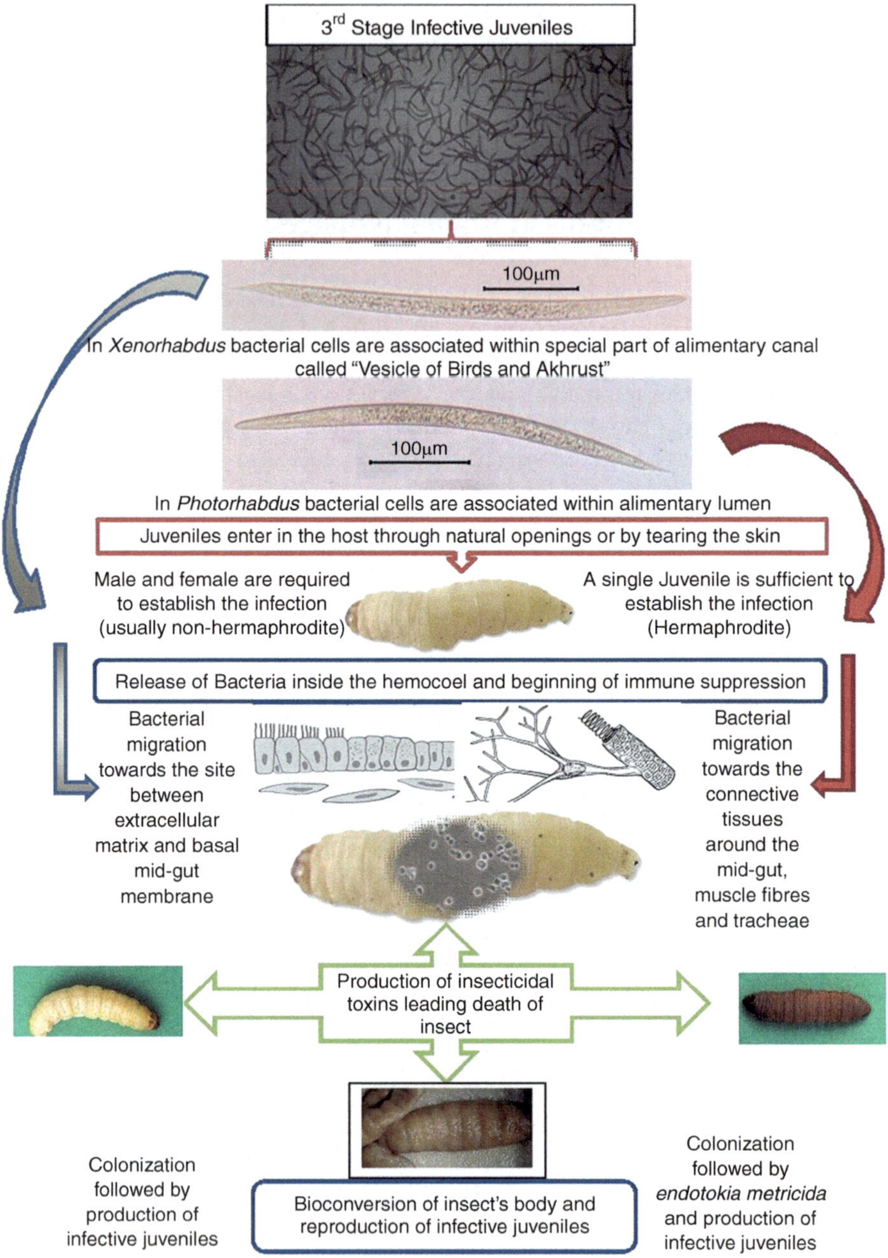

Fig. 11.1 A generalized life cycle of entomopathogenic nematodes

and humans, EPNs, as their name implies, only infect insects and kill them within a short period. EPNs live inside the body of their host, and so they are designated endoparasitic. EPNs are rhabditid nematodes belonging to families Steinernematidae Travassos, 1927, with genus *Steinernema* and Heterorhabditidae Poinar, 1976, having genus *Heterorhabditis*, members of which are also commercially produced and used in biological control. In 1994, Nguyen and Smart described a new genus *Neosteinernema* and added this genus to Steinernematidae family. The first EPN was described by Steiner in 1923 and was named *Aplectana kraussei* which was later renamed by Travassos (1927) as *Steinernema*. In continuation, Steiner in 1929 described one more nematode, *Neoaplectana glaseri*, and Glaser and Fox (1930) found that it infected Japanese beetle *Popillia japonica* (Newman). From that time, many more species have been acknowledged and have been used in field trials, starting with experiments to control *P. japonica*.

Entomopathogenic nematodes *Steinernema* and *Heterorhabditis* species are distributed to a large geographical area throughout the world. Species especially *S. kraussei*, *S. glaseri*, *S. feltiae*, *S. carpocapsae*, *H. bacteriophora*, *H. indica*, *H. megidis* and *H. zealandica* are known to have a cosmopolitan existence. The most adaptable species found in almost all warm and cold area of the different continents is *S. feltiae* common to Holarctic, Neotropic and Australian regions and recently also reported from the regions of Indo-Malaysian (Addis et al. 2011) and Afrotropical (Akyazi et al. 2012). In the case of other steinernematids, some of the species prefer temperate or torrid zone, or they can be found both in these zones. However, some of the results were found doubtful about the preference of a particular region as in the case of *S. kraussei* is a Holarctic species and its recovery from Neotropic in Colombia (Melo et al. 2009). On the other hand, *S. glaseri* and *S. carpocapsae* prefer Holarctic temperate zone but can also be found in torrid zone in Indo-Malaysian (India/Tamil Nadu) and Neoarctic (SE USA) regions (Seenivasan et al. 2012; Parkman and Smart 1996). In heterorhabditids, *H. bacteriophora* is the most widespread species common to both torrid and temperate zones, originally reported from Australian region. *H. indica* is found in all the tropical and subtropical areas, whereas the *H. megidis* is only in temperate zone of Holarctic. Interestingly, *H. zealandica*, originally described from New Zealand, was later found in north-eastern Europe and recently reported from north-eastern China, Florida and South Africa (Wang et al. 2014; Nguyen et al. 2007; Malan et al. 2011).

Over the past time, a great numerical increment in the number of new EPN species has been described all over the world. New lines present an inimitable combination of features and thus have a great potential for biological control of specific insect pests. Survey conducted throughout the world showed that steinernematids and heterorhabditids were common to all types of soil and more frequently inhabit agricultural and secondary forest ecosystems which provide the suitable conditions for insect's host populations (Půža et al. 2016). Recently, several surveys were conducted in Europe (Mráček et al. 2005), North America (Stock et al. 1999), South Africa, Ethiopian region (Malan et al. 2011), Southeast Asia, Indo-Malaysian region (Seenivasan et al. 2012) and tropical areas in Neotropic region (Barbosa-Negrisoli et al. 2010). Species of *Steinernema* were found to be more frequent as compared to *Heterorhabditis* where the number of *Steinernema* species reached more.

The global distribution of both the genera indicates that they were present when all land masses were combined as the Pangaea supercontinent. This theory was supported by a genetic study where *Heterorhabditis* was found as a sister group of vertebrate parasitic strongyloides, and both the groups arose independently from the free-living *Rhabditis* group (Kiontke et al. 2007). In 2011 (Poinar Jr 2011), the ancient age of *Heterorhabditis* clade was shown when a 100 million-year-old *Proheterorhabditis burmanicus* fossil was discovered from Early Cretaceous Burmese amber.

It is important to mention that the data accoutred on diversity of EPNs is highly influenced by wrong identification and doubts about several species such as *S. ritteri*, *S. rarum*, *S. scapterisci* and *S. riobrave* native to Southern and Northern America but later reported from Northeast China (Wang et al. 2014; Půža et al. 2016). Several steinernematid species claimed new, were either synonymized or considered as *species incuirendae* and *nomina nudum* due to inaccurate data (Hunt and Subbotin 2016).

In India, several EPNs were reported as new, namely, *S. masoodi*, *S. seemae*, *S. qazi* and *S. sayeedae*, but sufficient information about their origin were not provided (Ali et al. 2005, 2009; Ali and Azra 2011). However, in an attempt, Ali et al. (2010) provided the molecular data regarding the confirmation about the *S. masoodi*, *S. seemae* and *S. qazi*, but they failed as the molecular data did not belong to EPNs (Hunt and Subbotin 2016). Including these, *S. thermophilum* (Ganguly and Singh 2000), *S. meghalayensis* (Ganguly et al. 2005) and *S. dharanaii* (Kulkarni et al. 2012) reported by Indian scientists were not even found in citations in international literature and synonymized as junior synonyms of *S. abbasi*, *S. carpocapsae* and *S. hermaphroditum*, respectively (Hunt 2007; Hunt and Subbotin 2016). However, the data provided by Indian scientist rectified the ambiguities and helped to improve the quality (Bhat et al. 2017). Even after, due to inaccuracy in the data and improper identification, the number of EPN species reduces so far from more than 100 to 95 (Hunt and Subbotin 2016). Similarly, for *Heterorhabditis* list of 30 species reduced to 18 species only. List of all described species of *Steinernema* and *Heterorhabditis* presented in Tables 11.1 and 11.2 showed the authentic *Steinernema* and *Heterorhabditis*, respectively (synonymized species have not been included), with their worldwide location, host and authorities. List also included the species which are described recently (Grifaldo-Alcantara et al. 2017; Gorgadze et al. 2018; Stock et al., 2019). Figure 11.2 represents the year-wise discovery of all the EPNs along with the number of species synonymized. Species considered as junior synonym, *species incuirendae* and *nomina nuda* are not added in the list or in the graph.

11.4 Entomopathogenic Nematodes in Agriculture Insect Pest Management

To become effective and efficacious biological control agent, selection is based on host range, host finding or foraging strategy, tolerance of environmental factors and their effects on survival and efficacy (temperature, moisture, soil type, exposure to

Table 11.1 List of authentic *Steinernema* species identified worldwide with their respective authorities, year of identification, isolation source/host and location

S. No	Species	Authors	Year	Host	Location
1	*S. kraussei*	Steiner (1923)	1923	*Cephalcia abietis*	Germany, Europe
2	*S. glaseri*	Glaser and Fox (1930)	1930	*Popillia japonica*	Near Haddonfield, New Jersey
3	*S. feltiae*	Filipjev (1934)	1934	*Agrotis segetum*	Udmurt Republic, Ural region, Russia
4	*S. affine*	Bovien (1937), Wouts et al. (1982)	1937	*Phyla febrilis*	Denmark
5	*S. carpocapsae*	Weiser (1955)	1955	*Laspeyrasia pomonella*	Czechoslovakia
6	*S. intermedium*	Poinar (1985)	1985	Soil	Charleston, South Carolina, USA
7	*S. rarum*	de Doucet (1986), Mamiya (1988)	1986	Soil	Cordoba, Argentina
8	*S. kushidai*	Mamiya (1988)	1988	Soil	Shizuoka, Japan
9	*S. ritteri*	Douchet and Douchet (1990)	1990	Soil	Córdoba, Argentina
10	*S. scapterisci*	Nguyen and Smart (1990)	1990	*Scapteriscus vicinus*	Uruguay, South America
11	*S. caudatum*	Xu,Wang and Li (1991)	1991	Soil	China
12	*S. neocurtillae*	Nguyen and Smart (1992)	1992	*Neocurtilla hexadactyla*	Florida, USA
13	*S. longicaudum*	Shen and Wang (1992)	1992	Soil	Lailang, Shangdong province, China
14	*S. cubanum*	Mracek et al. (1994)	1994	Soil sample	Cuba
15	*S. riobrave*	Cabanillas et al. (1994)	1994	Soil	Weslaco, Texas, USA
16	*S. puertoricense*	Román and Figueroa (1994)	1994	Soil	Loiza, Puerto Rico
17	*S. bicornutum*	Tallosi et al. (1995)	1995	Soil	Serbia (formerly Yugoslavia)
18	*S. oregonense*	Liu and Berry (1996)	1996	Soil	Oregon USA

(continued)

Table 11.1 (continued)

S. No	Species	Authors	Year	Host	Location
19	*S. abbasi*	Elawad et al. (1997)	1997	Soil	Sultanate of Oman
20	*S. arenarium*	Artyukhovsky et al. (1997)	1997	*Melolontha hippocastani*	Russia
21	*S. ceratophorum*	Jian et al. (1997)	1997	Soil	North-eastern China
22	*S. monticolum*	Stock et al. (1997)	1997	Soil	Gyeongnam Province, Republic of Korea
23	*S. karii*	Waturu et al. (1997)	1997	Soil	Kirinyaga, Central Province, Kenya
24	*S. siamkayai*	Stock et al. (1998)	1998	Soil	Thailand
25	*S. tami*	Luc et al. (2000)	2000	Soil	Vietnam
26	*S. loci*	Phan et al. (2011)	2001	Soil	Vietnam
27	*S. sangi*	Phan et al. (2001)	2001	Soil	Vietnam
28	*S. thanhi*	Phan et al. (2001)	2001	Soil	Vietnam
29	*S. pakistanense*	Shahina et al. (2001)	2001	Soil	Karachi, Sindh Pakistan
30	*S. asiaticum*	Anis et al. (2002)	2002	Soil	Pakistan
31	*S. diaprepesi*	Nguyen et al. (2002)	2002	*Diaprepes abbreviates*	Florida
32	*S. anatoliense*	Hazir et al. (2003)	2003	Soil	Turkey
33	*S. scarabaei*	Stock and Koppenhofer (2003)	2003	*Anomala* (= *Exomala*) *orientalis* and *Popillia japonica*	New Jersey, USA
34	*S. weiseri*	Mrácek et al. (2003)	2003	Soil	Czech Republic, Europe
35	*S. apuliae*	Triggiani et al. (2004)	2004	Soil	Italy
36	S. *guangdongense*	Chen et al. (2004)	2004	Soil	China
37	*S. hermaphroditum*	Stock et al. (2004)	2004	Soil	Indonesia
38	*S. jollieti*	Spiridonov et al. (2004)	2004	Soil	Missouri valley near St. Louis, USA
39	*S. litorale*	Yoshida (2004)	2004	Soil	Japan

(continued)

Table 11.1 (continued)

S. No	Species	Authors	Year	Host	Location
40	*S. yirgalemense*	Nguyen et al. (2004)	2004	Soil	Yirgalem, Ethiopia
41	*S. aciari*	Qiu et al. (2005)	2005	Soil	China
42	*S. akhursti*	Qiu et al. (2006)	2005	Soil	China
43	*S. beddingi*	Qui et al. (2005)	2005	Soil	China
44	*S. robustispiculum*	Phan et al. (2005)	2005	Soil	Vietnam
45	*S. silvaticum*	Sturhan et al. (2005)	2005	Soil	Berlin, Germany, Europe
46	*S. ashiuense*	Phan et al. (2006)	2006	Soil	Japan
47	*S. backanense*	Phan et al. (2006)	2006	Soil	Vietnam
48	*S. cumgarense*	Phan et al. (2006)	2006	Soil	Vietnam
49	*S. eapokense*	Phan et al. (2006)	2006	Soil	Vietnam
50	*S. khoisanae*	Nguyen et al. (2006)	2006	Soil	South Africa
51	*S. leizhouense*	Nguyen et al. (2006)	2006	Soil	Southern China
52	*S. sasonense*	Phan et al. (2006)	2006	Soil	Vietnam
53	*S. sichuanense*	Mrabek et al. (2006)	2006	Soil	China
54	*S. hebeiense*	Chen et al. (2006)	2006	Soil	China
55	*S. costaricense*	Uribe-Lorio et al. (2007)	2007	Soil	Costa Rica
56	*S. puntauvense*	Uribe-Lorio et al. (2007)	2007	Soil	Costa Rica
57	*S. texanum*	Nguyen et al. (2007)	2007	Soil	Texas, USA
58	*S. cholashanense*	Nguyen et al. (2008)	2008	Soil	China
59	*S. colombiense*	López-Núñez et al. (2008)	2008	Soil	Colombia
60	*S. ichnusae*	Tarasco et al. (2008)	2008	Soil	Sardinia Island, Italy
61	*S. australe*	Edgington et al. (2009)	2009	Soil	Chile
62	*S. boemarei*	Lee et al. (2009)	2009	Soil	France

(continued)

Table 11.1 (continued)

S. No	Species	Authors	Year	Host	Location
63	*S. unicornum*	Edgington et al. (2009)	2009	Soil	Chile
64	*S. xueshanense*	Mracek et al. (2009)	2009	Soil	China
65	*S. brazilense*	Nguyen et al. (2010)	2010	Soil	Mato Grosso, Brazil
66	*S. schliemanni*	Spiridonov et al. (2010)	2010	*Osmoderma ceremita*	Germany, Europe
67	*S. minutum*	Maneesakorn et al. (2010)	2010	Soil	Thailand
68	*S. arasbaranense*	Nikdel et al. (2011)	2011	Soil	Iran
69	*S. citrae*	Malan et al. (2011)	2011	Soil	South Africa
70	*S. nepalense*	Khatri-Chhetri et al. (2011)	2011	Soil	Nepal
71	*S. surkhetense*	Khatri-Chhetri et al. (2011)	2011	Soil	Nepal
72	*S. lamjungense*	Khatri-Chhetri et al. (2011)	2011	Soil	Nepal
73	*S. phyllophagae*	Nguyen and Buss (2011)	2011	(*Phyllophaga* sp.)	Florida, USA
74	*S. pui*	Qui et al. (2011)	2011	Soil	Yunnan, China
75	*S. changbaiense*	Ma et al. (2012)	2012	Soil	China
76	*S. ethiopiense*	Tamiru et al. (2012)	2012	Soil	Ethiopia
77	*S. tielingense*	Juan et al. (2012)	2012	Soil	China
78	*S. xinbinense*	Juan et al. (2012)	2012	Soil	China
79	*S. cameroonense*	Ngo Kanga et al. (2012)	2012	Soil	Cameroon
80	*S. nyetense*	Kanga et al. (2012)	2012	Soil	Cameroon
81	*S. sacchari*	Nthenga et al. (2014)	2014	Soil	South Africa
82	*S. tophus*	Cimen et al. (2014)	2014	Soil	South Africa
83	*S. huense*	Phan et al. (2014)	2014	Soil	Vietnam
84	*S. poinari*	Mracek et al. (2014)	2014	Soil	Czech Republic
85	*S. innovationi*	Cimen et al. (2015)	2015	Soil	South Africa

(continued)

Table 11.1 (continued)

S. No	Species	Authors	Year	Host	Location
86	*S. jeffreyense*	Malan et al. (2015)	2015	Soil	South Africa
87	*S. papillatum*	San-Blas et al., (2015)	2015	Soil	Venezuela
88	*S. beitlechemi*	Puza et al. (2016)	2016	Soil	South Africa
89	*S. pwaniensis*	Puza et al. (2016)	2016	Soil	Tanzania
90	*S. fabii*	Abate et al. (2016)	2016	Soil	South Africa
91	*S. nguyeni*	Malam et al. (2016)	2016	Soil	South Africa
92	*S. biddulphi*	Cimen et al. (2016)	2016	Soil	South Africa
93	*S. ralatorei*	Grifaldo-Alcantara et al. (2017)	2017	Soil	Mexico
94	*S. borjomiense*	Gorgadze et al. (2018)	2018	*Oryctes nasicornis*	Georgia
95	*Steinernema khuongi*	Stock et al. (2019)	2018	–	Florida

Species considered as junior synonym, species incuerendae and nomun nudum, are given in the text
[a]Data is based on the available literature on the Internet up to 01 May 2018

ultraviolet light, salinity and organic content of soil, means of application, agrochemicals and others). In this context, EPN have great potential and positive attributes for biological control (Kaya and Gaugler 1993; Shapiro-Ilan and Grewal 2008). They have a wide host range where some of nematode species have been reported to infect dozens of insect species (Poinar Jr 1979; Klein 1990). They are also amenable to mass production using in vivo or in vitro methods (Shapiro-Ilan and Gaugler 2002; Shapiro-Ilan et al. 2014). Application of EPNs is safe for the environment, humans and other nontargeted organisms (Akhurst and Smith 2002; Ehlers 2005), and they are exempted for pesticide registration in many countries (Ehlers 2005) with few exception such as *S. scarabaei* (Koppenhöfer and Fuzy 2003a, b). Due to the development of morphological and behavioural changes, white grubs are one of the most difficult insects to control using EPNs (Klein et al. 2007). However, Japanese beetle *Popillia japonica* appears to be the most EPN-susceptible species infecting turf in the USA (Grewal et al. 2005; Klein et al. 2007). In a similar way, several other experiments were performed and reviewed on control tactics of larvae of other scarabids and lepidopterans (Klein 1990; Kaya and Gaugler 1993; Grewal et al. 2005; Lewis and Clarke 2012; Shapiro-Ilan et al. 2002, 2012). Weevils harbouring small fruit crops, ornamental plants and turf were also evaluated against

Table 11.2 List of authentic *Heterorhabditis* species identified worldwide with their respective authorities, year of identification, isolation source/host and location

S. No.	Species	Authors	Year	Host	Location
1	*H. bacteriophora*	Poinar (1976)	1975	*Heliothis punctigera*	South Australia
2	*H. megidis*	Poinar et al. (1987)	1987	*Popillia japonica*	Ohio, USA
3	*H. zealandica*	Poinar Jr. (1990)	1990	Soil	South Africa
4	*H. indica*	Poinar et al. (1992)	1992	*Scirpophaga excerptalis*	India
5	*H. hepialius*	Stock et al. (1996)	1996	*Hepialis californicus*	California
6	*H. marelatus*	Liu and Berry (1996)	1996	Soil	Oregon
7	*H. taysearae*	Shamseldean et al. (1996)	1996	Soil	Egypt
8	*H. downesi*	Stock et al. (2002)	2002		
9	*H. baujardi*	Phan et al. (2003)	2003	Soil	Vietnam
10	*H. mexicana*	Nguyen et al. (2004)	2004	Soil	Mexico
11	*H. amazonensis*	Andalo et al. (2006)	2006	Soil	Brazil
12	*H. floridensis*	Nguyen et al. (2006)	2006	Soil	Florida, USA
13	*H. georgiana*	Nguyen et al. (2008)	2008		Georgia
14	*H. safricana*	Malan et al. (2008)	2008	Soil	South Africa
15	*H. atacamensis*	Edgington et al. (2011)	2011	Soil	Chile
16	*H. beicherriana*	Xing-Yue et al. (2012)	2012	Soil	China
17	*H. rugaoensis*	Zhang et al. (2012)	2012	Soil	China
18	*H. noenieputensis*	Malan et al. (2014)	2014	Soil	South Africa

Species considered as junior synonym, species incuerendae and nomun nudum, are given in the text
[a]Data is based on the available literature on the Internet up to December 2016

the pathogenicity of several researchers. In an experiment, *Otiorhynchus sulcatus* (Curculionidae) larvae in strawberry plants were parasitized by *S. carpocapsae* and *H. megidis* in field conditions. In Ireland and Norway, *H. megidis* was proven to be a good candidate of controlling *O. sulcatus* larvae (Haukeland and Lola-Luz 2010). Higher pathogenicity of *S. scarabaei* was recorded against *P. japonica*, *Aonidiella orientalis*, *Rhizotrogus majalis*, *Cyclocephala borealis* and Asian garden beetle *Maladera castanea* (Cappaert and Koppenhöfer 2003; Koppenhöfer and Fuzy 2003a). *H. bacteriophora*, *H. megidis* and new isolates of *Heterorhabditis* spp. and *S. kushidai* were found highly pathogenic to several species of scarab in turf and lawns (Shapiro-Ilan et al. 2002; Grewal et al. 2004). *S. feltiae* and *Heterorhabditis* spp. isolates tested against fungus gnats *Lycoriella auripilla*, *L. mali*, *L. solani*, *Bradysia coprophila* and *B. difformis* were proven good biological control agent (Scheepmaker et al. 1998a, b; Jagdale et al. 2004, 2007; Jess et al. 2005; Tomalak et al. 2005; Grewal 2007). Due to their high efficacy against fungus gnats, commercially available products of *S. feltiae* are routinely used in the USA and Europe (Grewal and Georgis 1998; Georgis et al. 2006; Grewal 2007).

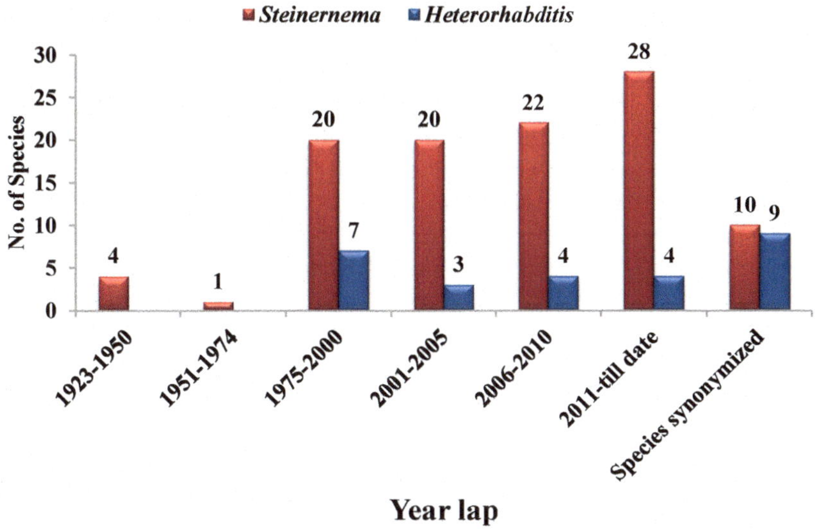

Fig. 11.2 Graphical presentation of year-wise identification of entomopathogenic nematode including the number of synonymized *Steinernema* and *Heterorhabditis* species

High larval mortality in field trials was recorded in *Diaprepes abbreviatus* (Coleoptera: Cucurlionidae) infesting citrus by *S. riobrave* by McCoy et al. (2002), Shapiro-Ilan et al. (2002) and Stuart et al. (2008). Cutworms (Lepidoptera: Noctuidae) (*Agrotis*, *Amathes*, *Noctua*, *Peridroma*, *Prodenia* spp.) control tactics using EPNs were found suitable when they were applied in turf and crops (Shapiro-Ilan et al. 2002; Ebssa and Koppenhöfer 2011). IJs of *S. carpocapsae* and *S. feltiae* used against codling moth, *Cydia pomonella* (Lepidoptera: Tortricidae), infesting worldwide pest of apple and other pome fruit, provide an excellent example of the successful use of EPNs for biological control of this pest (Unruh and Lacey 2001; Lacey et al. 2006). Some other targeted insect pests of EPN reported are filbert moth, *Melissopus latiferreanus* (Lepidoptera: Tortricidae) (Chambers et al. 2010); navel orangeworm, *Amyelois transitella* (Lepidoptera: Pyralidae) (Siegel et al. 2004, 2006); Oriental fruit moth, *Grapholita molesta* (Lepidoptera: Tortricidae) (Riga et al. 2006); peachtree borer *Synanthedon pictipes* (Lepidoptera: Sesiidae) (Shapiro-Ilan et al. 2010); and *S. litura* and *H. armigera* (Lepidoptera: Noctuidae) (Istkhar and Chaubey 2016, 2017).

The IJs of EPNs are compatible with other biological and chemical pesticides, fertilizers and soil amendments (Krishnayya and Grewal 2002). Integrated management of diamondback moth larvae, *Plutella xylostella* (Lepidoptera: Plutellidae), on watercress with *B. thuringiensis* and *S. carpocapsae* which produced 58% control was reported by Baur et al. (1998). Similar results were reported for *S. carpocapsae* and combination of the nematode and *B. thuringiensis* with a polymer by other workers (Schroer and Ehlers 2005; Schroer et al. 2005). Lello et al. (1996)

demonstrated that high-output hydraulic nozzles which deposited the greatest number of IJs onto foliage can produce up to 98% mortality in *P. xylostella*. Foliar application of nematodes has been successfully used to control the quarantine leaf-eating caterpillars on various crops (Choo et al. 1989; Rao et al. 1971); however, the results varied (Georgis et al. 2006). Foliar applications of *S. carpocapsae* against cabbageworm *Artogeia rapae* (Lepidoptera: Pieridae) under environmental conditions in Quebec did not provide good results (Bélair et al. 2003). However, *S. carpocapsae* and *S. feltiae* provided a better potential of control for leaf miners *Liriomyza trifolii* (Diptera: Agromyzidae), *Liriomyza huidobrensis*, *Tuta absoluta* and other leaf miner species in foliar application by several workers (Hara et al. 1993; LeBeck et al. 1993; Sher et al. 2000; Tomalak et al. 2005; Williams and Walters 2000; Batalla-Carrera et al. 2010).

At commercial level in Asia, Europe and North America, at least 13 different species, namely, *H. bacteriophora*, *H. indica*, *H. marelata*, *H. megidis*, *H. zealandica*, *S. carpocapsae*, *S. feltiae*, *S. glaseri*, *S. kushidai*, *S. kraussei*, *S. longicaudum*, *S. riobrave* and *S. scapterisci* (Lacey et al. 2015), have been commercialized.

11.5 Immune System of Insects

Vilmos and Kurucz (1998) reported that the immune system of insects exhibits a high degree of structural and functional similarity with the innate immune system of mammals. The quick responses of insects against invaders generally involve haemocytes and/or molecules which directed the interaction of pathogen-associated molecular patterns and pattern recognition receptors (PAMPs and PRRs) for neutralizing foreign substances (Medzhitov and Janeway 2002; Strand 2008; Fors et al. 2014). Hillyer (2016) indicated that the insects have developed sensitive mechanisms for detecting the presence of microbial infections and activating signalling pathways that control the production of molecules with antimicrobial activity. In this way, the cellular and humoral components of haemolymph form the essential components of immune system of insects. It has been described that haemocytes have essential roles in numerous physiological activities. The cellular immune defences of insect comprise haemocytes. Ratcliffe (1993) reported that haemocytes of insects function in a manner similar to that of phagocytes of human. Recent advances in chemistry and instrumentation have led to the recognition of six broad classes, viz. prohaemocytes, plasmatocytes, granulocytes, spherulocytes, adepohaemocytes and oenocytoids (Gupta 1994; Jones 1977). Six types of haemocytes were identified in *G. mellonella* by Boman and Hultmark (1987). Kalia et al. (2001) reported all six types of haemocytes from *H. armigera*. The classical work by haemocytes on coagulation has been studied by Gregoire (1955, 1957); on connective tissue synthesis by Wigglesworth (1955, 1956, 1973); on metamorphosis by Whitten (1964); on wound repair by Lai-Fook (1968); and on encapsulation by Salt (1970). Haemocyte-mediated encapsulation is an important part of the insect cellular immune defence system. Schmidt et al. (2001) reported that parasites such as parasitoids and nematodes are encapsulated by haemocytes. Inhibition of the immune defence system may be associated with changes in haemocyte population (Moreau et al. 2003; Ibrahim and Kim 2006; Istkhar and Chaubey 2018). Lavine and Strand (2002) reported plasmatocytes and granulocytes phagocytic in Lepidoptera. They also showed that cellular defence responses referred to haemocyte-mediated immune responses like phagocytosis, nodulation and encapsulation. Humoral defences were also reported by a number of workers which includes the production of antimicrobial peptides (Lowenberger 2001); reactive intermediates of oxygen and nitrogen (Bogdan et al. 2000); and prophenoloxidase (pro-PO) activating system that regulates coagulation or melanization of haemolymph (Gillespie et al. 1997; Kanost et al. 2004). Melanization is one of the important events in cellular defence mediated through phenoloxidase cascade. Pathogenesis of haemocytes considered as an index of melanization (Kalia et al. 2001). Bergin et al. (2005) reported that proteins involved in superoxide production homologous to the NADPH oxidase complex of human neutrophils help in phagocytosis. Some other humoral responses include cell adhesion molecule, and lysozymes and lectins have also been reported by Kanost et al. (2004).

11.6 Insect's Immune System and Nematode Defences

Kenney and Eleftherianos (2016) showed that the nematode parasites and their associated bacteria are outstanding pathogenic organisms for probing the insect host anti-nematode and antibacterial immune response. Cooper and Eleftherianos (2016) reported that the mechanisms of nematode immunomodulation display the variation of modulatory approaches developed by different types of parasitic nematodes to cripple the host immune response. Various active compounds in haemolymph along with prophenoloxidase (pro-PO) system mediate the insect host humoral defences (Hoffmann et al. 1996; Söderhall and Cerenius 1998). EPNs, *S. carpocapsae* showed the damaging strategy of host immune components by proteolytic secretions (Balasubramanian et al. 2010; Toubarro et al. 2013) that interfere with host immunological defences followed by hiding from host's immune system in *S. feltiae* and *S. glaseri* (Wang and Gaugler 1999; Brivio et al. 2004; Mastore and Brivio 2008). Wang and Gaugler (1999) reported that *S. glaseri* surface proteins defeat the host immune system of the Japanese beetle *Popillia japonica*, thereby protecting nematodes from encapsulation. Balasubramanian et al. (2010) purified a trypsin-like secreted protease from *S. carpocapsae* that suppresses the prophenoloxidase (pro-PO) in *G. mellonella*. Binda-Rossetti et al. (2016) demonstrated in their experiments with *S. carpocapsae* and *X. nematophila* that infection with live nematodes and bacteria can suppress the antibacterial peptide immune response of red palm weevil *Rhynchophorus ferrugineus*, but the inhibitory effect was not present when insects were injected with dead microorganisms.

Dowds and Peters (2002) reported that the bacteria and nematodes cooperate with each other to overwhelm the host's immune response, permitting the bacteria to multiply vegetatively. It has been reported that *S. carpocapsae* was able to suppress the immune response by secreting proteins, which may facilitate the release of their symbionts (Götz et al. 1981; Simões 1998). However, it was unknown whether similar proteins were produced by *Heterorhabditis* (Forst and Clarke 2002). Molecular study in different insect hosts revealed a divergent array of results for both *Steinernema* and *Heterorhabditis* species. For example, in *Manduca sexta*, *P. luminescens* cells secreted an antiphagocytic factor that permitted the bacterial cells to obstruct their own phagocytosis (Silva et al. 2002), whereas in *S. exigua*, *X. nematophila* cells were able to hamper nodule formation by inhibiting the eicosanoid biosynthetic pathway (Park and Kim 2000; Park et al. 2003). Additionally in *S. exigua* and *M. sexta*, *X. nematophila* inhibits transcription of insect genes encoding antimicrobial peptides (Ji and Kim 2004; Park et al. 2007). Production of lipopolysaccharide (LPS) was shown by both the genera i.e., *P. luminescens* and *X. nematophila* (Dunphy and Thurston 1990), where LPS of *X. nematophila* inhibits PO activity and in both systems the lipid A moiety of LPS was thought to be cytotoxic to haemocytes (Dunphy and Webster 1988, 1991). Certain serine proteases and serpins have been identified to regulate the activation of melanization in insects (Gulley et al. 2013; Veillard et al. 2016). Haemolysin activity was another similar activity shown by both genera (Brillard et al. 2001, 2002). At the end of growth cycle of bacteria, a variety of antimicrobial compound production was observed which showed their function in protection of cadaver from colonization by other microorganisms. Some of them were found closely related to antibacterial in *P. luminescens* and some against fungi and yeasts (Sharma et al. 2002; Boemare et al. 1997; Webster et al. 2002).

The symbiotic bacterium *Xenorhabdus* seems to be unrecognized and overcome the host immune system and produced a variety of toxins which interfere with the immune system and produced immunodeficiency (Dillman et al. 2012). Several other workers such as Wang et al. (1995), Forst et al. (1997), Jarosz (1998) and Owuama (2001) have proved that symbiotic bacteria released by EPN produce virulence factors that suppress host immune responses including toxin complexes, proteases, lipases and lipopolysaccharides. Ernst (2000) and Barbieri et al. (2002) have showed that bacteria have evolved numerous toxins and delivered type III effector molecules which can interfere with the actin cytoskeleton and inhibit phagocytosis. In a study, Brillard et al. (2001) reported that haemocyte monolayer from *S. littoralis* has shown two distinct haemolytic activities in supernatants from cultures of *X. nematophila*. Au et al. (2004) reported that *Photorhabdus* supernatants reduced haemocyte viability. The purified lipopolysaccharides (LPS), *Photorhabdus* protease fractions or *Xenorhabdus* lecithinase isomers showed no toxic effect following injection into the haemocoel (Bowen et al. 2000; Clarke and Dowds 1995; Thaler et al. 1998). A novel toxin complex identified from *P. luminescens* showed insecticidal activity when injected or given orally (Bowen et al. 1998). Another toxin complex a (Tca) purified by Blackburn et al. (1998) from *P. luminescens* has specific effect on the midgut epithelium of the insect *Manduca sexta*. Although *Xenorhabdus* and *Photorhabdus* bacteria showed similar lifestyles, the differences in their molecular defensive mechanisms have also been reported by some workers where they achieve successful host interactions (Goodrich-Blair and Clarke 2007). *Photorhabdus* used lipopolysaccharide (LPS) modification to resist the action of the host-derived AMPs (Bennett and Clarke 2005; Eleftherianos et al. 2006), but *X. nematophila* prevents induction of insect AMP expression altogether (Ji and Kim 2004; Park et al. 2007). The pathogenic effects of bacteria and the antibacterial defensive mechanisms have been documented very well in insects, and nowadays the nematode-associated defences are the object of recent studies.

11.7 Entomopathogenic Nematode Risk Assessments

Entomopathogenic nematodes efficiently control a variety of insect pests and are unusually safe biocontrol agents. The importance of these nematodes is that they are more specific and not a threat to the environment as expressed by chemical pesticides. Since the beginning of the application of the EPNs in insect pest control when they were applied against *Popillia japonica* (Glaser and Farrell 1935), no any inferior detrimental effect was observed on plants and animals or even to its user (Boemare et al. 1996; Akhurst and Smith 2002). Some of the studies showed that the symbiotic bacterial species of the genus *Photorhabdus* associated with *Heterorhabdus* spp. found to be opportunistic was reported from human, although they were non-symbiotic in nature (Farmer III et al. 1989; Peel et al. 1999). No reports exist that document any effect on humans by the symbiotic bacteria. A related non-symbiotic species, *Photorhabdus asymbiotica*, was reported five times from humans in the USA (Farmer III et al. 1989). Another group of non-symbiotic *Photorhabdus* is reported from five patients in Australia (Peel et al. 1999). From most of the patients, other human pathogenic bacteria were also recorded; thus, *Photorhabdus* spp. were opportunistic.

Scientific information on safety and possible EPN impact on environments is available where it was mentioned that nontarget mortality will be temporary and spatially restricted and will affect only part of the population (Bathon 1996). This can be supported by the study where it was mentioned that potential wide host range of >200 species targeted in the laboratory could not support in the fields (Buck and Bathon 1993; Koch and Bathon 1993; Bathon 1996). Introduction of EPNs from one area to another can execute the possible biological pollution being as exotic species. But this argument can be rejected as it is beneficial to the agroecosystem. In one study when exotic *S. riobrave* from Texas were introduced in North Carolina with endemic populations of *S. carpocapsae* and *H. bacteriophora*, the insect mortality was reported to reduce when soil samples were baited using *G. mellonella* (Barbercheck and Millar 2000). In an overall view, the importance and applicability of EPNs cannot be mistreated as they are safe for users and the environment and the benefits outweigh possible risks to nontarget organisms.

11.8 Conclusion

The prevailing use of chemical pesticides has generated several problems including pesticide resistance, outbreaks of secondary pests, safety risks for humans and domestic animals, contamination of groundwater and decrease in biodiversity among other environmental concerns. EPN are playing a result-oriented role in the management of insect pests in an effective way and in many cases have shown better performance over chemical and microbial control measures. If we focus to the last decade, more than half of the numbers of EPN species are described, and many more need to be identified and evaluated. Scientist are trying to find out more new species/ strains and trying to improve strains leveraged by genomics so that they could be targeted against insect pests in a better way. If we see, it was found that nematodes on one side are focused on commercial production, and their bacterial symbionts have now become the new focal point of research. These bacterial symbionts produce secondary metabolites having the potential for agriculture and medical fields. The identified toxic complexes are highly active against insect immune system with unusual and unique properties needs to be explored. Studies on EPN have experienced an exponential growth, and laboratories around the world are now working on different aspects of EPN and their bacterial symbionts. The immune system of insect is being studied at large scale nowadays, but their interactions with EPN are still a less studied area. As a future prospect, the study of the defences of insects and offences of EPN will provide more precise picture for the better management of insect pests. Identifying the defensive mechanisms of insects against the EPN and reducing them by any adjuvants or species of nematodes with higher pathogenicity and strong immunosuppressive capabilities could be helpful for future insect pest management program.

References

Adams B, Nguyen KB (2002) Taxonomy and systematics. In: Gaugler R (ed) Entomopathogenic nematology. CABI Publishing, Wallingford, pp 1–33

Addis T, Mulawarman M, Waeyenberge L, Moens M, Viaene N, Ehlers RU (2011) Identification and intraspecific variability of *Steinernema feltiae* strains from Cemoro Lawang village in eastern Java, Indonesia. Russ J Nematol 19:21–29

Akhurst R, Smith K (2002) Regulation and safety. In: Gaugler R (ed) Entomopathogenic nematology. CABI, Wallingford, pp 311–332

Akyazi F, Ansari MA, Ahmed BI, Crow WT, Mekete T (2012) First record of entomopathogenic nematodes (Steinernematidae and Heterorhabditidae) from Nigerian soil and their morphometrical and ribosomal DNA sequence analysis. Nematol Mediterr 40:95–100

Ali SS, Azra S (2011) *S. sayeedae* sp. n, a heat tolerant EPN from banana rhizosphere of Koshambi district, U.P. India. Trends Biosci 4:123–125

Ali SS, Ahmad R, Hussain MA, Pervez R (2005) Pest management in pulses through entomopathogenic nematodes. Indian Institute of Pulses Research, Kanpur, p 59

Ali SS, Shaheen A, Asif M, Akhtar MH (2009) *Steinernema qazi* sp. n. (Nematoda: Rhabditida: Steinernematidae) from Kanpur India. Trends Biosci 2(1):59–64

Ali SS, Riyaz A, Verma V, Azra S, Rashid P, Sobia A (2010) Molecular characterization of *Steinernema masoodi*, *S. seemae* and other Indian isolates of *Steinernema* spp. Trends Biosci 3:112–116

Au C, Dean P, Reynolds SE, Ffrench-Constant RH (2004) Effect of the insect pathogenic bacterium Photorhabdus on insect phagocytes. Cell Microbiol 6(1):89–95. https://doi.org/10.1046/j.1462-5822.2003.00345.x

Balasubramanian N, Toubarro D, Simoes N (2010) Biochemical study and in vitro immune suppression by a trypsin-like secreted protease from the nematode *Steinernema carpocapsae*. Parasite Immunol 32:165–175

Barbercheck ME, Millar LC (2000) Environmental impacts of entomopathogenic nematodes used for biological control in soil. In: Follett PA, Duan JJ (eds) Nontarget effects of biological control. Kluwer Academic, Dordrecht, NL, pp 287–308

Barbieri JT, Riese MJ, Aktories K (2002) Bacterial toxins that modify the actin cytoskeleton. Annu Rev Cell Dev Biol 18:315–344

Barbosa-Negrisoli CR, Garcia MS, Dolinski C, Negrisoli AS Jr, Bernardi D, dos Santos FJ (2010) Survey of entomopathogenic nematodes (Rhabditida: Heterorhabditidae, Steinernematidae) in Rio Grande do Sul State, Brazil. Nematol Bras 34:189–197

Batalla-Carrera L, Morton A, Garcia-del-Pino F (2010) Efficacy of nematodes against the tomato leafminer *Tuta absoluta* in laboratory and greenhouse conditions. BioControl 55:523–530

Bathon H (1996) Impact of entomopathogenic nematodes on non-target hosts. Biocontrol Sci Tech 6:421–434

Baur ME, Kaya HK, Tabashnik BE, Chilcutt CF (1998) Suppression of diamondback moth (Lepidoptera: Plutellidae) with an entomopathogenic nematode (Rhabditida: Steinernematidae) and *Bacillus thuringiensis* Berliner. J Econ Entomol 91:1089–1095

Bedding RA, Molyneux AS, Akhurst RJ (1983) *Heterorhabditis* spp., *Neoaplectana* spp., and *Steinernema kraussei*: interspecific and intraspecific differences in infectivity for insects. Exp Parasitol 55:249–257

Bélair G, Fournier Y, Dauphinais N (2003) Efficacy of Steinernematid nematodes against three insect pests of crucifers in Quebec. J Nematol 35:259–265

Bennett HPJ, Clarke DJ (2005) The pbg PE operon in *Photorhabdus luminescens* is required for pathogenicity and symbiosis. J Bacteriol 187:77–84

Bergin D, Reeves EP, Renwick J, Wientjes FB, Kavanagh K (2005) Superoxide production in *Galleria mellonella* hemocytes: identification of proteins homologous to the NADPH oxidase complex of human neutrophils. Infect Immun 73:4161–4170. https://doi.org/10.1128/IAI.73.7.4161-4170

Bhat AH, Istkhar, Chaubey AK, Půža V, San-Blas E (2017) First report and comparative study of *Steinernema surkhetense* (Rhabditida: Steinernematidae) and its symbiont bacteria from sub-continental India. J Nematol 49(1):92–102

Binda-Rossetti S, Mastore M, Protasoni M, Brivio MF (2016) Effects of an entomopathogen nematode on the immune response of the insect pest red palm weevil: focus on the host antimicrobial response. J Invertebr Pathol 133:110–119

Bird AF, Akhurst RJ (1983) The nature of the intestinal vesicle in nematodes of the family Steinernematidae. Int J Parasitol 13:599–606

Blackburn M, Golubeva E, Bowen D, Ffrench-Constant RH (1998) A novel insecticidal toxin from *Photorhabdus luminescens*, toxin complex a (Tca), and its histopathological effects on the midgut of *Manduca sexta*. Appl Environ Microbiol 64(8):3036–3041

Blaxter M, Denver DR (2012) The worm in the world and the world in the worm. BMC Biol 10:57. https://doi.org/10.1186/1741-7007-10-57

Boemare NE, Laumond C, Mauleon H (1996) The entomopathogenic nematode-bacterium complex: biology, life cycle and vertebrate safety. Biocontrol Sci Tech 6:333–346

Boemare N, Givaudan A, Brehelin M, Laumond C (1997) Symbiosis and pathogenicity of nematode-bacterium complexes. Symbiosis 22:21–45

Bogdan C, Rollinghoff M, Diefenbach A (2000) Reactive oxygen and reactive nitrogen intermediates in innate and specific immunity. Curr Opin Immunol 12:64–76

Boman G, Hultmark D (1987) Cell-free immunity in insects. Annu Rev Microbiol 41:103–126

Bowen D, Blackburn M, Rocheleau T, Grutzmacher C, Ffrench-Constant RH (2000) Secreted proteases from *Photorhabdus luminescens*: separation of the extracellular proteases from the insecticidal Tc toxin complexes. Insect Biochem Mol Biol 30:69–74

Bowen D, Rocheleau TA, Blackburn M, Andreev O, Golubeva E, Bhartia R, Ffrench-Constant RH (1998) Insecticidal toxins from the bacterium *Photorhabdus luminescens*. Science 280:2129–2132

Briese DT (1981) Resistance of insect species to microbial pathogens. In: Davidson EN (ed) Pathogenesis of invertebrate microbial diseases. Allanheld Osmum, Totowa, NJ, pp 511–545

Brillard J, Ribeiro C, Boemare N, Brehelin M, Givaudan A (2001) Two distinct hemolytic activities in *Xenorhabdus nematophila* are active against immunocompetent insect cells. Appl Environ Microbiol 67:2515–2525

Brillard J, Duchaud E, Boemare N, Kunst F, Givaudan A (2002) The Phl a hemolysin from the entomopathogenic bacterium *Photorhabdus luminescens* belongs to the two-partner secretion family of hemolysins. J Bacteriol 184:3871–3878

Brivio MF, Mastore M, Moro M (2004) The role of *Steinernemafeltiae* body-surface lipids in host-parasite immunological interactions. Mol Biochem Parasitol 135:111–121

Buck M, Bathon H (1993) Auswirkungen des Einsatzes entomopathogener Nematoden *Heterorhabditis* sp. im Freiland auf die Nichtzielfauna. 2. Teil: Diptera. Anz Schädlingskde Pflanzenschutz, Umweltschutz 66:84–88

Campbell JF, Gaugler R (1997) Inter-specific variation in entomopathogenic nematode foraging strategy: dichotomy or variation along a continuum? Fundam Appl Nematol 20:393–398

Campbell JF, Kaya HK (2002) Variation in entomopathogenic nematode (Steinernematidae and Heterorhabditidae) infective-stage jumping behaviour. Nematologica 4:471–482

Cappaert DL, Koppenhöfer AM (2003) *Steinernema scarabaei,* an entomopathogenic nematode for control of the European chafer. Biol Control 28:379–386

Castillo JC, Reynolds SE, Eleftherianos I (2011) Insect immune responses to nematode parasites. Trends Parasitol 27(12):537–547. https://doi.org/10.1016/j.pt.2011.09.001

Chambers U, Bruck DJ, Olsen J, Walton VM (2010) Control of overwintering filbertworm (Lepidoptera: Tortricidae) larvae with *Steinernema carpocapsae*. J Econ Entomol 103:416–422

Choo HY, Kaya HK, Burlando TM, Gaugler R (1989) Entomopathogenic nematodes: host-finding ability in the presence of plant roots. Environ Entomol 18:1136–1140

Ciche TA, Ensign JC (2003) For the insect pathogen *Photorhabdus luminescens,* which end of a nematode is out? Appl Environ Microbiol 69:1890–1897

Clarke DJ, Dowds BCA (1995) Virulence mechanisms of *Photorhabdus* Sp strain K122 toward wax moth larvae. J Invertebr Pathol 66:149–155

Cooper D, Eleftherianos I (2016) Parasitic nematode immunomodulatory strategies: recent advances and perspectives. Pathogens 5:58. https://doi.org/10.3390/pathogens5030058

Daborn PJ, Waterfield N, Blight MA, Ffrench-Constant RH (2001) Measuring virulence factor expression by the pathogenic bacterium *Photorhabdus luminescens* culture and during insect infection. J Bacteriol 183:5834–5839

Dhaliwal GS, Dhawan AK, Singh R (2007) Biodiversity and ecological agriculture: issues and perspectives. Indian J Ecol 34(2):100–109

Dillman AR, Chaston JM, Adams BJ, Ciche TA, Goodrich-Blair H, Stock SP, Sternberg PW (2012) An entomopathogenic nematode by any other name. PLoS Pathog 8(3):e1002527. https://doi.org/10.1371/journal.ppat.1002527

Dowds BCA, Peters A (2002) Virulence mechanisms. In: Gaugler R (ed) Entomopathogenic nematology. CABI Publishing, Wallingford, pp 79–98

Dunphy GB, Thurston GS (1990) Insect immunity. In: Gaugler R, Kaya HK (eds) Entomopathogenic nematodes in biological control. CRC Press, Boca Raton, FL, pp 301–323

Dunphy GB, Webster JM (1988) Virulence mechanisms of *Heterorhabditis heliothidis* and its bacterial associate, *Xenorhabdus luminescens*, in non-immune larvae of the greater wax moth, *Galleria mellonella*. Int J Parasitol 18:729–737

Dunphy GB, Webster JM (1991) Antihemocytic surface components of *Xenorhabdus nematophilus* var. Dutki and their modification by serum on nonimmune larvae of *Galleria mellonella*. J Invertebr Pathol 58:40–51

Ebssa L, Koppenhöfer AM (2011) Efficacy and persistence of entomopathogenic nematodes for black cutworm control in turfgrass. Biocontrol Sci Tech 21:779–796

Ehlers RU (2005) Forum on safety and regulation. In: Grewal PS, Ehlers RU, Shapiro-Ilan DI (eds) Nematodes as biocontrol agents. CABI, Wallingford, pp 107–114

Eleftherianos I, Millichap PJ, Ffrench-Constant RH, Reynolds SE (2006) RNAi suppression of recognition protein mediated immune responses in the tobacco hornworm *Manduca sexta* causes increased susceptibility to the insect pathogen *Photorhabdus*. Dev Comp Immunol 30:1099–1107

Endo BY, Nickle WR (1991) Ultrastructure of the intestinal epithelium, lumen and associated bacteria in *Heterorhabditis bacteriophora*. J Helminthol 58:202–212

Ernst JD (2000) Bacterial inhibition of phagocytosis. Cell Microbiol 2:379–386

Farmer JJ III, Jörgensen JH, Grimont PAD, Akhurst RJ, Poinar JGO, Ageron E, Pierce GV, Smith JA, Carter GP Wilson KL, Hickman-Brenner FW (1989) *Xenorhabdus luminescens* DNA hybridization group 5 from human clinical specimens. J Clin Microbiol 27:1594–1600

Ffrench-Constant R, Waterfield N, Daborn P, Joyce S, Bennett H, Au C Dowling A, Boundy S, Reynolds S, Clarke D (2003) *Photorhabdus*: towards a functional genomic analysis of a symbiont and pathogen. FEMS Microbiol Rev 26:433–456

Fors L, Markus R, Theopold U, Hambäck PA (2014) Differences in cellular immune competence explain parasitoid resistance for two coleopteran species. PLoS One 9(9):e108795. https://doi.org/10.1371/journal.pone.0108795

Forst S, Clarke D (2002) Bacteria-nematode symbiosis. In: Gaugler R (ed) Entomopathogenic nematology. CAB International, London, pp 57–77

Forst S, Dowds B, Boemare N, Stackebrandt E (1997) *Xenorhabdus* and *Photorhabdus* spp: bugs that kill bugs. Annu Rev Microbiol 51:47–72

Ganguly S, Singh LK (2000) *Steinernema thermophilum* sp. n. (Rhabditida: Steinernematidae) from India. Int J Nematol 10:183–191

Ganguly S, Rathour KS, Kumar S, Singh M (2005) *Steinernemameghalayensis* sp. n. (Rhabditida: Steinernematidae) from Northeastern Hilly Region of India. Ind J Nematol 41:83–97

Gaugler R, Bilgrami A (2004) Nematode behavior, pp 2–3

Georgis R, Koppenhöfer AM, Lacey LA, Belair G, Duncan LW, Grewal PS, Samish M, Tan L, Torr P, RWHM T (2006) Successes and failures in the use of parasitic nematodes for pest control. Biol Control 38:103–123

Gillespie JP, Kanost MR, Trenczek T (1997) Biological mediators of insect immunity. Annu Rev Entomol 42:611–643

Glaser RW, Farrell CC (1935) Field experiments with the Japanese beetle and its nematode parasite. J N Y Entomol Soc 43:345

Glaser RW, Fox H (1930) A nematode parasite of the Japanese beetle (*Popillia japonica* Newm.). Science 71:16–17

Goodrich-Blair H, Clarke DJ (2007) Mutualism and pathogenesis in *Xenorhabdus* and *Photorhabdus*: two roads to the same destination. Mol Microbiol 64:260–268

Gorgadze O, Fanelli E, Lortkhipanidze M, Troccoli A, Burjanadze M, Tarasco E, De Luca F (2018) *Steinernema borjomiense* n. sp. (Rhabditida: Steinernematidae), a new entomopathogenic nematode from Georgia. Nematology 20:1–17

Götz P, Boman A, Boman HG (1981) Interactions between insect immunity and an insect-pathogenic nematode with symbiotic bacteria. Proc R Soc Lond Ser B 212:333–350

Gregoire C (1955) Blood coagulation in arthropods. V. Studies on haemolymph coagulation in 420 species of insects. Arch Biol 66:103–148

Gregoire C (1957) Studies by phase contrast microscopy on distribution patterns of haemolymph coagulation in insects. Smithson Misc Collect 134:1–35

Grewal PS (2000) Enhanced ambient storage stability of an entomophathogenic nematode through Anhydrobiosis. Pest Manag Sci 56:401–406

Grewal PS (2007) Mushroom pests. In: Lacey LA, Kaya HK (eds) Field manual of techniques in invertebrate pathology: application and evaluation of pathogens for control of insects and other invertebrate pests, 2nd edn. Springer, Dordrecht, pp 457–461

Grewal PS, Georgis R (1998) Entomopathogenic nematodes. In: Hall FR, Menn JJ (eds) Biopesticides: use and delivery. Humana, Totowa, NJ, pp 271–299

Grewal PS, Power KT, Grewal SK, Suggars A, Haupricht S (2004) Enhanced consistency in biological control of white grubs (Coleoptera: Scarabaeidae) with new strains of entomopathogenic nematodes. Biol Control 30:73–82

Grewal PS, Koppenhöfer AM, Choo HY (2005) Lawn, turfgrass, and pasture applications. In: Grewal PS, Ehlers RU, Shapiro-Ilan DI (eds) Nematodes as biocontrol agents. CABI Publishing, Wallingford, pp 115–146

Grifaldo-Alcantara PF, Alatorre-Rosas R, Segura-León O, Hernandez-Rosas F (2017) *Steinernema ralatorei* n. sp. isolated from sugarcane areas at Veracruz, Mexico. Southwest Entomol 42 (1):171–190. https://doi.org/10.3958/059.042.0117

Gulley MM, Zhang X, Michel K (2013) The roles of serpins in mosquito immunology and physiology. J Insect Physiol 59:138–147

Gupta AP (1994) Insect haemocytes: classification and immunological function. In: Gujar GT (ed) Recent advances in insect physiology and toxicology. Agricole Publishing Academy, New Delhi, pp 106–206

Hall R (1995) Challenges and prospects of integrated pest management. In: Reuveni R (ed) Novel approaches to integrated pest management. Lewis, Boca Raton, FL, pp 1–19

Hara AH, Kaya HK, Gaugler R, LeBeck LM, Mello CL (1993) Entomopathogenic nematodes for biological control of the leafminer, *Liriomyza trifolii* (Dipt., Agromyzidae) (Rhabditida: Steinernematidae). Entomaphaga 38:359–369

Haukeland S, Lola-Luz T (2010) Efficacy of the entomopathogenic nematodes *Steinernema kraussei* and *Heterorhabditis megidis* against the black vine weevil *Otiorhynchus sulcatus* in open field-grown strawberry plants. Agric For Entomol 12:363–369

Hillyer JF (2016) Insect immunology and hematopoiesis. Dev Comp Immunol 58:102–118

Hoffmann JA, Reichhard JM, Hetru C (1996) Innate immunity in higher insects. Curr Opin Immunol 8:8–13

Hugot J, Baujard P, Morand S (2001) Biodiversity in helminths and nematodes as a field study: an overview. Nematology 3:199–208

Hunt DJ (2007) Overview of taxonomy and systematics. In: Nguyen KB, Hunt DJ (eds) Entomopathogenic nematodes: systematics, phylogeny and bacterial symbionts. Nematology monographs and perspectives, vol 5. Brill Publishing, Leiden, pp 27–57

Hunt DJ, Subbotin SA (2016) Taxonomy and systematics. In: Nguyen HB, Hunt DJ (eds) Advances in entomopathogenic nematode taxonomy and phylogeny. Brill Publishing, Leiden, pp 13–58

Ibrahim AMA, Kim Y (2006) Parasitism by *Cotesia plutellae* alters the hemocyte population and immunological function of the diamondback moth, *Plutella xylostella*. J Insect Physiol 52:943–950

Istkhar, Chaubey AK (2016) Potential of *Steinernema abbasi, Steinernema siamkayai and Heterorhabditis indica* for the bio-management of tobacco cutworm. Indian J Plant Prot 44 (3):322–328

Istkhar, Chaubey AK (2017) Impact of soil dwelling entomopathogenic nematodes, recovered from the soil of Uttar Pradesh (India), on cotton bollworm *Helicoverpa armigera* (Hübner) (Lepidoptera: Noctuidae). Appl Biol Res 19(1):63–70. https://doi.org/10.5958/0974-4517.2017.00009

Istkhar, Chaubey AK (2018) Challenging the larvae of *Helicoverpa armigera* and assessing the immune responses to nematode-bacterium complex. Phytoparasitica 46:75–87. https://doi.org/10.1007/s12600-018-0640-3

Jagdale GB, Casey ML, Grewal PS, Lindquist RK (2004) Application rate and timing, potting medium, and host plant effects on the efficacy of *Steinernema feltiae* against the fungus gnat, *Bradysia coprophila*, in floriculture. Biol Control 29:296–305

Jagdale GB, Casey ML, Cañas L, Grewal PS (2007) Effect of entomopathogenic nematode species, split application and potting medium on the control of the fungus gnat, *Bradysia difformis* (Diptera: Sciaridae), in the greenhouse at alternating cold and warm temperatures. Biol Control 43:23–30

Jarosz J (1998) Active resistance of entomophagous rhabditid *Heterorhabditis bacteriophora* to insect immunity. Parasitology 117:201–208

Jess S, Schweizer H, Kilpatrick M (2005) Mushroom applications. In: Grewal PS, Ehlers RU, Shapiro-Ilan DI (eds) Nematodes as biocontrol agents. CABI Publishing, Wallingford, pp 191–213

Ji D, Kim Y (2004) An entomopathogenic bacterium, *Xenorhabdus nematophila*, inhibits the expression of an antibacterial peptide, cecropin, of the beet armyworm, *Spodoptera exigua*. J Insect Physiol 50:489–496

Jones JC (1977) Circulatory system of insects. Thomas, Springfield, IL

Kalia V, Chaudhari S, Gujar GT (2001) Changes in haemolymph of American bollworm, *Helicoverpa armigera* (Hubner), infected with nucleopolyhedro virus. Indian J Exp Biol 39:1123–1129

Kanost MR, Jiang H, Yu XQ (2004) Innate immune responses of a lepidopteran insect, *Manduca sexta*. Immunol Rev 198:97–105

Kaya H, Gaugler R (1993) Entomopathogenic nematodes. Annu Rev Entomol 38:181–206

Kenney E, Eleftherianos I (2016) Entomopathogenic and plant pathogenic nematodes as opposing forces in agriculture. Int J Parasitol 46:13–19

Kiontke K, Barrière A, Kolotuev I, Podbilewicz B, Sommer R, Fitch DHA, Félix MA (2007) Trends, stasis, and drift in the evolution of nematode vulva development. Curr Biol 17:1925–1937

Klein MG (1990) Efficacy against soil-inhabiting insect pests. In: Gaugler R, Kaya HK (eds) Entomopathogenic nematodes in biological control. CRC, Boca Raton, FL, pp 195–214

Klein MG, Grewal PS, Jackson TA, Koppenhöfer AM (2007) Lawn, turf and grassland pests. In: Lacey LA, Kaya HK (eds) Field manual of techniques in invertebrate pathology: application and evaluation of pathogens for control of insects and other invertebrate pests, 2nd edn. Springer, Dordrecht, pp 655–675

Koch U, Bathon H (1993) Results of the outdoor application of entomopathogenic nematodes on nonobjective fauna 1 Coleoptera. Anz Schadlingsk Pflanz Umwelt 66:65–68

Koppenhöfer AM, Fuzy EM (2003a) *Steinernema scarabaei* for the control of white grubs. Biol Control 28:47–59

Koppenhöfer AM, Fuzy EM (2003b) Ecological characterization of *Steinernema scarabaei*, a scarab-adapted entomopathogenic nematode from New Jersey. J Invertebr Pathol 83:139–148

Krishnayya PV, Grewal PS (2002) Effect of neem and selected fungicides on viability and virulence of the entomopathogenic nematode *Steinernema feltiae*. Biocontrol Sci Tech 12:259–266

Kulkarni N, Rizvi AN, Kumar V, Paunikar S, Mishra VK (2012) Morphological and molecular characterization of *Steinernema dharanaii* sp. n.: a new entomopathogenic nematode from India. Indian J Trop Biodivers 20(2):107–116

Lacey LA, Arthurs SP, Unruh TR, Headrick H, Fritts R Jr (2006) Entomopathogenic nematodes for control of codling moth (Lepidoptera: Tortricidae) in apple and pear orchards: effect of nematode species and seasonal temperatures, adjuvants, application equipment and post-application irrigation. Biol Control 37:214–223

Lacey LA, Grzywacz DShapiro-Ilan DI, Frutos R, Brownbridge M, Goette MS (2015) Insect pathogens as biological control agents: back to the future. J Invertebr Pathol 132:1–41

Lai-Fook J (1968) The fine structure of wound repair in an insect *(Rhodnius prolixus)*. J Morphol 124:37–78

Lavine MD, Strand MR (2002) Insect hemocytes and their role in immunity. Insect Biochem Mol Biol 32:1295–1309

LeBeck LM, Gaugler R, Kaya HK, Hara AH, Johnson MW (1993) Host stage suitability of the leafminer *Liriomyza trifolii* (Diptera: Agromyzidae) to the entomopathogenic nematode *Steinernema carpocapsae* (Rhabditida: Steinernematidae). J Invertebr Pathol 62:58–63

Lello ER, Patel MN, Matthews GA, Wright DJ (1996) Application technology for entomopathogenic nematodes against foliar pests. Crop Prot 15:567–574

Lewis EE, Clarke DJ (2012) Nematode parasites and entomopathogens. In: Vega FE, Kaya HK (eds) Insect pathology, 2nd edn. Academic, San Diego, CA, pp 395–443

Lowenberger C (2001) Innate immune response of *Aedes aegypti*. Insect Biochem Mol Biol 31:219–229

Malan AP, Knoetze R, Moore SD (2011) Isolation and identification of entomopathogenic nematodes from citrus orchards in South Africa and their biocontrol potential against false codling moth. J Invertebr Pathol 108:115–125

Martens EC, Heungens K, Goodrich-Blair H (2003) Early colonization events in the mutualistic association between *Steinernema carpocapsae* nematodes and *Xenorhabdus nematophila* bacteria. J Bacteriol 185:3147–3154

Mastore M, Brivio MF (2008) Cuticular surface lipids are responsible for disguise properties of an entomoparasite against host cellular responses. Dev Comp Immunol 32(9):1050–1062

McCoy CW, Stuart RJ, Duncan LW, Nguyen K (2002) Field efficacy of two commercial preparations of entomopathogenic nematodes against larvae of *Diaprepesabbreviatus* (Coleoptera: Curculionidae) in alfisol type soil. Fla Entomol 85:537–544

Medzhitov R, Janeway CA (2002) Decoding the patterns of self and non self by the innate immune system. Science 296:298–300

Melo EL, Ortega CA, Susurluk A, Gaigl A, Bellotti AC (2009) Native entomopathogenic nematodes (Rhabditida) in four departments of Colombia. Rev Colomb Entomol 35:28–33

Moreau SJM, Eslin P, Giordanengo P, Doury G (2003) Comparative study of the strategies evolved by two parasitoids of the genus *Asobara* to avoid the immune response of the host, *Drosophila melanogaster*. Dev Comp Immunol 27:273–282

Mráček Z, Bečvář S, Kindlmann P, Jersáková J (2005) Habitat preference for entomopathogenic nematodes, their insect hosts and new faunistic records for the Czech Republic. Biol Control 34:27–37

Nguyen KB, Hunt DJ, Mracek Z (2007) Steinernematidae: species descriptions. In: Nguyen KB, Hunt DJ (eds) Entomopathogenic nematodes: systematics, phylogeny and bacterial symbionts, nematology monographs and perspectives. Brill, Boston, MA, pp 121–609

Owuama CI (2001) Entomopathogenic symbiotic bacteria, *Xenorhabdus* and *Photorhabdus* of nematodes. World J Microbiol Biotechnol 17:505–515

Park Y, Kim Y (2000) Eicosanoids rescue *Spodoptera exigua* infected with *Xenorhabdus nematophilus,* the symbiotic bacteria to the entomopathogenic nematode *Steinernema carpocapsae*. J Insect Physiol 46:1469–1476

Park Y, Kim Y, Putnam SM, Stanley DW (2003) The bacterium *Xenorhabdus nematophilus* depresses nodulation reactions to infection by inhibiting eicosanoid biosynthesis in tobacco hornworms, *Manduca sexta*. Arch Insect Biochem Physiol 52:71–80

Park Y, Herbert EE, Cowles CE, Cowles KN, Menard ML, Orchard SS, Goodrich-Blair H (2007) Clonal variation in *Xenorhabdus nematophila* virulence and suppression of Manduca sexta immunity. Cell Microbiol 9:645–656

Parkman JP, Smart GC (1996) Entomopathogenic nematodes, a case study: introduction of *Steinernema scapterisci* in Florida. Biocontrol Sci Tech 6:413–419

Peel MM, Alfredson DA, Gerrad JG, Davis JM, Robson JM, McDougall RJ, Scullie BL, Akhurst RJ (1999) Isolation, identification and molecular characterization of strains of *Photorhabdus luminescens* from infected humans in Australia. J Clin Microbiol 37:3647–3653

Pimentel D (2007) Area-wide pest management: environmental, economic and food issues. In: Vreysen MJB, Robinson AS, Hendrichs J (eds) Area-wide control of insect pests: from research to field implementation. Springer, Dordrecht, pp 35–47

Pimentel D (2009) Pesticides and pest control. In: Peshin R, Dhawan AK (eds) Integrated pest management: innovation-development, vol 1. Springer, Dordrecht, pp 83–88

Poinar GO Jr (1979) Nematodes for biological control of insects. CRC, Boca Raton, FL, p 277

Poinar GO Jr (1983) The natural history of nematodes. Prentice Hall, Engelwood Cliffs, NJ

Poinar GO Jr (2011) The evolutionary history of nematodes. Brill, Leiden, p 439

Poinar GO Jr (1990) Entomopathogenic nematodes in biological control. In: Gaugler R, Kaya HK (eds) Taxonomy and biology of Steinernematidae and Heterorhabditidae. CRC, Boca Raton, FL, pp 23–74

Poinar GO Jr (2003) Trends in the evolution of insect parasitism by nematodes as inferred from fossil evidence. J Nematol 35(2):129–132

Půža V, Mráček Z, Nermuť J (2016) Novelties in pest control by entomopathogenic and mollusc-parasitic nematodes. In: Gill H (ed) Integrated pest management (IPM): environmentally sound pest management. In Tech, Rijeka, pp 71–102. https://doi.org/10.5772/64578

Rajmohan N (1998) Pesticides resistance – a global scenario. Pestic World 3:34–40

Rao YR, Prakasa Rao PS, Verma A, Israel P (1971) Tests with an insect parasitic nematode DD-136 (Nematoda: Steinernematidae) against rice stem borer, *Tryporyza incertulas* walk. Indian J Entomol 33:215–217

Ratcliffe N (1985) Invertebrate immunity – a primer for the nonspecialist. Immunol Lett 10:253–270

Ratcliffe NA (1993) Cellular defense responses in insects: unresolved problems. In: Bechage NE, Thompson SN, Federice BA (eds) Parasites and pathogens of insects, vol 1. Academic, San Diego, CA, pp 579–604

Riga K, Lacey LA, Guerra N, Headrick HL (2006) Control of the oriental fruit moth, *Grapholita molesta*, using entomopathogenic nematodes in laboratory and bin assays. J Nematol 38:168–171

Salt G (1970) The cellular defense reactions of insects. Cambridge University Press, Cambridge

Scheepmaker JWA, Geels FP, Rutjens AJ, Smits PH, Van Griensven LJLD (1998a) Comparison of the efficacy of entomopathogenic nematodes for the biological control of the mushroom pests *Lycoriella auripila* (Sciaridae) and *Megaselia halterata* (Phoridae). Biocontrol Sci Tech 8:277–287

Scheepmaker JWA, Geels FP, Smits PH, van Griensven LJLD (1998b) Influence of *Steinernema feltiae* and diflubenzuron on yield and economics of the cultivated mushroom *Agaricus bisporus* in Dutch mushroom culture. Biocontrol Sci Tech 8:269–275

Schmidt O, Theopold U, Strand M (2001) Innate immunity and its evasion and suppression by hymenopteran endoparasitoids. BioEssay 23:344–351

Schroer S, Ehlers RU (2005) Foliar application of the entomopathogenic nematode *Steinernema carpocapsae* for biological control of diamondback moth larvae (*Plutella xylostella*). Biol Control 33:81–86

Schroer S, Sulistydanto D, Ehlers RU (2005) Control of *Plutella xylostella* using polymer-formulated *Steinernema carpocapsae* and *Bacillus thuringiensis* in cabbage fields. J Appl Entomol 129:128–204

Seenivasan N, Prabhu S, Makesh S, Sivakumar M (2012) Natural occurrence of entomopathogenic nematode species (Rhabditida. Steinernematidae and Heterorhabditidae) in cotton fields of Tamil Nadu, India. J Nat Hist 46:2829–2843

Segal D, Glazer I (2000) Genetics for improving biological control agents: the case of entomopathogenic nematodes. Crop Prot 19:685–689

Shapiro-Ilan DI, Gaugler R (2002) Production technology for entomopathogenic nematodes and their bacterial symbionts. J Ind Microbiol Biotechnol 28:137–146

Shapiro-Ilan DI, Grewal PS (2008) Entomopathogenic nematodes and insect management. In: Capinera JL (ed) Encyclopedia of entomology, 2nd edn. Springer, Dordrecht, pp 1336–1340

Shapiro-Ilan DI, Gouge DH, Koppenhöfer AM (2002) Factors affecting commercial success: case studies in cotton, turf and citrus. In: Gaugler R (ed) Entomopathogenic nematology. CABI Publishing, Wallingford, pp 333–355

Shapiro-Ilan DI, Cottrell TE, Mizell RFIII, Horton DL, Behleand RW, Dunlap CA (2010) Efficacy of *Steinernema carpocapsae* for control of the lesser peachtree borer, *Synanthedon pictipes*: improved aboveground suppression with a novel gel application. Biol Control 54:23–28

Shapiro-Ilan DI, Bruck DJ, Lacey LA (2012) Principles of epizootiology and microbial control. In: Vega FE, Kaya HK (eds) Insect pathology, 2nd edn. San Diego, CA, Academic, pp 29–72

Shapiro-Ilan DI, Han R, Qiu X (2014) Production of entomopathogenic nematodes. In: Morales-Ramos JA, Rojas MG, Shapiro-Ilan DI (eds) Mass production of beneficial organisms: invertebrates and entomopathogens. Academic, Amsterdam, pp 321–356

Sharma S, Waterfield N, Bowen D, Rocheleau T, Holland L, James R, Ffrench-Constant R (2002) The lumicins: novel bacteriocins from *Photorhabdus luminescens* with similarity to the uropathogenic-specific protein (USP) from uropathogenic *Escherichia coli*. FEMS Microbiol Lett 214:241–249

Sher RB, Parrella MP, Kaya HK (2000) Biological control of the Leafminer *Liriomyza trifolii* (burgess): implications for intraguild predation between *Diglyphus begini* Ashmead and *Steinernema carpocapsae* (Weiser). Biol Control 17:155–163

Siegel J, Lacey LA, Fritts R Jr, Higbee BS, Noble P (2004) Use of Steinernematid nematodes for post harvest control of navel orange worm (Lepidoptera: Pyralidae, Amyelois transitella) in fallen pistachios. Biol Control 30:410–417

Siegel J, Lacey LA, Higbee BS, Noble P, Fritts R Jr (2006) Effect of application rates and abiotic factors on *Steinernema carpocapsae* for control of overwintering navel orangeworm (Lepidoptera: Pyralidae, Amyelois transitella) in fallen pistachios. Biol Control 36:324–330

Silva CP, Waterfield NR, Daborn PJ, Dean P, Chilver T, Au CPY, Sharma S, Potter U, Reynolds SE, Ffrench-Constant RH (2002) Bacterial infection of a model insect: *Photorhabdus luminescens* and *Manduca sexta*. Cell Microbiol 4:329–339

Simões N (1998) Pathogenicity of the complex *Steinernema carpocapsae-Xenorhabdus nematophilus*: molecular aspects related with the virulence. In: Simões N, Boemare N, Ehlers RU (eds) Pathogenicity of entomopathogenic nematode versus insect defence mechanisms: impacts on selection of virulent strains. European Commission, Brussels, pp 73–83

Söderhall K, Cerenius L (1998) Role of prophenoloxidase-activating system in invertebrate immunity. Curr Opin Immunol 10:23–28

Spence KO, Lewis EE, Perry RN (2008) Host-finding and invasion by entomopathogenic and plant-parasitic nematodes: evaluating the ability of laboratory bioassays to predict. J Nematol 40 (2):93–98

Stock PS, Pryor BM, Kaya HK (1999) Distribution of entomopathogenic nematodes (Steinernematidae and Heterorhabditidae) in natural habitats in California. Biodivers Conserv 8:535–549

Stock SP, Campos-Herrera R, El Borai F, Duncan LW (2019) *Steinernema khuongi* n. sp. (Panagrolaimomorpha, Steinernematidae) a new entomopathogenic nematode species from Florida, USA. J Helminthol 93(2):226–241

Strand MR (2008) The insect cellular immune response. Insect Sci 15(1):1–14

Stuart RJ, El Borai FE, Duncan LW (2008) From augmentation to conservation of entomopathogenic nematodes: trophic cascades, habitat manipulation and enhanced biological control of Diaprepes abbreviatus root weevils in Florida citrus groves. J Nematol 40:73–84

Sudhaus W (1993) Die mittel symbiontischer Bakterien entomopathogenen Nematoden Gattungen *Heteror habditis* and *Steinernema* sind keine Schwestertaxa. Verh Dtsch Zool Ges 86:146

Sudhaus W (2008) Evolution of insect parasitism in rhabditid and diplogastrid nematodes. In: Makarov SE, Dimitrijevic RN (eds) Advances in arachnology and developmental biology. SASA, Belgrade, pp 143–161

Tabashnik BE, Finson N, Groeters FR, Moar WJ, Johnson MW, Luo K, Adang MJ (1994) Reversal of resistance to Bacillus thuringiensis in *Plutella xylostella*. Proc Natl Acad Sci U S A 91:4120–4124

Thaler JO, Duvic B, Givaudan A, Boemare N (1998) Isolation and entomotoxic properties of the *Xenorhabdus namatophilus* F1 Lecithinase. Appl Environ Microbiol 64:2367–2373

Tomalak M, Piggott S, Jagdale GB (2005) Glasshouse applications. In: Grewal PS, Ehlers RU, Shapiro-Ilan DI (eds) Nematodes as biological control agents. CABI Publishing, Wallingford, pp 147–166

Toubarro D, Avila MM, Montiel R, Simões N (2013) A pathogenic nematode targets recognition proteins to avoid insect defenses. PLoS One 8(9):e75691. https://doi.org/10.1371/journal.pone.0075691

Unruh TR, Lacey LA (2001) Control of codling moth, *Cydia pomonella* (Lepidoptera: Tortricidae) with *Steinernema carpocapsae*: effects of supplemental wetting and pupation site on infection rate. Biol Control 20:48–56

Veillard F, Troxler L, Reichhart JM (2016) *Drosophila melanogaster* clip-domain serine proteases: structure, function and regulation. Biochimie 122:255–269

Vilmos P, Kurucz E (1998) Insect immunity: evolutionary roots of the mammalian innate immune system. Immunol Lett 62:59–66

Wang Y, Gaugler R (1999) *Steinernema glaseri* surface coat protein suppresses the immune response of *Popillia japonica* (Coleoptera: Scarabaeidae) larvae. Biol Control 14:45–50

Wang Y, Campbell JF, Gaugler R (1995) Infection of entomopathogenic nematodes *Steinernema glaseri* and *Heterorhabditis bacteriophora* against *Popillia japonica* (Coleoptera, Scarabaeidae) larvae. J Invertebr Pathol 66:178–184

Wang H, Luan JB, Dong H, Qian HT, Cong B (2014) Natural occurrence of entomopathogenic nematodes in Liaoning (Northeast China). J Asia Pac Entomol 17:399–406

Webster JM, Chen G, Hu K, Li J (2002) Bacterial metabolites. In: Gaugler R (ed) Entomopathogenic nematology. CABI Publishing, Wallingford, pp 99–114

Whitten JM (1964) Haemocytes and the metamorphosing tissues in *Sarcophaga bullata, Drosophila melanogaster* and other cyclorrhaphous Diptera. J Insect Physiol 10:447–469

Wigglesworth VB (1955) The role of haemocytes in the growth and moulting of an insect – *Rhodnius prolixus* (Hemiptera). J Exp Biol 2:649–663

Wigglesworth VB (1956) The haemoeytes and connective tissue formation in an insect *Rhodnius prolixus* (Hemiptera). Q J Microbiol Sci 97:89–98

Wigglesworth VB (1973) Haemocytes and basement membrane formation in *Rhodnius*. J Insect Physiol 19:831–844

Williams EC, Walters KFA (2000) Foliar application of the entomopathogenic nematode *Steinernema feltiae* against leafminers on vegetables. Biocontrol Sci Tech 10:61–70

Chapter 12
Interaction Between Aromatic Oil Components and Bacterial Targets

Smaranika Pattnaik and Niranjan Behera

Abstract Aromatic and medicinal plants have been used as antimicrobial agents since time immemorial, though there has been a stark decline in the total quantitative use in recent times. Nonetheless they have been used, albeit with little or no precision in the knowledge of their modes of actions. Recent studies have indicated that different components of essential oils cause distinct types of injuries to microbial cells, each type of damage characteristic to one or more components of a particular essential oil. The damage to the microbial cells result from oxidative stress, protein dysfunction or membrane impairment. The modern interdisciplinary research has been successful in furthering our comprehension of the various chemotypes of essential oils as well as improving our insight on designing active compounds for use as antimicrobial agents and as alternatives to antibiotics. Here, we have briefly reviewed the chemical principles that underlie the antibacterial activity of some promising essential oils. We have also discussed the pros and cons of preferring compounds for specific microbial targeting. Further, we have emphasized on the possible steps to catalogue and leverage this uncharted fraction of the study of antibacterial properties of essential oils.

12.1 Introduction

Plants are reservoir of biologically active compounds with therapeutic properties has been well established (Egamberdieva et al. 2016) due to their reliable pharmacological actions and affordability to common people which makes them effective in control of various diseases (Vikram et al. 2014). The identification and scientific exploration of phytochemicals is still a fascinating field in therapeutic world as

S. Pattnaik (✉)
Department of Biotechnology and Bioinformatics, Sambalpur University, Burla, Odisha, India

N. Behera
School of Life Sciences, Sambalpur University, Burla, Odisha, India

A. Varma et al. (eds.), *Plant Biotic Interactions*,
https://doi.org/10.1007/978-3-030-26657-8_12

demonstrated by the recent paradigms of the anticancer drug epothilone, the immune modulator rapamycin or the proteasome inhibitor salinosporamide, to name but a few of hundreds of possible examples (Dickschat 2011). The pharmacological or clinical characterization of several known plant-derived compounds like the taxanes, artemisinin derivatives or boswellic acids, tacrolimus, ascomycin, betulinic acid and shikonin was significantly broadened (Heilmann 2010). Unexpected functions of known secondary metabolites are continuously being unraveled and are fulfilling some of the needs of present-day medicine and show great promise for the future (Demain and Vaishnav 2011). As a result of increasing interest in ethnic and traditional phytotherapeutics (Zielińska and Matkowski 2014), medicinal plants are the most important source of life-saving drugs for the majority of the world's population (Siahsar et al. 2011).

For a number of these plant drugs, controlled clinical trials have proven their efficacy and can thus be prescribed in evidence-based medicine (Bellamy et al. 1992; Ramawat and Goyal 2009; Van Wyk et al. 2008; Rates 2001; Schulz et al. 2002; Katiyar et al. 2012; Wink 2015). For several hundred medicinal plants, reported from corners of the world, monographs have been published, in which the therapeutic evidence has been synchronized like The German Commission E, the European Pharmacopoeia (PhEur), Proceedings from a symposium held at The Norwegian Academy of Science and Letter (2008), the European Scientific Cooperative on Phytotherapy (ESCOP), the World Health Organization (WHO monographs), and the European Medicines Agency-Herbal Medicinal Products. More recently, the term "medicinal and aromatic plants" (MAPs) has been used in a slightly broader sense, distinguishing the fragrant (aromatic, ethereal) ingredients containing group of medicinal plants (Mathe 2015).

12.1.1 Phytochemical Profile of Medicinal Plants

A distinct phytochemical entity of a medicinal plant is the phytochemical profile of natural products. Phytochemical constituents of medicinal plants encompass a diverse space of chemical scaffolds which are used for rational design of novel drugs (Mohanraj et al. 2018).

Phytochemicals vary in their chemistry but can be divided into hydrophilic and hydrophobic compounds, of which a wide variety of polyphenolic and terpenoid compounds, as well as alkaloids, carbohydrates and non-protein amino acids, invoke special interest (Acamovic and Brooker 2005). The Ministry of Environment, Forest and Climate Change, Government of India, has identified and documented over 9500 plant species considering their importance in the pharmaceutical industry (https://archive.india.gov.in). In addition, India has been considered as a treasure house of valuable medicinal and aromatic plant species.

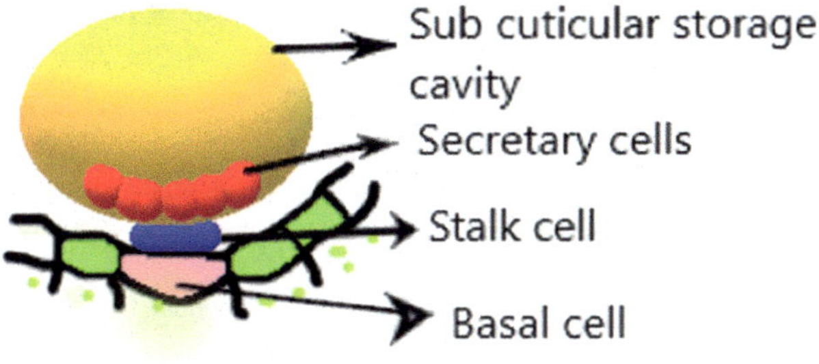

Fig. 12.1 The sub-cuticular oil storage cavity of a typical glandular cell (After Turner and Croteau 2004)

12.2 Phytochemical Profile of Medicinal and Aromatic Plants (MAPs)

The medicinal and aromatic plants (MAPs) are the biosynthetic factories of essential oils. The essential oils are biosynthesized in specialized cells types, such as osmophores, glandular trichomes and ducts and cavities, present on different parts of these plants (Rehman and Asif Hanif 2016). Figure 12.1 represents a pictorial sketch of a typical oil gland structure and its associated cells. Interest in the use of essential oils has enjoyed a massive revival in recent decades. As their name implies, the essential oils are volatile in nature. They differ entirely both physically and chemically from fixed oils. The major chemical components of essential oils include terpenes, esters, aldehydes, ketones, alcohols, phenols and oxides, which are volatile and may produce characteristic odour. Table 12.1 enlists some selected plants and their active constituents. This is pertinent to mention that different varieties of the same species may have different chemotypes (different chemical composition of the same plant species as a result of different harvesting methods or locations) and thus different types of effects (Pattnaik et al. 1995a, b; Pattnaik et al. 1997).

12.3 Phytochemical Profile of Essential Oils

In general, essential oils like other organic compounds can be subdivided into two distinct groups of chemical constituents: the hydrocarbons which are made up almost exclusively of terpenes (monoterpenes, sesquiterpenes and diterpenes) and the oxygenated compounds which are mainly esters, alcohols, ketones, alcohols, phenols and oxides (https://www.abundanthealth4u.com). Due to their structural relationship within the same chemical group, essential oil components are known to easily convert into each other by oxidation, isomerization, cyclization or dehydrogenation reactions, triggered either enzymatically or chemically (Turek and Stingzing 2013). In Fig. 12.2, there is depiction of various classes of essential oil components which are the key players in making an essential oil therapeutically active.

Table 12.1 The chemotypes of selected aromatic and medicinal plants

Composition	Examples of plants	Active constituents
Terpenes/ sesquiterpenes	*Pinus* sp., *Juniperus communis*	Terpenes, pinene, camphene, sesquiterpenes, cadiene
Alcohols	*Coriandrum sativum, Pelargonium graveolens, Cymbopogon martinii, Santalum album*	Linalool, geraniol, citronellol, santol
Esters and alcohols	*Lavandula officinalis, Rosmarinus officinalis, Mentha piperita*	Linalool, linanyl acetate, ethyl pentyl ketone, borneol, cineole, bornyl acetate, menthol, methyl acetate
Aldehydes	*Cinnamomum verum, Cinnamomum cassia, Citrus limon, Cymbopogon citrates, Eucalyptus citriodora*	Cinnamic aldehyde, euginol, citral, limonene, citronellol, citronellal
Ketones	*Mentha spicata, Carum carvi, Anethum graveolens, Salvia officinalis, Artemisia absinthium*	Carvone, limonene, thujone, camphor, cineole, thujyl alcohol, azuline
Phenols	*Cinnamomum verum, Syzygium aromaticum, Thymus vulgaris, Trachyspermum ammi*	Euginol, vanilin, thymol
Esters	*Pimpinella anisum, Foeniculum vulgare, Eucalyptus globulus*	Anethol, chavicol, cineole, apiole, myristin

12.4 Antibacterial Action of Essential Oils on Sensitive Bacterial Cells

With a surge in lethal infections and the antibiotic-resistant microorganisms, there has been dearth of the availability of proper antibiotics for the health treatment around the world. Often a desired antibiotic is going to be obsolete in curbing a pathogen, as a result of which, there has been a growing concern among the pharmaceutical industries for designing and production of antibiotics. This has necessitated an approach for the treatment of pathogenic microorganisms with alternative medicines. In this context, phytochemicals hold promise, as many of such have been reported to exhibit antimicrobial/antivirulent activities.

Essential oils are fundamentally different being volatile, organic chemicals that are distilled from various parts of plants, including seeds, bark, leaves, stems, roots, flowers and fruit. These are highly concentrated, beneficial end products of the plant's metabolism, communication and defence response. Essential oil constituents are lipid soluble and comparatively smaller molecules and have ability to easily penetrate the skin and travel throughout the body in a matter of minutes when applied topically. They can penetrate the blood–brain barrier and access the central nervous system. They have been observed to interact with various receptor sites in the brain, such as those for GABA and glutamate (Aoshima and Hamamoto 1999: Elisabetsky et al. 1999). Many are known to possess significant antimicrobial activity against a wide range of microorganisms. Elucidation of the mechanism of action of these compounds may lead to identification of new antibiotic targets. Such

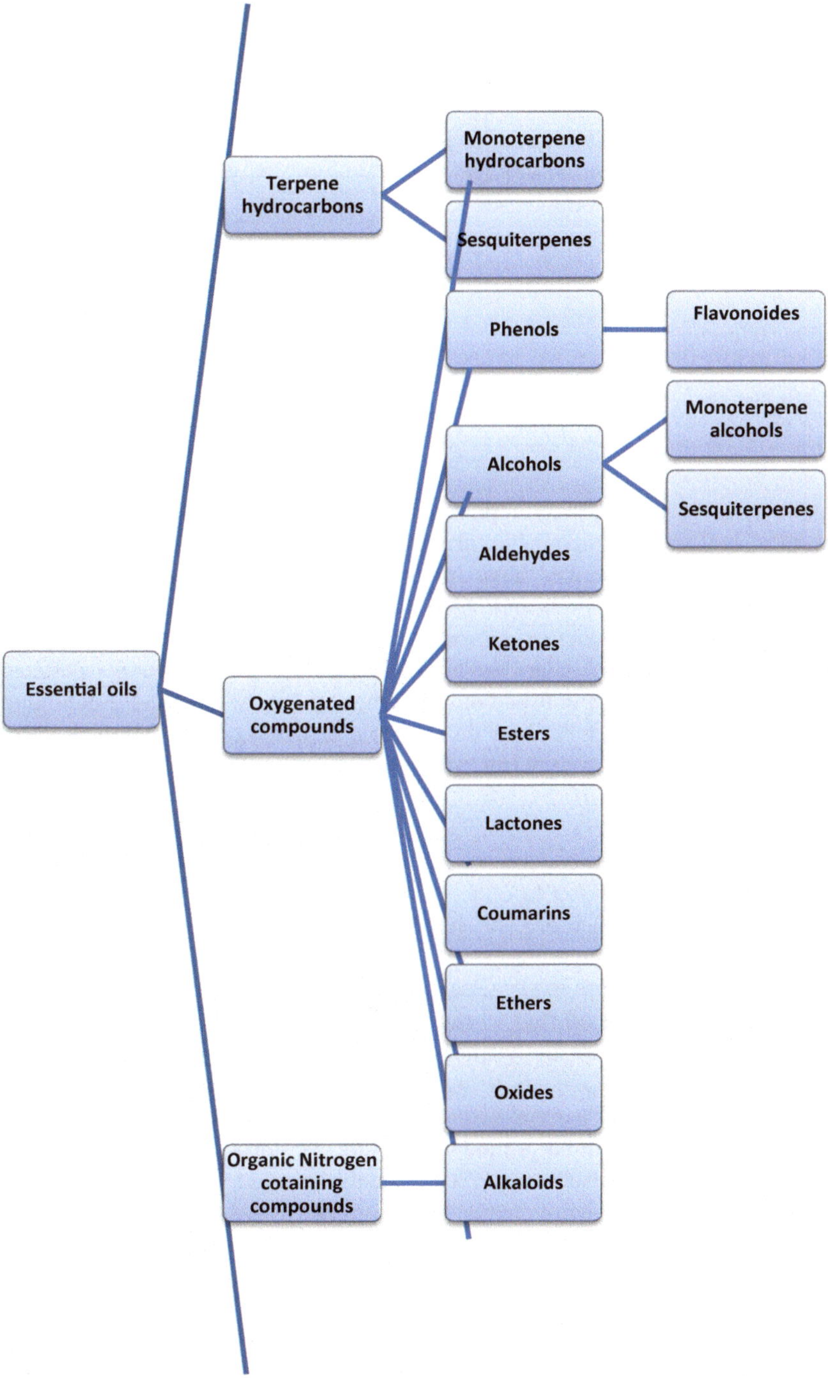

Fig. 12.2 The chemical components making the essential oils therapeutically active

Table 12.2 Essential oil components and chemotypes

Names of compounds	Class of chemotype
1. Geraniol	A monoterpene alcohol
2. Citronellol	A monoterpene alcohol
3. Menthol	Cyclic terpene alcohol
4. Linalool	Terpene alcohol
5. Thymol	Monoterpene phenol
6. Eucalyptol (1,8 cineole)	Cyclic ether and a monoterpene
7. Euginol	Phenylpropropene (allylbenzene)
8. Citral a (geranial)	Aliphatic terpene aldehyde
9. Citral b (neral)	Aliphatic terpene aldehyde
10. Citronellal	Aliphatic terpene aldehyde
11. D-Limonene	Cyclic monoterpene
12. β-Pinene	Monoterpene

targets, once identified, may represent biosynthetic or regulatory pathways not currently inhibited by available drugs (Boire 2013). Target-identification and mechanism-of-action studies have important roles in small-molecule probe and drug discovery. Whether it is an FDA-approved drug with an unknown mechanism of action, a natural product or a novel small molecule identified from a laboratory screen, researchers must understand how a molecule of interest exerts its effects. As a plethora of reports are published on EOs emphasizing its remarkable antimicrobial properties, this chapter will focus on the topic of mode of interaction between the active constituents of essential oils with putative cellular targets of some medically important bacterial strains.

Among the various essential oil components identified so far, terpene alcohols, terpene aldehydes and monoterpenes (Table 12.2) are observed to be explored more scientifically in comparison to other compounds. Hence, this communication will highlight the possible mode of interaction between these components with antibacterial targets.

From various literature works (Pattnaik et al. 1995a, b, 1997, 2010, 2017), it was found that the essential oil (EO) components enlisted in Table 12.2 have sufficient druggability, druglikeness and leadlikeness required for a prodrug (Pattnaik 2018). Therefore, the EO components being smaller in size and lipophilic in nature can enter inside a sensitive bacterial cell and engage with target molecules.

Figure 12.3 depicts about a model, showing the absorption of EO components with staphylococcal cells (Pattnaik 2018). In this context, the nodes of this chapter are the key essential oil components labelled by their main target or mode of action like cell rupture, cell lysis, membrane blebbing, loss of nucleic acid and cytoplasmic inclusions, filamentation, spheroplast formation, presence of degenerate chromosome, capsular degradation, biofilm degradation, etc. There is a critical analysis, discussion, explanation and more over implementation of hypotheses on the possible mode of action of efficient essential oils and their components against susceptible microbial strains. For the purpose of findings of other researchers as well as self-studied results are included in this report.

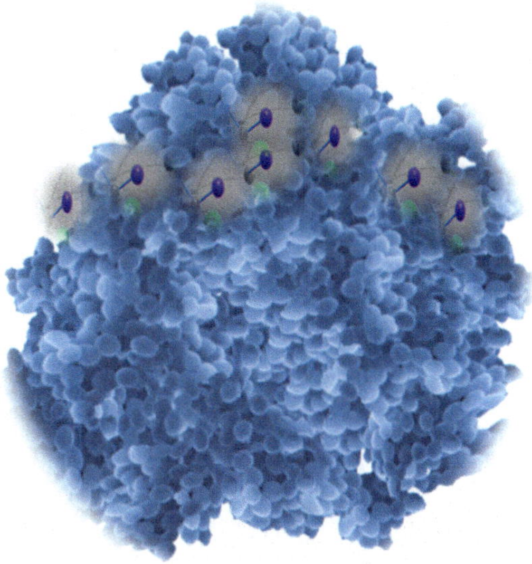

Fig. 12.3 A model showing the interaction of essential oil components with staphylococcal cells

12.5 Effect of EO Components on the Bacterial Morphology

Here is a presentation of finding of an in vitro study carried out to decipher the mechanism of action of linalool against cells of *Escherichia coli*. The cells were grown in nutrient agar media and were subjected to quantitative assay with various concentrations of linalool using a "tube dilution method" (Pattnaik et al. 1995a, b). The cells grown in subinhibitory concentration (SIC) of linalool were visualized under electron microscope (Central Instrument Facility, North East Hill University, Shillong), and the images are depicted in Fig. 12.4. The normal rod-shaped *E. coli* cells were observed to be 1–1.5 μm in length (a), while there was presence of swollen, L-shaped, polymorphs and more over round-shaped cells (b, c and d) in smeared slides.

In addition, Fig. 12.5 is displaying an injured cell with severe cellular punctures and disruptions. Likewise, Fig. 12.6 is the SEM image of *Escherichia coli* cells in presence of linalool. There is illustration of degraded bacterial cells (indicated with arrows) along with normal cells.

The mentioned piece of work is quite supportive about the observations made by other authors. Based upon these reports, it may be hypothesized that the essential oils and/or its major or minor components have a modulating effect on susceptible bacterial cells. They may cause bacterial membrane damage, blebbing and more over leakage of nuclear material along with cytoplasmic contents (Fig. 12.7) provoking the bacterial cell to die. The component eucalyptol could have either degraded or removed the cell wall of *E. coli* cells; thus there was presence of large spheroplasts (Fig. 12.8) instead of normal rods. Therefore, the mode activity of eucalyptol may be comparable with β-lactam antibiotics.

As per the report of Jones et al. (2013), pathogenic and non-pathogenic foodborne bacteria had the ability to form filaments in microbiological growth media and foods

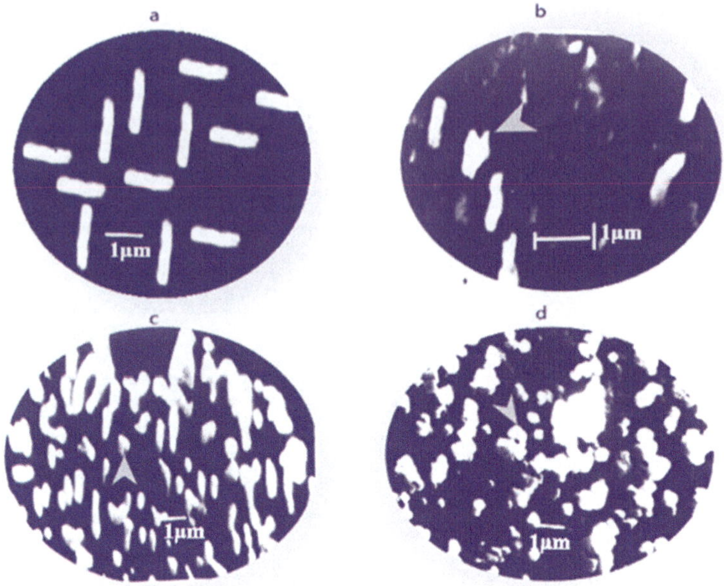

Fig. 12.4 Cellular deformity in *Escherichia coli* cells in presence of SIC of linalool

Fig. 12.5 Cellular puncture and cell wall disruption in *Escherichia coli* cell in presence of SIC of linalool

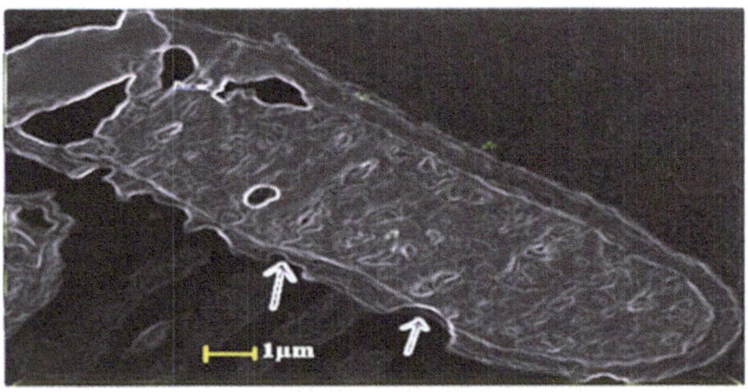

after prolonged exposure to sublethal stress or marginal growth conditions. In many cases, nucleoids were evenly spaced throughout the filamentous cells, but septa are not visible, indicating that there was a blockage in the early steps of cell division but the mechanism behind filament formation is not clear. The formation of filamentous cells appeared to be a reversible stress response. When filamentous cells were exposed to more favourable growth conditions, filaments divide rapidly into a number of individual cells. Further, they had added that the filament formation under a number of sublethal stresses might be linked to a reduced energy state of bacterial cells.

At this point, some interesting images taken from the optical (Division of Microbiology, Silchar Medical College, Silchar, Assam, Laboratory of Medical Microbiology, School of Life Sciences, Sambalpur University, Sambalpur) as well as scanning electron microscopic studies (NEHU, Shillong, Meghalaya) are given here to substantiate the observations reported by other researchers.

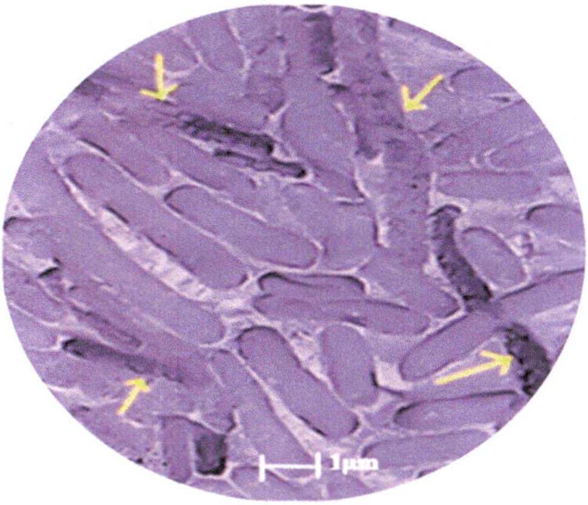

Fig. 12.6 The SEM image of *E. coli* cells in presence of subinhibitory concentration (SIC) citronellol. NB: The arrows are indicating the cellular degradation

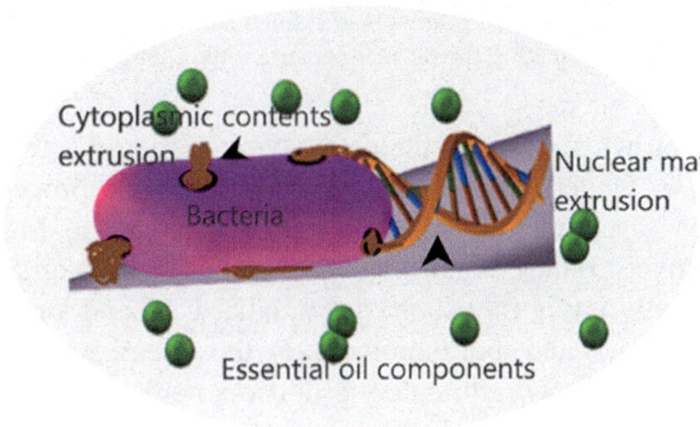

Fig. 12.7 A diagram showing the leakage of cellular contents along with nucleic acid due to effect of essential oil components

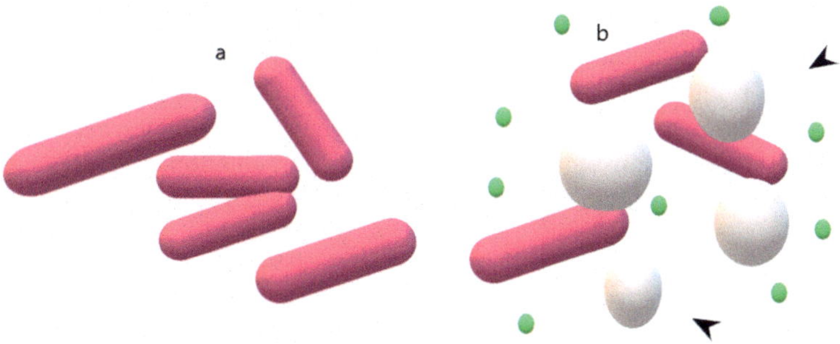

Fig. 12.8 A hypothetical view of the conversion of rod-shaped *E. coli* cells into spheroplasts in presence of essential oil components

In the Fig. 12.9, it was seen that the normal *Bacillus subtilis* cells were of 1–1.5 μm in length as observed under optical microscope at a magnification of 1000X (a). While in the presence of SIC of cinnamaldehyde, the cells were observed to be much elongated. The possible mechanism of action of cinnamaldehyde may be explained here. The Gram-positive *Bacillus subtilis* exhibits an additional mode of

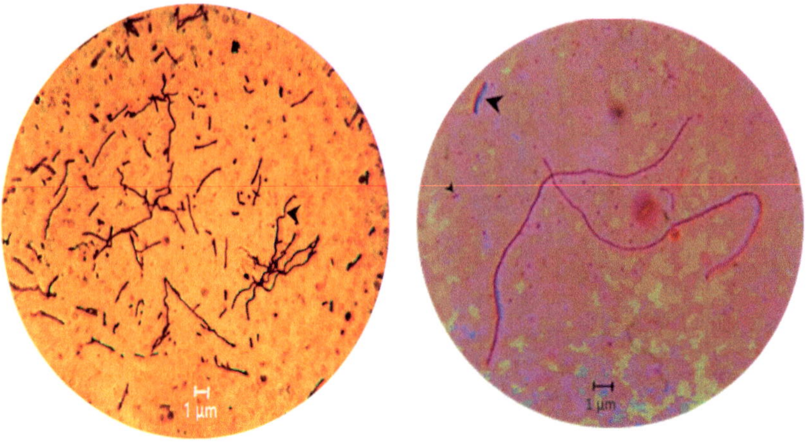

Fig. 12.9 Formation of filaments in stained *Bacillus subtilis* cells in presence of cinnamaldehyde. NB: Arrows indicating the normal bacteria with a length of 1.5 µm

division when it enters the pathway to sporulate regulated by the sporulating gene, *SpoOM*. Under such conditions, it undergoes a process of asymmetric division in which a septum is formed at a polar (Latch and Margolin 1997) in presence of mitomycin rather than a medial position. The polar septum divides the cell asymmetrically into a fore spore (the smaller cell) and a mother cell (Kemp et al. 2002). Cell division and sporulation require the assembly of a septum to initiate the cascade of events that will ultimately produce a new cell: a daughter cell in the case of cell division and a spore in the case of sporulation. Vega-Cabrera et al. (2017) had provided evidence that *SpoOM* interacts with cytoskeletal proteins involved in cell division. Therefore, it may be hypothesized that cinnamaldehyde, as an effector molecule, could have interacted with the putative SpoOM protein, thereby regulating the septum formation between the forespore and mother cell. As a result there was neither sporulation nor septum formation between the daughter and mother cell. Figure 12.10 also demonstrates the induction of filamentation in *E. coli* cells, when exposed to citral, a component of lemongrass oil.

12.6 Mechanism of Induction of Filamentation in Bacillary Form of Bacteria

Shape is a selectable feature that aids survival. Young (2006) had spelled out the physical, environmental and biological forces that favour different bacterial morphologies and which, therefore, contribute to natural selection. Specifically, cell shape is driven by eight general considerations: nutrient access, cell division and segregation, attachment to surfaces, passive dispersal, active motility, polar differentiation, the need to escape predators and the advantages of cellular differentiation. Elongating increases a cell's uptake-proficient surface without changing its surface-to-volume ratio

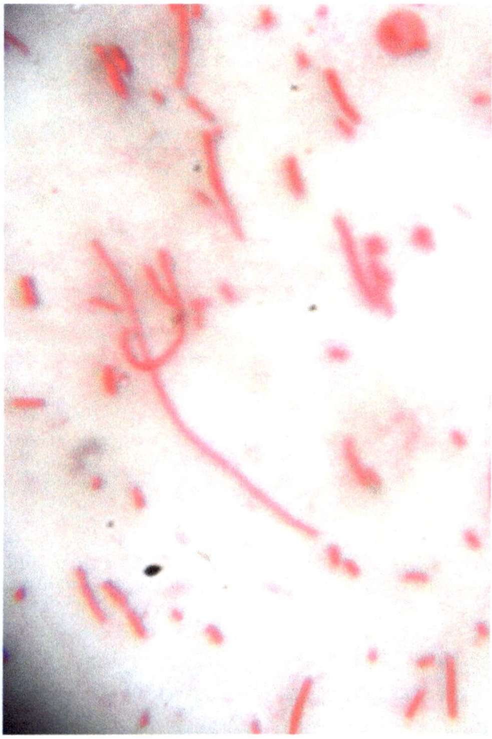

Fig. 12.10 Showing the induction of *E. coli* cells (negatively stained) in presence of citral. A typical elongated form is measured as 127 μm in comparison to the length of a normal *E. coli* cell (1–1.5 μm)

appreciably. Filamentation may have beneficial effects on cells, because there is increase of total surface area to absorb more nutrition during stress condition. Additional benefit may be keeping itself in direct contact with solid medium.

Surveys of populations of bacterial pathogens have revealed a clear link between environmental stress mediated by the immune system and antibiotics and the prevalence of mutators (Torres-Barcelo et al. 2013). It reports that *sufI* (cell division protein renamed as FtsP) is essential for the viability of *E. coli* cells subjected to various forms of stress, including oxidative stress and DNA damage. The *sufI* mutant also exhibits *sulA*-independent filamentation. Cell division protein that is required for growth during stress conditions might have involved in protecting or stabilizing the divisional assembly under conditions of stress.

FtsZ is one of several proteins essential for cell division in *Escherichia coli* and acts at a very early stage of cytokinesis before mid-cell constriction is visible. FtsZ is a GTPase (Sun et al. 2014) and a tubulin homologue (Vedyaykin et al. 2016). FtsZ is the key protein of bacterial cell division and an emergent target for new antibiotics. It is a filament-forming GTPase and a structural homologue of eukaryotic tubulin (Schaffner-Barbero et al. 2012).

Keeping this view in mind and listing the in vitro-based self-observations, it may be suggested that the essential oil components being lipophilic in nature can enter into a susceptible bacterial cell and are able to interact with the pilot protein FtsZ, responsible

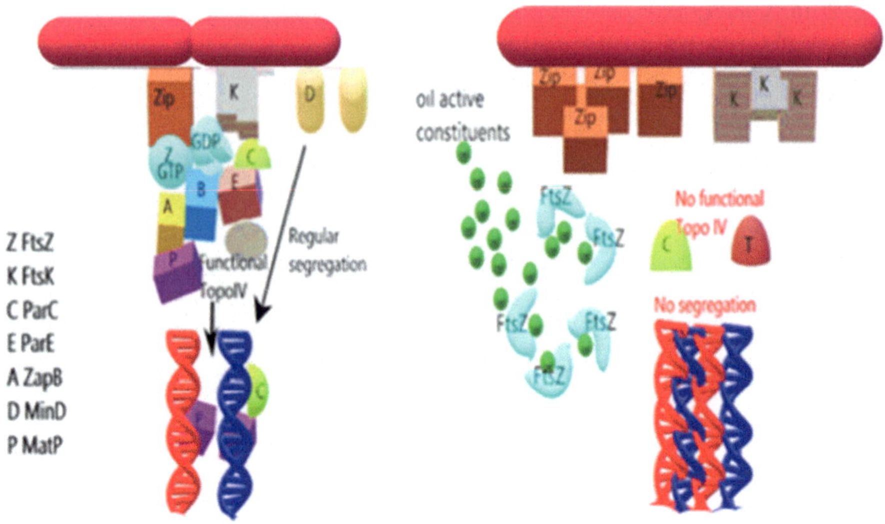

Fig. 12.11 The schematic representation of possible mode of action of essential oil components to cause filamentation by interacting with FtsZ proteins

for cell septa formation. Further, it may be added here that prior to cell division, a bacterial cell is ready to lay down a FtsZ ring with the help of a team of proteins, namely, ftsK, ParC, ParE, ZapB, MinD and MatP, which constitute a functional topoisomerase IV. As a result, there is a uniform segregation of DNA molecules to daughter cells after the separation phase. By taking active acquisition of information from the primary observations regarding induction of filamentation of *E. coli* as well as *B. subtilis* cells in presence of respective essential oil components, citral a and cinnamaldehyde, menthol, etc., an assumption can be made here. There may be possibility of interaction between active constituents of essential oils and FtsZ proteins leading to dessolution of team work of proteins responsible for septa formation. This event may lead to production of a non-functional topoisomerase IV, and finally there may not be any segregation of doubled-up DNA molecules. And there is induction of multi-chromosomal filament forms which are observable in culture media. Figure 12.11 is the schematic representation of possible mode of action of essential oil components to cause filamentation by interacting with FtsZ protein.

Research is in progress to evaluate the antibacterial efficacy of an array of essential oil and its components against coccus form of bacterial strains having clinical consequences. Results of self-studied experiments, had inferred about drastic effect of essential oil like lemongrass, peppermint, palmarosa, eucalyptus, lantana, patchouli, citronella, kewda, etc., oil had remarkable bacteria killing effect on cells of various strains of *Staphylococcus aureus*, *Staphylococcus saprophyticus* and more *Staphylococcus haemolyticus*. Figure 12.12 is showing bacteria killing effect of thymol against the cells of *Staphylococcus haemolyticus*.

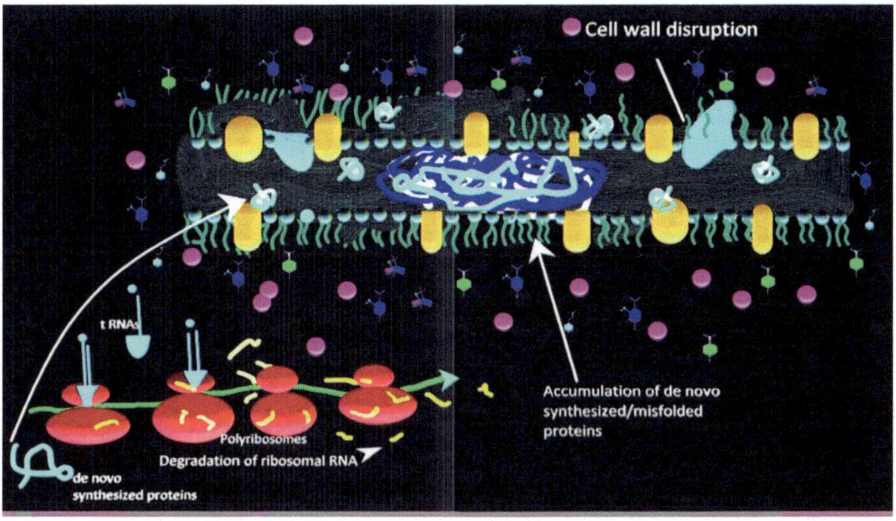

Fig. 12.12 The putative role of essential oil components for accumulation of de novo synthesized and/or misfolded proteins and degradation of ribosomal RNA

12.7 Modulation of EO Components Towards Accumulation of Misfolded Proteins and Degradation of RNA

There is an intense interest in exploring roles of heat shock proteins in pathogen biology to exploit pathogen chaperones as drug targets (Neckers and Tatu 2008). Only folded proteins are functional; protein folding in bacteria is tightly controlled genetically and transcriptionally and at the protein sequence level (Sabate et al. 2010). Outer membrane proteins were upregulated in treated cells; indeed the bacterial envelope stress response is trigged by the accumulation of misfolded outer membrane proteins (Maguire et al. 2002). However, bacteria have evolved sophisticated quality control systems consisting primarily of chaperones and proteases that exert multiple activities such as (1) de novo folding of newly synthesized proteins, (2) preventing aggregation of unfolded proteins, (3) removing of terminally misfolded proteins by degradation and (4) resolubilizing protein aggregates for subsequent refolding or degradation (Mogk et al. 2011). Results demonstrated rRNA degradation by the carvacrol/cinnamaldehyde mixture, and this effect was due to carvacrol.

Figure 12.12 is the graphic representation of putative role of essential oil components for accumulation of de novo synthesized or misfolded proteins in the cell. The origin of misfolded proteins may be due to action of essential oil components; hence there is every possibility that these misfolded proteins would accumulate in the cell. Because of accumulation of more and more number of proteins inside, cell may disturb the regular function of de novo synthesized proteins. In addition, as reported by these authors, there is also a possibility of degradation of RNA due to activity of essential oil components.

12.8 Conclusion and Future Prospective

In a nutshell, it may be concluded that the EO components have enough potentiality to enter a bacterial cell and can interact with target molecules, thus inhibiting bacterial growth. Although a handful of research work is in progress to decipher the antimicrobial action of EO components on microbes of clinical interest, still more in-depth studies are required to place EO components at par with chemotherapeutic agents. The glaring lacuna behind several studies lies in consistency of the data. The reason behind may be due to insufficient correlation between in vitro and in vivo assays. In this debate, the emerging molecular docking studies between the drug and target component are expected to provide conclusive evidence about the antimicrobial efficacy of essential oils.

Acknowledgement This chapter is part of DSc thesis (2018) submitted to Sambalpur University, Odisha, India.

References

Acamovic T, Brooker JD (2005) Biochemistry of plant secondary metabolites and their effects in animals. Proc Nutr Soc 64:403–412

Aoshima H, Hamamoto K (1999) Potentiation of GABA receptors expressed in *Xenopus oocytes* by perfume and Phytoncid. Biosci Biotechnol Biochem 63(4):743–748

Bellamy W, Takase M, Wakabayashi H, Kawase K, Tomita M (1992) Antibacterial spectrum of Lactoferricin B, a potent bactericidal peptide derived from the N-terminal region of bovine Lactoferrin. J Appl Bacteriol 73(6):472–479

Boire NA (2013) Essential oils and future antibiotics: new weapons against emerging 'superbugs'? J Anc Dis Prev Rem 01(02):1–5

Demain AL, Vaishnav P (2011) Natural products for cancer chemotherapy. Microb Biotechnol 4 (6):687–699

Dickschat JS (2011) Biosynthesis and function of secondary metabolites. Beilstein J Org Chem 7:1620–1621

Egamberdieva D, Mamedov N, Ovidi E, Tiezzi A, Craker L (2016) Phytochemical and pharmacological properties of medicinal plants from Uzbekistan: a review. J Med Act Plants 5(2):59–75

Elisabetsky E, Brum LFS, Souza DO (1999) Anticonvulsant properties of linalool in glutamate-related seizure models. Phytomedicine 6(2):107–113

Heilmann J (2010) New medical applications of plant secondary metabolites. In: Wink M (ed) Annual plant reviews, vol 39, (Functions and biotechnology of plant secondary metabolites, 2nd edn. Wiley Blackwell, Oxford

Jones TH, Vail KM, McMullen LM (2013) Filament formation by foodborne bacteria under sublethal stress. Int J Food Microbiol 165(2):97–110

Katiyar C, Gupta KS, Katiyar S (2012) Drug discovery from plant sources: an integrated approach. Ayu 33(1):10–19

Kemp JT, Driks A, Losick R (2002) Fts A mutants of *Bacillus subtilis* impaired in sporulation. J Bacteriol 184(14):3856–3863

Latch JN, Margolin W (1997) Generation of buds, swellings, and branches instead of filaments after blocking the cell cycle of *Rhizobium meliloti*. J Bacteriol 179:2373–2381

Maguire M, Coates ARM, Henderson B (2002) Chaperonin 60 unfolds its secrets of cellular communication. Cell Stress Chaperones 7(4):317–329

Mathe A (2015) Introduction, utilization/significance of medicinal and aromatic plants. In: Mathe A (ed) Medicinal and aromatic plants of world. Springer, Dordrecht, pp 1–12

Mogk A, Huber D, Bukau B (2011) Integrating protein homeostasis strategies in prokaryotes. Cold Spring Harb Perspect Biol 3(4):1–19

Mohanraj K, Karthikeyan BS, Vivek-Ananth RP, Bharath Chand RP, Aparna SR, Mangalapandi P, Samal A (2018) IMPPAT: a curated database of Indian Medicinal Plants, Phytochemistry and Therapeutics. Sci Rep 8:4329

Neckers L, Tatu U (2008) Molecular chaperones in pathogen virulence: emerging new targets for therapy. Cell Host Microbe 4(6):519–527

Pattnaik S (2018) Plant derived essential oils are the inducers of stress in bacterial cells of clinical relevance. In: Rakshit A, Tripathi VK, Chandola VK, Singh A, Sekhar S, Sarkar DR (eds) Innovative approach of integrated resource management. New Delhi Publishers, New Delhi, pp 163–173

Pattnaik S, Subramanyam VR, Kole CR, Sahoo S (1995a) Antibacterial activity of essential oils from Cymbopogon: inter- and intra-specific differences. Microbios 84(341):239–245

Pattnaik S, Subramanyam VR, Rath CC (1995b) Effect of essential oils on the viability and morphology of *Escherichia coli* (SP-11). Microbios 84(340):195–199

Pattnaik S, Subramanyam VR, Bapaji M, Kole CR (1997) Antibacterial and antifungal activity of aromatic constituents of essential oils. Microbios 89(358):39–46

Pattnaik S, Padhan DK, Jana GK (2010) Evaluation of cinnamon oil, peppermint oil, cardamom oil & Orange oil as antimicrobial agents. J Pharm Res 3(2):414–416

Pattnaik S, Behera SK, Mohapatra N (2017) Homology modeling of FtsZ protein from virulent bacterial strains and its interaction with eucalyptol: an In silico approach for therapeutics. Bioinformatics 1:24870

Ramawat KG, Goyal S (2009) Indian herbal drugs scenario in global perspectives. In: Ramawat KG, Merillon JM (eds) Bioactive compounds and medicinal plants. Springer, Heidelberg, p 323

Rates SM (2001) Plants as source of drugs. Toxicon 39(5):603–613

Rehman R, Asif Hanif M (2016) Biosynthetic factories of essential oils: the aromatic plants. Nat Prod Chem Res 04(04):227

Sabate R, De Groot NS, Ventura S (2010) Protein folding and aggregation in bacteria. Cell Mol Life Sci 67(16):2695–2715

Schaffner-Barbero C, Martín-Fontecha M, Chacón P, Andreu JM (2012) Targeting the assembly of bacterial cell division protein FtsZ with small molecules. ACS Chem Biol 7(2):269–277

Schulz B, Boyle C, Draeger S, Ro A, Krohn K (2002) Endophytic fungi: a source of novel biologically active secondary metabolites. Mycol Res 106(September):996–1004

Siahsar B, Rahimi M, Tavassoli A, Raissi A (2011) Application of biotechnology in production of medicinal plants 1. Am J Agric Environ Sci 11(3):439–444

Sun N, Chan FY, Lu YJ, Neves MAC, Lui HK, Wang Y et al (2014) Rational design of berberine-based FtsZ inhibitors with broad-spectrum antibacterial activity. PLoS One 9(5):e97514

Torres-Barcelo C, Cabot G, Oliver A, Buckling A, MacLean RC (2013) A trade-off between oxidative stress resistance and DNA repair plays a role in the evolution of elevated mutation rates in bacteria. Proc R Soc B Biol Sci 280(1757). https://doi.org/10.1098/rspb.2013.0007

Turek C, Stingzing FC (2013) Stability of essential oils: a review. Compr Rev Food Sci Food Saf 12:40–51

Turner GW, Croteau R (2004) Organization of monoterpene biosynthesis in Mentha. Immunocytochemical localizations of geranyl diphosphate synthase, limonene-6-hydroxylase, isopiperitenol dehydrogenase, and pulegone reductase. Plant Physiol 136(4):4215–4227

Van Wyk BE, de Wet H, Van Heerden FR (2008) An ethnobotanical survey of medicinal plants in the southeastern Karoo, South Africa. S Afr J Bot 74(4):696–704

Vedyaykin AD, Vishnyakov IE, Polinovshaya VS, Khodorkovskii MA, Sabansev AV (2016) New insights into FTsZ rearrangements during the cell division of *Escherichia coli* from single molecule localization microscopy of fixed cells. Microbiologyopen 5(3):378–386

Vega-Cabrera LA et al (2017) Analysis of Spo0M function in *Bacillus subtilis*. PLoS One 12 (2):1–24. https://doi.org/10.1371/journal.pone.0172737

Vikram P, Chiruvella KK, Ripain IHA, Arifullah M (2014) A recent review on phytochemical constituents and medicinal properties of Kesum (*Polygonum minus* Huds.). Asian Pac J Trop Biomed 4(6):430–435

Wink M (2015) Modes of action of herbal medicines and plant secondary metabolites. Medicines 2 (3):251–286

Young KD (2006) The selective value of bacterial shape. Microbiol Mol Biol Rev 70(3):600–703

Zielińska S, Matkowski A (2014) Phytochemistry and bioactivity of aromatic and medicinal plants from the genus Agastache (Lamiaceae). Phytochem Rev 13(2):391–416

Chapter 13
Enhancement of Active Constituents of Medicinal Plants Through the Use of Microbes

Charu Gupta and Dhan Prakash

Abstract The traditional system of medicine plays a pivotal role in primary health care for the prevention and treatment of various ailments. Plants are an important source of biologically active compounds with therapeutic properties. These active compounds are commonly called as phytoconstituents or phytochemicals and are of significant importance in the efficacy of medicinal plants. These are mainly produced in those plants which are under stressful conditions. The amount of phytochemicals varies from plant to plant, and their concentration depends on location, local climate, and types of soil. Medicinal plants are normally collected from their natural and wild habitat, but now due to their growing market demand by the pharmaceutical industries, they are also cultivated. However, the cultivated plants may have lower efficacy due to the presence of lower amounts of active ingredients. This chapter focuses on the role and interactions of microbes with medicinal plants, impact on their functional efficacy, and possible synergistic role in quantitative enhancement of bioactive phytoconstituents.

13.1 Introduction

It is believed that about 7600 plants are used in traditional health-care system and are very popular mainly in rural and tribal areas of India. The plant-based customary medication systems play a crucial role in the well-being of masses, in particular the economically weaker segment of the society. Nowadays, the medicinal herbal plants, herbal health-care products, drugs, dietary supplements, nutraceuticals, cosmeceuticals, and functional foods are in great demand, and their market comprises more than 50% of all the allopathic medicines in medical use. The sources of these herbal drugs are either directly or indirectly from natural sources such as florae, faunas, microbes, and molds.

According to an estimate of the National Medicinal Plant Board (NMPB), the business of Indian herbal industry may rise up to an order of Rs. 80–90 billion by

C. Gupta (✉) · D. Prakash
Amity Institute of Herbal Research & Studies, Amity University Uttar Pradesh, Noida, Uttar Pradesh, India

A. Varma et al. (eds.), *Plant Biotic Interactions*,
https://doi.org/10.1007/978-3-030-26657-8_13

2020. Besides this, India is promoting the Traditional System of Indian Medicine as a part of AYUSH (Ayurveda, Yoga, Unani, Siddha, and Homeopathy) mandate in healthiness upkeep segment through overall links (Sarker and Nahar 2007).

Plants are a pool of naturally dynamic complexes with healing properties and have been used globally to deal with several diseases. Successful cultivation of medicinal plants depends on biological and geographic factors which can control the composition of secondary metabolites, essential oil, and the usage of organic nourishments and pesticides should be avoided. Innovative biotechnological methodologies such as the usage of root-associated beneficial microbes which are able to stimulate plant development and nutrient uptake and defend florae against numerous soil-borne pathogens can benefit florae to acclimatize to a number of ecological stresses and have vital importance in medicinal plant research. It is very significant to test and develop the important medicinal plants for their active phytochemical constituents. The active phytoconstituents are of significant importance in the efficacy of medicinal plants. The active pharmaceutical ingredient (API) content varies in the concentration with location, local climate, and types of soil. Chemically, the phytochemicals include alkaloids, tannins, flavonoids, phenol, phytosterols, and terpenes that are present in active extracts (Palombo 2006; Van Wyk and Wink 2004; Prasad et al. 2008). These active constituents may be present in any part of the plant including leaves, seeds, rhizomes (roots), bark, stem, flowers, and fruits (Doughari et al. 2009). The plants produce a wide variety of phytoconstituents to survive and grow in their natural habitat and to overcome any stressful environmental conditions such as salinity, water deficiency, high temperature, low temperature, and nutrient limitations to name a few (Wu et al. 2007). These complexes defend plants against bacterial contaminations or infestations by pests. These phytoconstituents also provide protection to the plants from the attack of wild animals (Kaufman et al. 1999).

The phytochemicals are generally categorized under plant secondary metabolites and are valuable source of pharmaceuticals and nutraceuticals. Therefore, it is essential to promote the cultivation of medicinal plants (Phillipson 2001). But the successful cultivation of medicinal plants rests on both biological and geographical factors which can either enhance or decrease the amount of active phytoconstituents including essential oil composition and its yield, in case of aromatic plants (Juliani et al. 2006). It is also imperative to shun the usage of synthetic fertilizers and pesticides in the farming of plants since they are naturally consumed without being further treated after harvest (Banchio et al. 2008). There are various studies carried out in the past that show a positive impact of microorganisms on the enrichment of bioactive constituents in medicinal and aromatic florae.

13.2 Microorganisms and Plant Growth

Intensive cropping and current agricultural practices in India have resulted in severe nutrient deficiency of soil. Though various methods are being employed to manage nutrient loss in agricultural lands, all are costly and practically difficult as soil pH and chemical fixation by soil minerals determine their quantity of availability. To improve their mineralization and solubilization, microbial facilitated nutrient supervision is an eco-friendly and cost-effective method. This can be further explained by taking an example of phosphorus (P), which is an important component of soil. It serves as macronutrient for plants and plays key functions in their growth and progress (Sharma et al. 2013). Despite its abundance in agricultural soil, phosphorus has a growth-restrictive feature because of its slow diffusion and high fixation rate in soil (Miller et al. 2010). Microorganisms regulate P mineralization by various modes of action such as production of extracellular phosphatases, secretion of organic acids, and production of siderophores to enhance phosphorus availability by chelating cations such as iron, aluminum, or calcium that are intricate in the development of insoluble phytates (Singh et al. 2011). Endophytes possess the potential to mineralize complex and insoluble forms of macro- and micronutrients by secreting extracellular hydrolytic enzymes and enhance the accessibility of nutrients to florae. Several reports are available on the phosphate-solubilizing potential of endophytic bacteria and their positive effect on herb development (Oteino et al. 2015). Microbes mediate the nutrient cycling in soil through the secretion of extracellular enzymes such as amylase, protease, lipase, cellulase, chitinase, phosphatase, urease; thereby playing a significant role in mineralization of organic compounds (Das and Varma 2011). The activity of soil enzymes has been used as an indicator of soil value and strength (Badiane et al. 2001).

The impact of endophytes in plant development, nutrient accessibility, produce, and value of medicinal plants is validated by many researchers, and there are growing interests in the use of endophytes for the cultivation of medicinal plants without using synthetic agrochemicals (Keû et al. 2015). An extensive diversity of endophytic bacteria and fungi has been recognized that have great importance in plant nutrient acquirement and secondary metabolite modification of medicinal plants (Tiwari et al. 2010; Qadri et al. 2013).

The microflora present in rhizosphere has been widely considered as they are advantageous for growth and health of plants. They majorly belong to the group of nitrogen-fixing bacteria, mycorrhizal fungi, plant growth-promoting rhizobacteria (PGPR), biocontrol microorganisms, mycoparasitic fungi, and protozoa. Rhizosphere microbes stimulate plant growth and defend plants from pathogen attack by different modes of action (Lugtenberg and Kamilova 2009). Some of them are mainly addition of biofertilizers, stimulation of root growth, rhizo-remediation, control of abiotic stress, and disease control. These mechanisms are commonly observed in rhizobacteria belonging to the group *Proteobacteria* and *Firmicutes*, which includes *Pseudomonas* and *Bacillus*; fungi from the group *Deuteromycetes*, which includes *Trichoderma* and *Gliocladium*; and fungi from the order *Sebacinales*, which includes *Piriformospora* (Kogel et al. 2006; Qiang et al. 2012).

Similarly, rhizosphere microbes can also assist in the uptake of specific trace elements such as iron. Although iron is abundant in soil, under neutral to alkaline conditions it is present in the insoluble ferric oxide form, which is not accessible for microbial growth. In order to enhance the bioavailability and solubility of iron, many bacteria secrete a specific class of chemicals commonly known as siderophores (Hider and Kong 2010). On the plant side, the solubility of inorganic iron in the rhizosphere is increased by secreting phyto-siderophores that are subsequently transported back into the root tissue by a specific uptake system (Walker and Connolly 2008).

In rice cultivation, the solubility of iron is increased through the production of siderophores by fluorescent pseudomonads, which stimulate iron nutrition not only for Graminaceous plants but also for dicotyledonous plant species (Shirley et al. 2011). Besides this, a fungal siderophore rhizoferrin, secreted by fungus *Rhizopus arrhizus*, was found to be a proficient carrier of iron to plants with an efficiency comparable to that of synthetic chelating agents (Yehuda et al. 2000). Some of Rhizobacteria such as *Bacillus subtilis* GB03 are also able to activate the plant's own iron acquirement machinery (Zhang et al. 2009).

There are three other mechanisms to explain the microbial activity on enhancing plant growth: firstly by influencing the hormonal signaling of plants, secondly by repelling or outcompeting pathogenic microbial strains, and lastly by increasing the bioavailability of soil-borne nutrients (Jacoby et al. 2017).

13.3 Correlation Between Microbes and Plant Growth

The microflora that intensively colonizes the surface of root and its surroundings constitutes its rhizoplane and rhizosphere, respectively (Lugtenberg et al. 2001). The rhizosphere microorganisms are typically of two types: first, that establish the symbiotic association with plants and secondly, that are free living in the soil and root (Berg et al. 2013). Scientific studies have shown that beneficial rhizobacteria also known as plant growth-promoting rhizobacteria (PGPR) has the potential to increase the root and shoot growth, has better seed germination, increases nutrient uptake and yield, increases the plant stress tolerance, and is also able to control various plant pathogens and diseases (Jabborova et al. 2013).

Some notable examples of plant growth-promoting rhizobacteria are *Pseudomonas*, *Rhizobium*, *Azospirillum*, *Azotobacter*, *Bacillus*, *Bradyrhizobium*, *Burkholderia*, *Arthrobacter*, *Cellulomonas*, *Clostridium*, *Enterobacter*, *Flavobacterium*, *Micrococcus*, *Paenibacillus*, *Sinorhizobium*, and *Serratia* (Egamberdieva 2011; Egamberdieva et al. 2013). Besides bacteria, there are certain plant growth-promoting fungi—arbuscular mycorrhizal fungi (AM fungi)—that also improve plant growth and development through increasing the supply of mineral nutrients to plants, especially phosphorus (Hameed et al. 2014). They play a crucial role in controlling toxicity induced by salt stress, thus regularizing the uptake mechanism in plants by providing vital nutrients.

It was found that the application of various PGPR microbes to several plants stimulates and increases the content of secondary metabolites including total phenols, tannins, alkaloids, and lycopene and their antioxidant activity (Elango 2004). Furthermore, using combinations of PGPR and AM fungi on various plants also develops synergistic interactions that further increase their growth yield. This application also functions by increasing the uptake of vital mineral nutrients such as P, N, and K and some other minerals (Egamberdieva et al. 2010).

Several scientific studies have proved that rhizosphere microbes play a key role in the development of plant growth of medicinally important plants and increase the concentration of phytochemical bioactive constituents which are broadly used for curing several diseases (Bharti et al. 2013; Teixeira da Silva and Egamberdieva 2013). In another study, the effects of biofertilizers on growth, fruit yield, and oil composition of fennel plants were studied. A mixture of biofertilizer comprising of *Azotobacter chroococcum*, *Azospirillum lipoferum*, and *Bacillus megaterium* was applied mixed with 50% of the recommended dosage of NPK (chemical fertilizers). The main observations were increased vegetative growth in the form of plant height, branch number, fresh herb, and dry weight per plant as compared to the chemical fertilizer treatments alone (Mahfouz and Sharaf-Eldin 2007). In yet another study, the effects of vesicular-arbuscular (VA) mycorrhizal fungi (*Glomus mosseae* and *G. epigaeum*) on the growth, uptake of elements, and production of bioactive compounds in Chinese medicinal herb, *Datura stramonium*, under different concentrations showed that both species of VA mycorrhizal fungi improved the growth of *Datura stramonium*, increased the uptake of different elements, and enhanced the concentrations of hyoscyamine and hyoscine, the active phytochemicals (Geitang and Honggang 1989). Further, the role of endophytic bacteria in enhancing the nutrient level in soil was explored, and also its effect on growth of *Withania somnifera* was assessed. In this case, the growth-promoting potential of endophytic bacterial strain PSE-1 was explored which was isolated from the roots of *W. somnifera* and identified as *Pseudomonas* sp. on the basis of 16S rRNA gene sequencing. Results obtained suggested that endophytic bacteria *Pseudomonas* sp. plays a significant role in improving the nutrient level in soil and displayed positive effects on plant growth (Singh and Arora 2016). Furthermore, some of the endophytic fungi such as *Aspergillus* sp., *Fusarium* sp., and *Ramularia* sp. were isolated from *Rumex gmelini* Turcz (RGT). All of these three strains were observed to produce some similar bioactive secondary metabolites of their host. However, the ability to produce bioactive compounds decreased considerably when these fungi were cultured alone for a long time and was even difficult to recover. In order to obtain more bioactive secondary metabolites, the co-culture of tissue culture seedlings of *Rumex gmelini* Turcz and its endophytic fungi was established and RGT seedling was selected as a producer. Amongst these fungi, *Aspergillus* sp. showed the most noteworthy improvement on bioactive compound accumulation in RGT seedlings. These results showed that co-culture with endophytic fungi can significantly augment bioactive secondary metabolite production of RGT seedlings (Ding et al. 2018).

13.4 Enhancement of Phytoconstituents Through Application of Microbes

There are several research studies that showed the positive impact of application of plant growth-promoting bacteria (PGPB) and arbuscular mycorrhizal (AM) fungi on plant growth, nutrient uptake, and increase in the production of secondary metabolites such as phenols, flavonoids, alkaloids, saponin, and tannins present in medicinal plants. These studies have been carried out on *Catharanthus roseus*, *Origanum majorana*, *Matricaria chamomilla*, *Ocimum basilicum*, *Salvia miltiorrhiza*, *Mentha arvensis*, and *Withania somnifera* (Egamberdieva and Teixeira da Silva 2015). The synergistic combinations of some selected medicinal/aromatic plants and the beneficial effect of PGPB and arbuscular mycorrhizal (AM) fungi are compiled in Table 13.1 (Egamberdieva and Teixeira da Silva 2015).

In a research by Banchio et al. (2009) on *Ocimum basilicum* (sweet basil), it was found that when the plant was inoculated with *Bacillus subtilis* GB03 in its roots, the content of their two essential oils (terpineol and eugenol) increased up to tenfold. Similarly, the plant growth and essential oil content of *Ocimum* spp. improved after plants were inoculated with *Glomus fasciculatum* and *Azotobacter chroococcum*, *Pseudomonas putida* and *A. chroococcum*, and the AM fungus, *Glomus mosseae* (Egamberdieva and Teixeira da Silva 2015). Similarly, better essential oil contents in the shoots of *Origanum* sp. and *Pelargonium* species after inoculation with the AM fungus *Glomus mosseae* were also reported. Related results were also observed by Gupta et al. (2002) when *Mentha arvensis* was inoculated with AM fungus *Glomus fasciculatum*. The plant height was increased along with shoot growth and essential oil content. Similarly, PGPR *Pseudomonas fluorescens* and *Azospirillum brasilense* increased the biosynthesis of the major essential oil components up to 70% and total phenolic content in *Tagetes minuta* (Cappellari et al. 2013). In yet other study, the highest carvone content (63.22%) and the lowest contents of limonene (25.16%) in essential oil of *Anethum graveolens* L. were attained after the treatment of *Pseudomonas putida* combined with vermicompost (Tajpoor et al. 2013). Another study by Bahadori et al. (2013) concluded that when *Thymus daenensis* was co-inoculated with *G. mosseae* and *Bacillus subtilis*, it resulted in a 75% increase in shoot/root dry weight and a 117% increase in plant yield and also increased the essential oil yield by 93% as compared to uninoculated controls.

Further, Karagiannidis et al. (2012) showed that there is an increase in essential oil content in aromatic plants such as *Santolina chamaecyparissus*, *Salvia officinalis*, *Lavandula angustifolia*, *Geranium dissectum*, and *Origanum dictamnus* by 28.75, 55.56, 56.95, 53.63, and 55.24% respectively, when inoculated with AM fungus *Glomus lamellosum*. Similar results were also observed by Geneva et al. (2010) where essential oil content, bornyl acetate, 1,8-cineole, and α- and β-thujones of *Salvia officinalis* were improved when root was inoculated with *Glomus intraradices* (Geneva et al. 2010). In yet another experiment, inoculation of *Anethum graveolens* with AMF *Glomus macrocarpum* and *Glomus fasciculatum* markedly improved limonene and α-phellandrene content (Kapoor et al. 2002).

Withania somnifera (ashwagandha) is an important medicinal plant that is used in the treatment of cancer and nervous disorders as it contains a therapeutically active withanolide, withaferin A. When ashwagandha was inoculated with the bacterial composition comprising a mixture of *Azospirillum*, *Azotobacter chroococcum*, *Pseudomonas fluorescens*, and *Bacillus megaterium*, the plant height, root length, and withaferin content were significantly improved (Rajasekar and Elango 2011). Similarly, *Coleus forskohlii* (family Lamiaceae) is widely used to relieve coughs,

Table 13.1 Some selected synergistic combinations of medicinal/aromatic plants and the beneficial effect of PGPB and arbuscular mycorrhizal (AM) fungi on their respective phytochemical(s)

PGPB/VM fungi	Medicinal/ aromatic plant	Phytochemical(s) enhanced
Trichoderma harzianum + *Glomus mosseae*	*Andrographis paniculata*	Andrographolide
Glomus macrocarpum + *Glomus fasciculatum*	*Anethum graveolens*	Limonene, α-phellandrene
Pseudomonas putida	*Anethum graveolens*	Carvone, limonene
Glomus macrocarpum + *Glomus fasciculatum*	*Artemisia annua*	Artemisinin
Glomus fasciculatum	*Coleus forskohlii*	Forskolin
Glomus fasciculatum + *Pseudomonas monteilii*	*Coleus forskohlii*	Forskolin
Azospirillum brasilense + *Pseudomonas fluorescens*	*Catharanthus roseus*	Ajmalicine
Glomus lamellosum	*Geranium dissectum*	Essential oil
Glomus aggregatum + *Trichoderma harzianum* + *Bacillus coagulans*	*Glycyrrhiza glabra*	Phenols, ortho-dihydroxy phenols, tannins, flavonoids, alkaloids
Glomus lamellosum	*Lavandula angustifolia*	Essential oil
Glomus intraradices + *Glomus etunicatum*	*Lonicera confuse*	Chlorogenic acid
Glomus fasciculatum	*Mentha arvensis*	Essential oil
Glomus fasciculatum + *Azotobacter chroococcum*	*Ocimum spp.*	Essential oil
Pseudomonas putida + *Azotobacter chroococcum*	*Ocimum basilicum*	Essential oil
Bacillus subtilis	*Ocimum basilicum*	Terpineol, eugenol
Glomus mosseae	*Ocimum basilicum*	Essential oil
Pseudomonas fluorescens + *Bradyrhizobium* sp.	*Origanum majorana*	Terpinen-4-ol, cis-sabinene hydrate, trans-sabinene hydrate, α-terpineol
Azospirillum + *Azotobacter chroococcum* + *Pseudomonas fluorescens* + *Bacillus megaterium*	*Withania somnifera*	Withaferin A
Pseudomonas fluorescens + *Azospirillum brasilense*	*Tagetes minuta*	Essential oil, phenolic content
Bacillus megaterium + *Azospirillum* sp. + AM fungi	*Stevia rebaudiana*	Stevioside

eczemas, skin infections, tumors, glaucoma, and cardiac problems and to treat certain types of cancers (Kavitha et al. 2010). Its roots contain a labdane diterpene compound known as forskolin. It was observed that forskolin content was significantly improved by up to 25% on inoculation with the AM fungus *Glomus fasciculatum* (Sailo and Bagyaraj 2005) and improved further after collective inoculation of *G. fasciculatum* and *Pseudomonas monteilii* (Singh et al. 2012).

Stevia rebaudiana is a medicinal plant that is used as a source of natural sweeteners due to the presence of bioactive phytochemical, steviol glycosides. The plant has been reported for hypotensive and heart tonic actions (Ferri et al. 2006). Gupta et al. (2011) reported that when *S. rebaudiana* was inoculated with a consortium of phosphorus-solubilizing bacteria (PSB) such as *Burkholderia gladioli* MTCC 10216, *B. gladioli* MTCC 10217, *Enterobacter aerogenes* MTCC 10208, and *Serratia marcescens* MTCC 10238, the plant showed better root and shoot biomass and stevioside and rebaudioside-A contents (291 and 575%, respectively) on a whole-plant basis as compared to the control plants. Similar results were also reported by Das and Dang (2010) on collective inoculation of *Bacillus megaterium, Azospirillum* sp., and AM fungi on stevioside content of *S. rebaudiana.*

Likewise, *Artemisia annua* L. (Asteraceae) or annual wormwood is an important medicinal plant in Chinese customary medication and is a source of an antimalarial compound "artemisinin." Kapoor et al. (2007) reported that when the plant was root-inoculated with two AM fungi, *Glomus macrocarpum* and *Glomus fasciculatum*, it improved plant growth and artemisinin content. Karthikeyan et al. (2009) reported a rise in the production of terpenoid indole alkaloids (ajmalicine) in *Catharanthus roseus* when inoculated with *Azospirillum brasilense* and *Pseudomonas fluorescens*. Arpana and Bagyaraj (2007) reported that *Glomus mosseae* and *Trichoderma harzianum* increased plant root length, improved shoot growth, and increased dry weight, phosphorus uptake, and andrographolide content in *Andrographis paniculata* as compared to uninoculated plants. Selvaraj and Sumithra (2011) observed that the AM fungi such as *Glomus aggregatum*, *Trichoderma harzianum*, and *Bacillus coagulans* enhanced the plant biomass and polyphenolic compounds such as glycyrrhizin in licorice (*Glycyrrhiza glabra*). Shi et al. (2013) proved growth improvement and chlorogenic acid content in flowers of *Lonicera confusa*, a traditional Chinese medicine herb that is used for treating cold, flu, and acute fever, on inoculation with *Glomus intraradices*. All these studies demonstrate and confirm the role of microorganisms in improving the amount of phytochemical constituents and essential oil content in medicinal plants.

13.5 Role of Rhizosphere Bacteria in Ancillary Plant Growth Under Biotic Stress

The microflora present in a plant's rhizosphere provides the major protection to plant roots against attack by soil-borne pathogens. There are various members of rhizosphere microflora that can antagonize soil-borne pathogens before and during primary infection and during secondary spread on and even in root tissues. The key mechanisms through which rhizosphere microorganisms get rid of plant pathogens are antibiosis; competition for trace elements, nutrients, and microsites; parasitism; interference with quorum sensing affecting virulence; and induced systemic resistance (Mendes et al. 2013).

All rhizobacteria produce secondary metabolites that inhibit the growth or activity of competing microorganisms. Rhizosphere fungi such as *Trichoderma* species are inexhaustible producers of antibiotic metabolites and has therefore considerable attention as a source of biocontrol agent for plant protection (Brakhage and Schroeckh 2011; Druzhinina et al. 2011). Most fungal and bacterial biocontrol strains produce more than one antibiotic compound with overlapping or different degrees of antimicrobial activity. For example, bacteriocins such as agrocin 84 produced by *Agrobacterium radiobacter* (Kim et al. 2006) exhibit antibiotic activities against closely related genera, whereas many polyketide and non-ribosomal peptide antibiotics exhibit broad-spectrum activities against some plant pathogens (Raaijmakers et al. 2010).

Rhizosphere microorganisms also possess the ability to assist the uptake of specific trace elements such as iron. Some rhizosphere bacteria employ a variety of mechanisms to regulate intracellular iron concentrations by secretion of siderophores (Hider and Kong 2010). Also, rhizoferrin, a fungal siderophore produced by *Rhizopus arrhizus*, was found to be a proficient carrier of iron to plants.

Volatile organic compounds (VOCs) are also amongst the important metabolites produced by rhizosphere microorganisms. These are small molecules (<300 Da) with high vapor pressures and have the ability to diffuse through the water- and gas-filled pores in soil (Insam and Seewald 2010). Various bacterial species including *Stenotrophomonas maltophilia, Serratia plymuthica, Pseudomonas trivialis, P. fluorescens, B. subtilis,* and *Burkholderia cepacia* produce VOCs that impede mycelial growth of fungal plant pathogens (Jamalizadeh et al. 2010). VOCs can also induce systemic resistance in plants (Han et al. 2006) and promote plant growth (Bailly and Weisskopf 2012). Members of the rhizosphere microbiome can also modulate the plant immune system (Zamioudis and Pieterse 2012). The systemic resistance response induced in plants by beneficial rhizobacteria is mainly regulated by the phytohormones such as jasmonic acid and ethylene (Zamioudis and Pieterse 2012).

13.6 Role of Rhizosphere Bacteria in Supporting Plant Growth Under Abiotic Stress

Rhizosphere microflora adds to the ability of some plant species to survive under extreme conditions (Jorquera et al. 2012) such as under conditions of flooding, the bacteria has been shown to sustain the plant growth (Grichko and Glick 2001). In yet another study, the rhizospheric microflora from wheat plants grown in a saline zone were isolated, and it was found that out of 130 rhizobacterial isolates, only

24 bacterial isolates were tolerant to relatively high levels (8%) of sodium chloride salt (Upadhyay et al. 2009). All of the 24 salt-tolerant isolates produced indole-3-acetic acid, 10 isolates solubilized phosphorus, eight produced siderophores, six produced gibberellin, and two isolates contained the nifH gene, indicating their potential for nitrogen fixation. The dominant bacterial genus isolated under these conditions was *Bacillus* sp. (Upadhyay et al. 2009). Halotolerant bacterial strains were also isolated from halophytic plant species found in coastal soils in Korea (Siddikee et al. 2010). New halotolerant diazotrophic bacteria embracing indole acetic acid production, phosphate solubilization, and 1-aminocyclopropane-1-carboxylic acid (ACC) deaminase activity were also isolated from roots of the extreme halophyte *Salicornia brachiate* (Jha et al. 2012).

Several mechanisms are discussed by which microorganisms may alter plant physiological response under saline stress. They affect the water homeostasis by osmolyte accumulation, affect plant energetics by modulating the source–sink relationships, affect root uptake of toxic ions and nutrients by altering host physiology, modify physical barriers around the roots, or directly reduce foliar accumulation of toxic ions and also affect the crop's salt tolerance by altering hormonal root–shoot signaling.

Burkholderia phytofirmans increased grapevine root growth and physiological activity at temperatures down to 4 °C (Barka et al. 2006). When co-inoculated with a mixture of *Bradyrhizobium japonicum* and *Serratia proteamaculans,* they stimulated the soybean growth at 15 °C, the temperature at which soybean nodule infection and nitrogen fixation are normally inhibited. In another study, plants treated with a 2,4-diacetylphloroglucinol (DAPG)-producing *P. fluorescens* strain showed their role in pathogen control, which can also act to ameliorate abiotic stress factors (Raudales et al. 2009). Thus, members of the rhizosphere microbiome can lighten biotic and abiotic stresses on plants.

13.7 Conclusion and Future Prospects

Thus from the foregoing discussion, it is quite pertinent to state that plant rhizosphere-associated microorganisms play a significant role in promoting plant growth, biomass, and nutrient uptake, provide protection against both biotic and abiotic stress and against plant pathogens, and also enhance the concentration of bioactive phytochemical constituents and essential oil in both medicinal and aromatic plants, respectively, under various climatic conditions. The root inoculation with agriculturally important microorganisms offers an attractive alternative to replace the use of chemical fertilizers, pesticides, and other supplements for cultivation of both food and non-food plants. However, more studies are required to investigate the possible mechanisms by which bacteria increase phytochemical constituents in medicinally important plants at the tissue, cell, or molecular level. These studies can also be used to develop genetically modified and disease/drought/saline resistant plants along with high nutritive value such as those rich in vitamin A, carotene, and/or iron.

Acknowledgments The authors gratefully acknowledge Dr. Ashok K. Chauhan, Founder President, Amity Group of Institutions, and Mr. Atul Chauhan, Chancellor, Amity University UP, Noida, for the encouragement, research facilities, and financial support.

References

Arpana J, Bagyaraj DJ (2007) Response of kalmegh to an arbuscular mycorrhizal fungus and a plant growth promoting rhizo-microorganism at two levels of phosphorus fertilizers. Am-Euras J Agric Environ Sci 2:33–38

Badiane NNY, Chotte JL, Patê E, Masse D, Rouland D (2001) Use of soil enzymes activities to monitor soil quality in natural and improve fallows in semi-arid tropical regions. Appl Soil Ecol 18:229–238

Bahadori F, Ashorabadi ES, Mirza M, Matinizade M, Abdosi V (2013) Improved growth, essential oil yield and quality in *Thymus daenensis* Celak on mycorrhizal and plant growth promoting rhizobacteria inoculation. Int J Agron Plant Prod 4:3384–3391

Bailly A, Weisskopf L (2012) The modulating effect of bacterial volatiles on plant growth. Plant Signal Behav 7:1–7

Banchio E, Bogino PC, Zygadlo J, Giordano W (2008) Plant growth promoting rhizobacteria improves growth and essential oil yield in *Origanum majorana* L. Biochem Syst Ecol 36:766–771

Banchio E, Xie X, Zhang H, Pare PW (2009) Soil bacteria elevate essential oil accumulation and emissions in sweet basil. J Agric Food Chem 57:653–657

Barka EA, Nowak J, Clement C (2006) Enhancement of chilling resistance of inoculated grapevine plantlets with a plant growth-promoting rhizobacterium, *Burkholderia phytofirmans* strain PsJN. Appl Environ Microbiol 72:7246–7252

Berg G, Alavi M, Schmidt CS, Zachow C, Egamberdieva D, Kamilova F, Lugtenberg B (2013) Biocontrol and osmoprotection for plants under saline conditions. In: de Bruijn FJ (ed) Molecular microbial ecology of the rhizosphere. Wiley-Blackwell, Hoboken, NJ

Bharti N, Yadav D, Barnawal D, Maji D, Kalra A (2013) *Exiguobacterium oxidotolerans*, a halotolerant plant growth promoting rhizobacteria, improves yield and content of secondary metabolites in *Bacopa monnieri* (L.) Pennell under primary and secondary salt stress. World J Microbiol Biotechnol 29:379–387

Brakhage AA, Schroeckh V (2011) Fungal secondary metabolites – strategies to activate silent gene clusters. Fungal Genet Biol 48:15–22

Cappellari LR, Santoro MV, Nievas F, Giordano W, Banchio E (2013) Increase of secondary metabolite content in marigold by inoculation with plant growth-promoting rhizobacteria. Appl Soil Ecol 70:16–22

Das K, Dang R (2010) Influence of biofertilizers on stevioside content in *Stevia rebaudiana* grown in acidic soil condition. Arch Appl Sci Res 4:44–49

Das SK, Varma A (2011) Role of enzymes in maintaining soil health. In: Shukla G, Varma A (eds) Soil enzymology, Soil biology, vol 22. Springer, Heidelberg

Ding C-H, Wang Q-B, Guo S, Wang Z-y (2018) The improvement of bioactive secondary metabolites accumulation in *Rumex gmelini* Turcz through co-culture with endophytic fungi. Braz J Microbiol 49(2):362–369

Doughari JH, Human IS, Bennade S, Ndakidemi PA (2009) Phytochemicals as chemotherapeutic agents and antioxidants: possible solution to the control of antibiotic resistant verocytotoxin producing bacteria. J Med Plant Res 3:839–848

Druzhinina IS, Seidl-Seiboth V, Herrera-Estrella A et al (2011) Trichoderma: the genomics of opportunistic success. Nat Rev Microbiol 9:896

Egamberdieva D (2011) Role of microorganisms in nitrogen cycling in soils. In: Miransari M (ed) Soil nutrients. Nova Science, New York, pp 159–176

Egamberdieva D, Teixeira da Silva JA (2015) Chapter 14: medicinal plants and PGPR: a new frontier for phytochemicals. In: Egamberdieva D, Shrivastava S, Varma A (eds) Plant-growth-promoting rhizobacteria (PGPR) and medicinal plants. Springer, New York

Egamberdieva D, Berg G, Lindström K, Räsänen LA (2013) Alleviation of salt stress of symbiotic *Galega officinalis* L. (goat's rue) by co-inoculation of rhizobium with root colonizing *Pseudomonas*. Plant Soil 369(1):453–465

Egamberdieva D, Berg G, Lindstrom K, Rasanen L (2010) Root colonizing *Pseudomonas* spp. improve growth and symbiosis performance of fodder galega (*Galega orientalis* LAM) grown in potting soil. Eur J Soil Biol 46:269–272

Elango KV (2004) Studies on the effect of native AM fungi and PGPR's on growth and productivity of *Gloriosa superba* L. Ph.D. thesis, Bharathidasan University, Tiruchrappalli, Tamil Nadu, India

Ferri LA, Alves-Do-Prado W, Yamada SS, Gazola S, Batista MR, Bazotte RB (2006) Investigation of the antihypertensive effect of oral crude stevioside in patients with mild essential hypertension. Phytother Res 20:732–737

Geitang W, Honggang W (1989) Effects of VA Mycorrhizal fungi on growth, nutrient uptake and effective compounds in Chinese medicinal herb *Datura stramonium* L. Sci Agric Sin 5:1989–1905

Geneva MP, Stancheva IV, Boychinova MM, Mincheva NH, Yonova PA (2010) Effects of foliar fertilization and arbuscular mycorrhizal colonization on *Salvia officinalis* L. growth, antioxidant capacity and essential oil composition. J Sci Food Agric 90:696–702

Grichko VP, Glick BR (2001) Amelioration of flooding stress by ACC deaminase-containing plant growth-promoting bacteria. Plant Physiol Biochem 39:11–17

Gupta M, Bisht S, Singh B, Gulati A, Tewari R (2011) Enhanced biomass and steviol glycosides in *Stevia rebaudiana* treated with phosphate-solubilizing bacteria and rock phosphate. Plant Growth Regul 65:449–457

Gupta ML, Prasad A, Ram M, Kuma S (2002) Effect of the vesicular-arbuscular mycorrhizal (VAM) fungus *Glomus fasciculatum* on the essential oil yield related characters and nutrient acquisition in the crops of different cultivars of menthol mint (*Mentha arvensis*) under field conditions. Bioresour Technol 81:77–79

Hameed A, Egamberdieva D, Abd-Allah EF, Hashem A, Kumar A, Ahmad P (2014) Salinity stress and arbuscular mycorrhizal symbiosis in plants. In: Miransari M (ed) Use of microbes for the alleviation of soil stresses. Springer, New York, pp 139–159

Han SH, Lee SJ, Moon JH et al (2006) GacS-dependent production of 2,3-butanediol by Pseudomonas chlororaphis O6 is a major determinant for eliciting systemic resistance against *Erwinia carotovora* but not against *Pseudomonas syringae pv. tabaci* in tobacco. Mol Plant-Microbe Interact 19:924–930

Hider RC, Kong X (2010) Chemistry and biology of siderophores. Nat Prod Rep 27:637–657

Insam H, Seewald MSA (2010) Volatile organic compounds (VOCs) in soils. Biol Fertil Soils 46:199–213

Jabborova D, Egamberdieva D, Räsänen L, Liao H (2013) Salt tolerant *Pseudomonas* strain improved growth, nodulation and nutrient uptake of soybean grown under hydroponic salt stress condition. In: XVII international plant nutrition colloquium and boron satellite meeting proceedings book, Istanbul, Turkey, pp 260–261

Jacoby R, Peukert M, Succurro A, Koprivova A, Kopriva S (2017) The role of soil microorganisms in plant mineral nutrition—current knowledge and future directions. Front Plant Sci 8:1617

Jamalizadeh M, Etebarian HR, Aminian H et al (2010) Biological control of *Botrytis mali* on apple fruit by use of *Bacillus* bacteria, isolated from the rhizosphere of wheat. Arch Phytopathol Plant Protect 43:1836–1845

Jha B, Gontia I, Hartmann A (2012) The roots of the halophyte *Salicornia brachiata* are a source of new halotolerant diazotrophic bacteria with plant growth-promoting potential. Plant Soil 356:265–277

Jorquera MA, Shaharoona B, Nadeem SM, de la Luz Mora M, Crowley DE (2012) Plant growth-promoting rhizobacteria associated with ancient clones of creosote bush (*Larrea tridentata*). Microb Ecol 64:1008–1017

Juliani HR, Kapteyn J, Jones D, Koroch AR, Wang M, Charles D, Simon JE (2006) Application of near-infrared spectroscopy in quality control and determination of adulteration of African essential oils. Phytochem Anal 17:121–128

Kapoor R, Chaudhary V, Bhatnagar AK (2007) Effects of arbuscular mycorrhiza and phosphorus application on artemisinin concentration in *Artemisia annua* L. Mycorrhiza 17:581–587

Kapoor R, Giri B, Mukerji KG (2002) *Glomus macrocarpum* a potential bioinoculant to improve essential oil quality and concentration in dill (*Anethum graveolens* L.) and carum (*Trachyspermum amni* (Linn.) Sprague). World J Microbiol Biotechnol 18:459–463

Karagiannidis N, Thomidis T, Lazari D, Panou-Filotheou E, Karagiannidou C (2012) Response of three mint and two oregano species to *Glomus etunicatum* inoculation. Aust J Crop Sci 6:164–169

Karthikeyan B, Joe MM, Jaleel CA (2009) Response of some medicinal plants to vesicular arbuscular mycorrhizal inoculations. J Sci Res 1:381–386

Kaufman PB, Cseke LJ, Warber S, Duke JA, Brielmann HL (1999) Natural products from plants. CRC, Boca Raton, FL

Kavitha C, Rajamani K, Vadivel E (2010) *Coleus forskohlii*: a comprehensive review on morphology, phytochemistry and pharmacological aspects. J Med Plant Res 4:278–285

Keû A, Slimene IB, Karkouch I, Rihouey C, Azaeiz S, Bejaoui M, Belaid R, Cosette P, Jouenne T, Limam F (2015) Characterization of endophytic *Bacillus* strains from tomato plants (*Lycopersicon esculentum*) displaying antifungal activity against *Botrytis cinerea* Pers. World J Microbiol Biotechnol 31:1967–1976

Kim JG, Park BK, Kim SU et al (2006) Bases of biocontrol: sequence predicts synthesis and mode of action of agrocin 84, the Trojan Horse antibiotic that controls crown gall. Proc Natl Acad Sci U S A 103:8846–8851

Kogel KH, Franken P, Hückelhoven R (2006) Endophyte or parasite – what decides? Curr Opin Plant Biol 9:358–363

Lugtenberg B, Kamilova F (2009) Plant-growth-promoting rhizobacteria. Annu Rev Microbiol 63:541–556

Lugtenberg BJ, Dekkers L, Bloemberg GV (2001) Molecular determinants of rhizosphere colonization by *Pseudomonas*. Annu Rev Phytopathol 39:461–490

Mahfouz SA, Sharaf-Eldin MA (2007) Effect of mineral vs. biofertilizer on growth, yield and essential oil content of fennel (*Foeniculum vulgare* mill.). Int Agrophys 21:361–366

Mendes R, Garbeva P, Raaijmakers JM (2013) The rhizosphere microbiome: significance of plant beneficial, plant pathogenic, and human pathogenic microorganisms. FEMS Microbiol Rev 37:634–663

Miller SH, Browne P, Prigent-Cambaret C, Combes-Meynet E, Morrissey JP, O'Gara F (2010) Biochemical and genomic comparison of inorganic phosphate solubilisation in *Pseudomonas* species. Environ Microbiol Rep 2:403–411

Oteino N, Lally RD, Kiwanuka S, Lloyd A, Ryan D, Germaine KJ, Dowling DN (2015) Plant growth promotion induced by phosphate solubilizing endophytic *Pseudomonas* isolates. Front Microbiol 6:745

Palombo EA (2006) Phytochemicals from traditional medicinal plants used in the treatment of diarrhoea: modes of action and effects on intestinal function. Phytother Res 20:717–724

Phillipson JD (2001) Phytochemistry and medicinal plants. Phytochemistry 56:237–243

Prasad R, Sharma M, Kamal S, Rai MK, Rawat AKS, Pushpangdan P, Varma A (2008) Interaction of *Piriformospora indica* with medicinal plants. In: Varma A (ed) Mycorrhiza. Springer, Heidelberg, pp 655–678

Qadri M, Johri S, Shah BA, Khajuria A, Sidiq T, Lattoo SK, Abdin MZ, Riaz-Ul-Hussain S (2013) Identification and bioactive potential of endophytic fungi isolated from selected plants of the Western Himalayas. Springer Plus 2:8

Qiang X, Weiss M, Kogel KH, Schafer P (2012) *Piriformospora indica* a mutualistic basidiomycete with an exceptionally large plant host range. Mol Plant Pathol 13:508–518

Raaijmakers JM, de Bruijn I, Nybroe O, Ongena M (2010) Natural functions of lipopeptides from *Bacillus* and *Pseudomonas*: more than surfactants and antibiotics. FEMS Microbiol Rev 34:1037–1062

Rajasekar S, Elango R (2011) Effect of microbial consortium on plant growth and improvement of alkaloid content in *Withania somnifera* (Ashwagandha). Curr Bot 2:27–30

Raudales RE, Stone E, McSpadden Gardener BB (2009) Seed treatment with 2,4-diacetylphloroglucinol-producing pseudomonads improves crop health in low-pH soils by altering patterns of nutrient uptake. Phytopathology 99:506–511

Sailo GL, Bagyaraj DJ (2005) Influence of different AM-fungi on the growth, nutrition and forskolin content of *Coleus forskohlii*. Mycol Res 109:795–798

Sarker SD, Nahar L (2007) Chemistry for pharmacy students general, organic and natural product chemistry. Wiley, Chichester, pp 283–359

Selvaraj T, Sumithra P (2011) Effect of *Glomus aggregatum* and plant growth promoting rhizo-microorganisms on growth, nutrition and content of secondary metabolites in *Glycyrrhiza glabra* L. Indian J Appl Pure Biol 26:283–290

Sharma SB, Sayyed RZ, Trivedi MH, Gobi TA (2013) Phosphate solubilizing microbes: sustainable approach for managing phosphorus deficiency in agricultural soils. Springer Plus 2:587

Shi AD, Li Q, Huang JG, Yuan L (2013) Influence of arbuscular mycorrhizal fungi on growth, mineral nutrition and chlorogenic acid content of *Lonicera confusa* seedlings under field conditions. Pedosphere 23:333–339

Shirley M, Avoscan L, Bernaud E, Vansuyt G, Lemanceau P (2011) Comparison of iron acquisition from Fe-pyoverdine by strategy I and strategy II plants. Botany 89:731–735

Siddikee M, Chauhan P, Anandham R, Han GH, Sa T (2010) Isolation, characterization, and use for plant growth promotion under salt stress, of ACC deaminase-producing halotolerant bacteria derived from coastal soil. J Microbiol Biotechnol 20:1577–1584

Singh B, Kunze G, Satyanarayana T (2011) Developments in biochemical aspects and biotechnological applications of microbial phytases. Biotechnol Mol Biol Rev 6:69–87

Singh R, Soni SK, Kalra A (2012) Synergy between *Glomus fasciculatum* and a beneficial *Pseudomonas* in reducing root diseases and improving yield and forskolin content in *Coleus forskohlii* Briq under organic field conditions. Mycorrhiza 23:35–44

Singh R, Arora NK (2016) Growth enhancement of medicinal plant *Withania somnifera* using phosphate solubilizing endophytic bacteria *Pseudomonas* sp. as bioinoculant. Inter J Sci Technol Soci 2(1&2):13–18

Tajpoor N, Moradi R, Zaeim AN (2013) Effects of various fertilizers on quantity and quality of dill (*Anethum graveolens* L.) essential oil. Int J Agric Crop Sci 6:1334–1341

Teixeira da Silva JA, Egamberdieva D (2013) Plant-growth promoting rhizobacteria and medicinal plants. In: Govil JN, Bhattacharya S (eds) Recent progress in medicinal plants, essential oils III and phytopharmacology, vol 38. Studium, Houston, TX, pp 26–42

Tiwari R, Kalra A, Darokar MP, Chandra M, Aggarwal N, Singh AK, Khanuja SPS (2010) Endophytic bacteria from *Ocimum sanctum* and their yield enhancing capabilities. Curr Microbiol 60:167–171

Upadhyay SK, Singh DP, Saikia R (2009) Genetic diversity of plant growth promoting rhizobacteria isolated from rhizospheric soil of wheat under saline condition. Curr Microbiol 59:489–496

Van Wyk BE, Wink M (2004) Medicinal plants of the world. Briza, Pretoria, pp 54–56

Walker EL, Connolly EL (2008) Time to pump iron: iron-deficiency-signaling mechanisms of higher plants. Curr Opin Plant Biol 11:530–535

Wu JY, Ng J, Shi M, Wu SJ (2007) Enhanced secondary metabolite (tanshinone) production of *Salvia miltiorrhiza* hairy roots in a novel root bacteria co-culture process. Appl Microbiol Biotechnol 77:543–550

Yehuda Z, Shenker M, Hadar Y, Chen YN (2000) Remedy of chlorosis induced by iron deficiency in plants with the fungal siderophore rhizoferrin. J Plant Nutr 23:1991–2006

Zamioudis C, Pieterse CMJ (2012) Modulation of host immunity by beneficial microbes. Mol Plant-Microbe Interact 25:139–150

Zhang H, Sun Y, Xie X, Kim MS, Dowd SE, Pare PW (2009) A soil bacterium regulates plant acquisition of iron via deficiency-inducible mechanisms. Plant J 58:568–577

Chapter 14
Effect of *Agnihotra* Ash on Drug-Resistant *Escherichia coli* in Water

Reshma Tuladhar, Bijaya Laxmi Maharjan, Supriya Sharma, Anjana Singh, and Ulrich Berk

Abstract *Agnihotra* ash is the product of *Agnihotra* or *Homa* fire performed in a copper pyramid exactly at the time of sunrise and sunset using rice, dried cow dung, and clarified butter. *Agnihotra* was known to enhance growth and physiology of plants and provide beneficial effects on animals. The energy emanating from *Agnihotra* disseminates a powerful healing effect in the surroundings and the energy is locked in the resultant ash. This ash is used to treat various ailments and known to possess antibacterial properties. *Agnihotra* ash has been used to purify water and reduce the load of bacteria in water. Multidrug-resistant *Escherichia* coli were found to be reduced in the water sample when treated with *Agnihotra* ash. Thus, this can be an area to explore for an alternative to mitigate the water pollution problem.

14.1 Introduction

Availability of safe drinking water is fundamental to leading a healthy life. Contaminated water is the source of numerous diseases in humans with developing countries being the most vulnerable. The recent rise in drug-resistant bacteria in the water sources has been a serious concern (Chen et al. 2017; Diwan et al. 2010). While numerous cost-effective methods of treating contaminated water are being developed, the use of the ancient method of *Agnihotra* has been almost neglected.

Agnihotra is a *Homa* fire performed in a copper pyramid of predetermined size and shape, tuned to the biorhythm of sunrise/sunset. It comes from the ancient *Vedic* sciences of bioenergy, medicine, agriculture, and climate engineering. According to the ancient tradition, energies emanating from the sun extend all the way to the earth and produce a flood effect at those coordinates where the sun is said to rise. It is believed that the flood enlivens and purifies everything in its path while destroying impurities (Berk 2016).

R. Tuladhar · B. L. Maharjan · S. Sharma · A. Singh (✉)
Central Department of Microbiology, Tribhuvan University, Kirtipur, Kathmandu, Nepal

U. Berk
Deutsche Gesellschaft für Homa-Therapie, Haldenhof, Mühlingen, Germany

A. Varma et al. (eds.), *Plant Biotic Interactions*,
https://doi.org/10.1007/978-3-030-26657-8_14

Although the practice of *Agnihotra* has been assumed to be older than Indian tradition, at present time it has been comprehended as a part of the Indian tradition or religion and not much as a part of science (Berk 2016). Typically, people perform *Agnihotra* in their home to reduce stress, treat different kinds of ailments, and boost the spiritual power (Berk 2016; Golechha et al. 1991). However, public attention was drawn following some positive consequences on agriculture application (Berk 2016). Germination of seed and plant growth was enhanced by *Agnihotra* ash and *Agnihotra* fumes in several experiments (Abhang et al. 2015; Pathade and Abhang 2014). Plant growth was increased with the ash used in the experiment from *Agnihotra* performed during sunrise and sunset compared to other times of the day (Pathade and Abhang 2014; Rameshwar et al. 2009). Similarly, *Agnihotra* performed in a copper vessel chanting *mantra* and using unpolished rice and clarified butter from cow resulted in better plant growth compared to *Agnihotra* performed in a steel vessel, without *mantra*, and using polished rice and clarified butter from buffalo (Pathade and Abhang 2014).

The fertility of the soil was found to increase with the addition of *Agnihotra* ash in the soil (Berde et al. 2015). The load of nitrogen-fixing bacteria and phosphate-solubilizing bacteria was higher in the soil when supplemented with *Agnihotra* ash and incubated for a week (Berde et al. 2015). In addition, it stimulated photosynthesis and respiration in the plant (Abhang et al. 2015; Berde et al. 2015).

Furthermore, *Agnihotra* purified atmosphere has a positive effect on human and animals. *Agnihotra* atmosphere and ash improved the health of cows and has benefitted apiculture, pisciculture, and breeding of animals (Berk 2016). The neurophysiological effect of *Agnihotra* on the human body and brain has been associated with the well-being of humans and has a potency to cure alcoholism when used as adjunct (Golechha et al. 1991; Nair 2016). The fumes liberated from the fire have helped protect the upper respiratory tract and lungs from bacterial infection (Bansal et al. 2015). Researchers have mentioned about the use of *Agnihotra* against pathogenic bacteria (Nautiyal et al. 2007; Pathade and Abhang 2014). With the increase in antibiotic-resistant bacteria at an alarming rate, any strategy to impede the detrimental effect of these bacteria will be beneficial for mankind.

14.2 Prerequisite for *Agnihotra*

While correct timing for performing *Agnihotra* is essential, the choice of precise material is equally indispensable for an effective consequence. The shape, size, and material of the vessel are very important. A pyramid-shaped copper container which is 14.5 cm × 14.5 cm at the top and 5.25 cm × 5.25 cm at the bottom with 6.5 cm height is prescribed (Paranjpe 1989).

The vibration and resonance being affected by size, selection of accurate pyramid dimension is important. The pyramid shape has special indication since ancient time. Although the exact meaning pertaining to the shape is yet to be unraveled, it is believed that pyramids have the ability to capture the cosmic energy from the surroundings (Kumar et al. 2005). As for the material of pyramid, copper and gold are the only choice. Copper and gold have been known to possess antimicrobial and healing properties (Nie et al. 2010; Rai et al. 2010; Zhou et al. 2018).

Agnihotra fire is an important element for transforming solid and liquid substance into gaseous form. Specific materials are required for fire preparation which includes the cow dung patties properly smeared with cow's ghee. Cow dung is flattened into pancake-like patties and dried in the sun. *Agnihotra* fire is ignited from this dried cow dung. These materials are offered to fire while chanting *Agnihotra mantras*. In ancient cultures, cow dung has been believed to possess medicinal value.

The clarified butter (ghee) used for *Agnihotra* fire should be derived purely from cow's milk devoid of any additives. It is prepared by heating cow milk in low heat allowing water to evaporate. After the water has evaporated, the milk solid appearing white in color will rise to the top. When passed through a fine strainer, clarified butter is collected leaving the solid behind. Thus prepared clarified butter is considered to have medicinal property and can be stored for a long period of time. The clarified butter is believed to act as a carrier agent for subtle energies which lock up powerful energy (Berk 2016).

14.3 Preparation of the Agnihotra Fire

The preparation is made by placing a flat piece of dried cow dung at the bottom of the copper pyramid and loosely arranging pieces of dried cow dung over it to allow air to pass freely in the pyramid. A small amount of cow ghee (clarified butter) is applied on a piece of cow dung, which is lit, and *Agnihotra* fire is prepared a few minutes before sunrise or sunset.

An additional material required for *Agnihotra* fire is rice. For this purpose, unpolished and unbroken brown rice is essential since the subtle energy structure around the material is disturbed if the rice is broken. It is recommended to add a few grains of rice in order to not have a big effect on the temperature of the flame. The temperature of the flame will be between 600 °C and 800 °C (Berk 2016).

A vital component of *Agnihotra* is *mantra* which is the sound uttered by the person performing it and is known to activate vibrations that exist in the surroundings and create a certain atmosphere of effects. The energy from *Agnihotra* is emanated not only from the fire but the rhythm and *mantra* also create subtle energies which are thrust into the atmosphere by fire (Paranjpe 1989). When specific *mantras* are uttered at specific times of sunrise/sunset, resonance takes place in the pyramid to which is attributed all the powerful healing effects in the surroundings (Berk 2016).

14.4 *Agnihotra* Ash and Control Ash

Agnihotra ash is the product of *Agnihotra Homa* fire. *Agnihotra* ash is collected at the end of the Homa process after it has totally cooled down. The ash is sieved and kept in a natural material container like clay or glass. However, use of metal or plastic containers is prohibited. It is believed that properly performed *Agnihotra* with perfect timing and correct materials and following the right process will result in the ash that has potential benefit (Berk 2016).

Experiments have been carried out to see the beneficial effect of *Agnihotra*. In order to prove its effectiveness, comparisons have been made with control ash. Control ash is prepared with the same ingredients used for *Agnihotra* ash but without *mantras*, not the specific timing of *Agnihotra* and not the copper pyramid vessel. Any kind of ash has a positive effect on plants and is traditionally used in the gardens (Berk 2016). Ashes were found to exhibit some antibacterial activity and reduced the bacterial population in the wastewater (Ivankovic et al. 2014). Thus, in order to demonstrate the power of *Agnihotra* ash, it is necessary to carry the experiment simultaneously with control ash and see the difference in the potency.

14.5 Configuration of Energy Transformation in *Agnihotra*

The sun is the foundation of the energy for *Agnihotra*, the copper pyramid is the generator, and the fire is the turbine (Berk 2016). The importance of the time of the day for performing *Agnihotra* is that during morning at the time of sunrise, the energies, electricity, and ethers are attracted to the pyramid in its shape and at sunset these energies are thrust out in the same shape (Berk 2016). The torrent of subtle energies at sunrise carries music with it and the morning *Agnihotra mantra* is the essence of that music.

When the rice mixed with clarified butter is offered to the fire prepared in the prescribed copper pyramid with mantra uttered, a channel is created through the atmosphere (Berk 2016). With tremendous amount of energy drawn around the *Agnihotra* pyramid, a magnetic-type field is created which neutralizes negative energies and reinforces the positive. The energy emanating from the *Agnihotra* pyramid is believed to benefit plants, animals, and humans (Berk 2016). The effect of *Agnihotra* is not only limited during the period of performance since the energy is locked in the resultant ash collected after the flame had died. This ash is used to treat various ailments (Berk 2016).

14.6 Effect on Microorganisms

Numerous experiments have been carried out to see the effect of *Agnihotra* and *Agnihotra* ash on microorganisms associated with air, water, and soil (Abhang et al. 2015; Gerlecka 1985; Mondkar 1982; Pathade and Abhang 2014; Sharma et al. 2011). Experiments on microorganisms are relatively easier since the result can be obtained quicker compared to experiments on plants and animals. Besides, it can be performed at low cost without the need of large space.

The beneficial effect of *Agnihotra* on the reduction of pathogenic bacteria in air was reported by Mondkar in 1982. Smoke emanating from burning wood and medicinal herbs caused reduction of over 94% of aerial bacterial population within one hour and disinfected the air within 24 h. Bactericidal potential of the medicinal smoke treatment was demonstrated by the absence of pathogenic bacteria such as *Corynebacterium urealyticum*, *Enterobacter aerogenes*, *Kocuria rosea*, *Pseudomonas syringae* pv. *persicae*, and *Staphylococcus lentus* in the room for 30 days (Nautiyal et al. 2007).

Change in the virulence capacity of *Staphylococcus aureus*, *Klebsiella pneumonia*, and *Pseudomonas aeruginosa* with the effect of *Agnihotra* ash was investigated by Pathade and Abhang (2014). *Agnihotra* ash, though known to have a detrimental effect on pathogenic bacteria, does not have a negative effect on beneficial bacteria. Several researchers have studied the beneficial effect of *Agnihotra* on non-pathogenic bacteria (Berde et al. 2015).

Volatilized aliphatic acids enhance the germicidal activity of the smoke (Huang et al. 2011; Shilling et al. 2013). Lauric acid was found to possess antibacterial activity against *Propionibacterium acnes* (Nakatsuji et al. 2009). Organic acids released from volatilization of fatty acid esters such as caprylic acid ethyl ester, decanoic acid ethyl ester, behenic acid ethyl ester, and lauric acid ethyl ester (Table 14.1) enhance the beneficial effects of the *Agnihotra* smoke (Nair 2016). Similarly, compound 2,4-bis(tert-butyl)-phenol, identified in *Agnihotra* ash, has

Table 14.1 Major compounds identified in the *Agnihotra* sample based on similarity search in GC-MS library

S. N.	Name of the compounds	Retention time (min)
1	Isoborneol (exo-2-hydroxy-1,7,7- trimethyloborane)	05.00
2	Borneol (endo-1,7,7-trimethyl-bicyclo[2.2.1]heptan-2-ol)	05.10
3	1-Dodecene	5.30
4	Ethyl octanoate (caprylic acid ethyl ester)	5.35
5	n-Tridecane	5.40
6	n-Octanal (caprylic aldehyde)	5.50
7	Endo-isocamphonone	5.60
8	1-Tetradecene	8.00
9	Ethyl decanoate (decanoic acid ethyl ester)	8.09
10	Hexadecane	8.15
11	Elemene	8.20
12	2.4-bis(tert-butyl)-phenol	9.60
13	Dodecanoic acid (lauric acid)	10.17
14	Ethyl dodecanoate (lauric acid ethyl ester)	10.60
15	Ethyl dodecanoate (behenic acid ethyl ester)	12.80
16	1-hexadecanol	16.70
17	Lirioresinol beta-dimethyl ether (1H,3H- furo[3,4-c]furan,tetrahydro-1,4-bis(3,4,5-trimethoxyphenyl)	25.90

Adapted from Nair (2016)

inhibitory effect in the spore germination and hyphal growth of agriculturally important fungi *Fusarium oxysporum* (Dharni et al. 2014; Nair 2016). Antibacterial/germicidal effect has been linked with the compounds listed in Table 14.1 (Nair 2016).

14.7 *Agnihotra* for Water Purification

Agnihotra when performed regularly was found to purify the water. It is has been assumed that regular performance of *Agnihotra* creates an energy field which helps purify water (Berk and Sharma 2015).

The present experiment was conducted in sterile water inoculated with multidrug-resistant (MDR) *Escherichia coli* isolated from contaminated water. Nine sterile bottles with sterile water were inoculated with MDR *E. coli* suspension making bacterial load of 1.5×10^{8}cfu/ml. In the first set of three bottles, *Agnihotra* ash was added, followed by control ash in the second set of three bottles, and the third set of bottles was control (which did not contain any form of ash). The bottles were incubated at room temperature. Following 24 h of incubation, each of the bottles from each set was used for bacterial count and load estimation by the membrane filter method and Most Probable Number (MPN) method. A similar experiment was performed on the third and fifth day of incubation using the remainder of the incubated bottles. The load of bacteria was reduced remarkably on the fifth day with *Agnihotra* ash treatment (Fig. 14.1). The reduction of bacterial load was

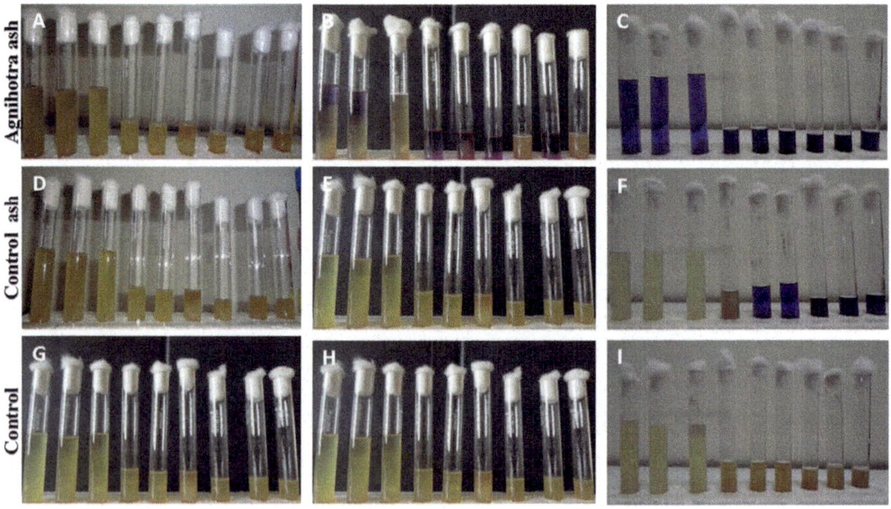

Fig. 14.1 Estimation of *E. coli* in *Agnihotra* ash-treated water by MPN method. All the tubes were inoculated with *E. coli* and treated with different parameters as: (**a**) *Agnihotra* ash for 24 h, (**b**) *Agnihotra* ash for 3 days, (**c**) *Agnihotra* ash for 5 days, (**d**) control ash for 24 h, (**e**) control ash for 3 days, (**f**) control ash for 5 days, (**g**) control for 24 h, (**h**) control for 3 days, and (**i**) control for 5 days

Table 14.2 Estimation of coliform number in water at different days following treatment of *Agnihotra* ash, control ash, and sample without any ash

Treatment	1 day (MPN/100 ml)	3 days (MPN/100 ml)	5 days (MPN/100 ml)
Agnihotra ash	2400	210	<3
Control ash	2400	2400	93
Control	2400	2400	2400

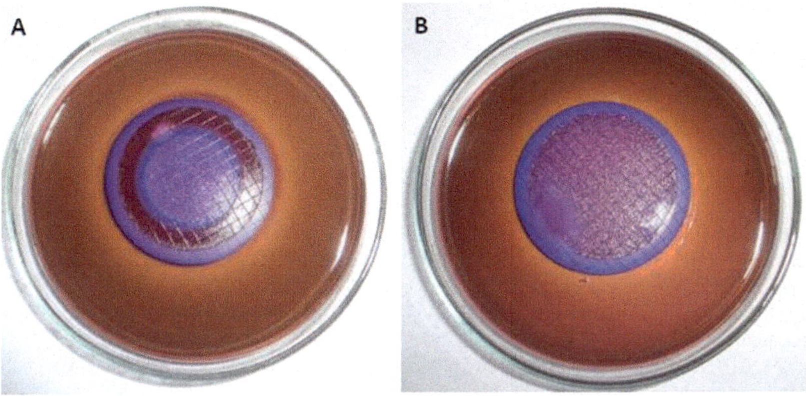

Fig. 14.2 MDR *E. coli* colonies from water samples on membrane filter. (**a**) Water treated with *Agnihotra* ash, (**b**) water treated with control ash

observed on the third day of incubation in *Agnihotra* ash, while there was no reduction in number in control ash at this time period. The result of control ash was identical to the control experiment on the third day. However, there was some reduction in control ash on the fifth day of incubation, which was still less compared to that seen in the *Agnihotra*-treated ash (Fig. 14.1, Table 14.2).

Antibacterial property of ash was demonstrated by the reduction in bacterial load in control ash on the fifth day or treatment compared to control experiment which is devoid of ash (Table 14.2). This is attributed to some antibacterial activity associated with ash (Ivankovic et al. 2014). But the efficacy of *Agnihotra* ash over any kind of ash has been demonstrated by the reduction of bacterial load starting from the third day of treatment with significant reduction on the fifth day compared to control ash (Fig. 14.1, Table 14.2). Similarly, a few colonies of *E. coli* were seen in *Agnihotra* ash compared to control ash (Fig. 14.2).

In the light of reduced bacterial load in *Agnihotra* ash-treated water, the beneficial effect of *Agnihotra* in the reduction of pathogenic bacteria has a potential in the treatment of contaminated water. However, more rigorous experiments are needed to be done in both laboratory conditions and natural water bodies to validate the claim that *Agnihotra* can be an effective alternative for water treatment.

14.8 Conclusion

If the tremendous benefit of *Agnihotra* can be utilized for purification of water bodies to reduce the number of harmful bacteria, the problems associated with bacterial contamination of water bodies can be significantly reduced. Not only that this method will be economically sustainable but most importantly devoid of any side effect on human and environment, it is imperative that more research should be done to confirm its implementation.

References

Abhang P, Manasi P, Moghe P (2015) Beneficial effects of Agnihotra on environment and agriculture. Int J Agric Sci 5:111–120

Bansal P, Kaur R, Gupta V, Kumar S, Kaur R (2015) Is there any scientific basis of *hawan* to be used in epilepsy-prevention/cure? J Epilepsy Res 5:33–45

Berde C, Kulkarni A, Potphode A, Gaikwad A, Gaikwad S (2015) Application of Agnihotra ash for enhancing soil fertility. IJESRT 4:2546–2551

Berk U (2016) Suggested experiments with Agnihotra and homa therapy: what has been done and what can be done. Haldenhof, Mühlingen

Berk U, Sharma S (2015) Effect of Agnihotra energy field on water purification. Indian J Tradit Knowl 1(1):63–68

Chen Z, Yu D, He S, Ye H, Zhang L, Wen Y, Zhang W, Shu L, Chen S (2017) Prevalence of antibiotic-resistant *Escherichia coli* in drinking water sources in Hangzhou City. Front Microbiol 8(1133). https://doi.org/10.3389/fmicb.2017.01133

Dharni S, Maurya A, Samad A, Srivastava S, Sharma A, Patra D (2014) Purification, characterization, and in vitro activity of 2,4-di-tert-butylphenol from *Pseudomonas monteilii* PsF84: conformational and molecular docking studies. J Agric Food Chem 62:6138–6146

Diwan V, Tamhankar AJ, Khandal RK, Sen S, Aggarwal M, Marothi Y, Iyer RV, Sundblad-Tonderski K, Stalsby-Lundborg C (2010) Antibiotics and antibiotic-resistant bacteria in waters associated with a hospital in Ujjain, India. BMC Public Health 10(1):414. https://doi.org/10.1186/1471-2458-10-414

Gerlecka E (1985) Report on Agnihotra healing in Gdansk, Poland. Satsang 13:10–11

Golechha GR, Sethi IC, Deshpande M, Rani U (1991) Agnihotra in the treatment of alcoholism. Indian J Psychiatry 33:44–47

Huang C, Alimova Y, Myers T, Ebersole J (2011) Short- and medium- chain fatty acids exhibit antimicrobial activity for oral microorganisms. Arch Oral Biol 56:650–654

Ivankovic T, Hrenovic J, Itskos G, Koukouzas N, Kovacevic D, Milenkovic J (2014) Alkaline disinfection of urban wastewater and landfill leachate by wood fly ash. Arh Hig Rada Toksikol 65(4):365–375

Kumar IR, Swamy NVC, Nagendra HR (2005) Effect of pyramids on microorganisms. Indian J Tradit Knowl 4:373–379

Mondkar AD (1982) Agnihotra and microbes, a laboratory experience. Satsang 9:20

Nair RR (2016) Agnihotra Yajna: a prototype of south Asian traditional medical knowledge. J Acupunct Meridian Stud 10(2):143–150

Nakatsuji T, Kao M, Fang J, Zouboulis C, Zhang L, Gallo R (2009) Antimicrobial property of lauric acid against *Propionibacterium acnes*: its therapeutic potential for inflammatory acne vulgaris. J Invest Dermatol 129:2480–2488

Nautiyal CS, Chauhan PS, Nene YL (2007) Medicinal smoke reduces airborne bacteria. J Ethnopharmacol 114(3):446–451

Nie Y, Kalapos C, Nie X, Murphy M, Hussein R, Zhang J (2010) Superhydrophilicity and antibacterial property of a Cu-dotted oxide coating surface. Ann Clin Microbiol Antimicrob 9:25. https://doi.org/10.1186/1476-0711-9-25

Paranjpe VV (1989) Homa therapy – our last chance? Fivefold Path, Madison, VA

Pathade GR, Abhang P (2014) Scientific study of Vedic knowledge Agnihotra. Bharatiya Bouddhik Sampada Q Sci Res J Vijnana Bharati 43:18–27

Rai A, Prabhune A, Perry CC (2010) Antibiotic mediated synthesis of gold nanoparticles with potent antimicrobial activity and their application in antimicrobial coatings. J Mater Chem 20:6789–6798

Rameshwar R, Punam P, Atul D (2009) Research on Homa organic farming in Palampur. In: Berk U and Johnson B (eds) Brainstorming Conference – Bringing Homa organic farming into mainstream Indian agriculture system, Dhule 2009

Sharma S, Sengupta T, Sunar K (2011) Somayag influences on different parameters of Narmada River. Indian Res Commun:55–56

Shilling M, Matt L, Rubin E, Visitacion M, Haller N, Grey S (2013) Antimicrobial effects of virgin coconut oil and its medium-chain fatty acids on *Clostridium difficile*. J Med Food 16:1079–1085

Zhou Q, Kang H, Bielec M, Wu X, Cheng Q, Wei W, Dai H (2018) Influence of different divalent ions cross-linking sodium alginate-polyacrylamide hydrogels on antibacterial properties and wound healing. Carbohydr Polym 197:292–304

Chapter 15
Plant Microbe Interface: The Plant Antimicrobial Peptides

S. Manivannan and P. Umadevi

Abstract Antimicrobial peptides (AMPs) are known to play important roles in plant development and stress tolerance. The AMPs are involved in the defense reaction of innate plant immunity and are known to increase the transcription level in response to abiotic or biotic stress factors. There are numerous reports available on the structure and the in vitro efficiency of antimicrobial peptides against phytopathogens. This chapter provides the available pipelines to identify the plant AMPs and its role in mediating the defense in plants in particular.

15.1 Introduction to Plant AMPs

Antimicrobial peptides (AMP) are small peptides, size ranging from 2 to 9 kDa with broad-spectrum antimicrobial activity. AMPs have been isolated from various plant parts, viz., leaves, stems, roots, flowers, and seeds. They are constitutively expressed or regulated upon stress (Nawrot et al. 2014; Umadevi et al. 2018) and often tissue specific and were proved to act against phytopathogens. The major plant AMPs are thionin (PR-13 family), defensin (PR-12 family), hevein-like peptide, knottin, α-hairpinin, lipid transfer protein (PR-14 family), snakins which are cysteine-rich peptides (Park et al. 2000), and cyclotides, the circular peptides (Gillon et al. 2008). Classification of plant AMP families is largely based on their Cys motifs which exhibit a characteristic Cys pattern with a defined number of non-Cys residues between the two neighboring Cys. The cysteine-rich peptides are the prominently identified AMPs in plants, while the description of cysteine-free AMPs from plants is very limited (Egorov et al. 2005; Silva et al. 2012; Zipfel 2009; Umadevi et al. 2018).

S. Manivannan (✉)
Department of Horticulture, Sikkim University, Gangtok, Sikkim, India
e-mail: smanivanan@cus.ac.in

P. Umadevi
Division of Crop Improvement and Biotechnology, ICAR-Indian Institute of Spices Research, Kozhikode, Kerala, India

A. Varma et al. (eds.), *Plant Biotic Interactions*,
https://doi.org/10.1007/978-3-030-26657-8_15

15.2 Identification of AMPs from Plants

The presence of variable numbers of cysteine residues are the signature tags for each group of AMP. Hence these tags are used to identify the AMPs from plants (Table 15.1).

Cation exchange chromatography-based identification of AMPs from onion seeds was demonstrated. The basic fraction of the total protein was purified using S-sepharose high-performance column (Cammue et al. 1992). The EST-based study identified the Sp-AMP family: Sp-AMP1 (79aa) and Sp-AMPs 2–4 (105aa) in the cDNA library of root tissue infected with the root rot fungus *Heterobasidion annosum* from *Pinus sylvestris* (Scots pine) (Asiegbu et al. 2003). Ke et al. (2015) used ESTs including those generated from *B. napus* 187,272 cDNA libraries of seeds, 35,788 pathogen-challenged leaves, and 644,998 deposited in the public databases, as a model, to perform in silico identification against APD2 database. They found 606 peptides that were orthologous to different AMP families.

A lot of genes encoding antimicrobial peptides have been discovered in many plants due to the development of transcriptome in many plants. The annotation of transcriptomes yielded many transcripts coding for the AMPs. The specific pipeline of AMP search using transcriptome data has also been developed. AMP mining in transcriptomes of the wheat species *Triticum kiharae* Dorof. et Migusch, highly resistant to pathogens, was done (Kovtun et al. 2018). 563 unique putative AMPS (pAMPs) were identified by a hidden Markov model based on conserved cysteine motifs characteristic to different AMP families, viz., defensins, thionins, hevein-like peptides, LTPs, snakins, and cysteine-rich peptides with novel cysteine motifs. Other cysteine-rich peptides with defensive role, such as Bowman-Birk inhibitors and proteinase inhibitors II, as well as RALF peptides involved in defense signaling were identified, together with other CRPs, such as MEG Ae1 and pollen Ole e1. For the first time, three types of hevein-like peptides that differ in the number of cysteine residues and the length of the C terminal prodomain were discovered in wheat. Slavokhotova et al. (2015) predicted more than 160 putative AMPs in lyme grass

Table 15.1 AMP type and the signature tags

AMP type	Signature tags in a protein/transcript
Thionine	8 Cys residues near the N-terminus
Defensin	8–10 Cys residues at C terminal
Hevein	8 Cys residues as continuous sequence of cysteine (–CC–) near the center of the molecule
Knottin	6 Cys residues as continuous sequence of cysteine (–CC–) near the center of the molecule
Lipid transfer protein	8 cysteine residues involved in 4 disulfide bridges and a consensus amino acid signature
Cyclotides	Cyclotide precursors with endoplasmic reticulum (ER) signal peptide, an N-terminal propeptide (NTPP) with 46–68 residues of variable sequence, and a C terminal hydrophobic propeptide of 3–7 residues

seedlings from transcriptome. The identified AMPs of seven families include defensins, thionins, hevein-like peptides, snakins, cyclotide, alfa-hairpinins, and LTPs.

In all of the above techniques, the major parameter for the identification of AMP from plants relied on the cysteine-rich motifs/sequence in a peptide. The recent advancement in bioinformatics and also in proteomics allowed the newer ways of identifying the peptides from plants. Ramada et al. (2017) developed a pipeline to find out the intragenic AMPs (IAMPs) from internal fragments of various plant proteins. This hypothesis-driven search strategy allowed the selection of all putative encrypted antimicrobial peptides based upon their physicochemical properties that were iteratively selected by an in-house computer program named Kamal. This program enabled the in silico processing of protein sequences for specific physicochemical parameters (net charge, average hydrophobicity, and hydrophobicity) capable of identifying putative antimicrobial peptide fragments inside larger polypeptide chains.

The label-free proteomics strategy successfully described the AMP signature profile of black pepper both constitutive (Table 15.2) and induced upon *Phytophthora capsici* infection (Umadevi et al. 2018) (Figs. 15.1 and 15.2).

The peptide data from antimicrobial peptide (APD) database was used to query the peptide ion spectra of black pepper using Progenesis QI software. Out of 24 only 5 peptides had the match in PhytAMP database. The search in CAMPR3, the multiorganism database, showed the identity for all 16 non-plant AMPs of black pepper origin. They are supposed to be homologs of animal/insect AMP signatures. The presence of conserved domains in the identified peptides revealed the label-free proteomics approach as a reliable and quick method to identify AMPs even from complex samples. This study enabled the identification of 13 cysteine-free AMPs from plants for the first time which were otherwise the underrepresented group from other methods (Table 15.3).

15.3 Antimicrobial Peptides in Plant Defense Mechanism

The involvement of AMPs in plant development and signaling (Salas et al. 2014), symbiosis (Wang et al. 2017), and metal tolerance (Mirouze et al. 2006) had been demonstrated. The plant defense mechanism to pathogens is a complex process. The endogenous plant peptides, the AMPs which are active against the pathogens, are one of the essential components in it. Though there are enormous reports on the direct effect of AMPs (Fig. 15.3) against phytopathogens, we focus on some elegant research on the involvement of AMPs in plant defense mechanism. Huffaker et al. (2006) isolated a peptide with 23 amino acids (*At*Pep1) from the Arabidopsis leaves. The defense gene PROPEP1 transcription was found to be activated by AtPep1 peptide. The gene was expressed at low levels in all tissues but the overexpression *PROPEP1* conferred resistance to root pathogen *Pythium irregulare*. The cumulative results indicated that *At*Pep1 requires H_2O_2 for the transcription of the defensin gene in Arabidopsis against pathogen. Changes in expression levels of the six annotated *PROPEP* gene family members (*PROPEP1 to 6*) in response to pathogens

Table 15.2 Some representative AMPs from black pepper

APD database ID	Phyt AMP database ID	AMP type	Signature peptide	Molecular weight	Isoelectric point
AP00308 AP00307 AP00489 AP00225	PHYT00044	Defensin (LCR76)	AGLQFPVGR	4.3 kDa	12.41
AP00936	PHYT00125	Lipid transfer protein (Pa_LTP_1)	ITGACGLNGR	3.3 kDa	8.54
AP01541	PHYT00089	Thionin (thionin-2.4)	AVLDILK	2.3 kDa	7.25
AP01373	PHYT00062	Thionin (denclatoxin-B)	KICLDPDAPK	7.4 kDa	8.78
AP01917	PHYT00225	Hevein family (EAFP1)	FSVFGK	4.1 kDa	10.78
AP00433	PHYT00002	Defensin (antifungal protein AX1)	RNGGVCIPIR	5.4 kDa	12.23
AP01559	PHYT00182	Cyclotide (cycloviolacin-O7)	AQSGKTAICKCYVKVCPR	5.8 kDa	8.76
AP02274	PHYT00201	Cyclotide (kalata B5)	FISGLIGGLK	3.06	5.87
AP01597	PHYT00227	Hevein family (Ee-CBP)	CAPKMKQIGTCGMPQVKCCK	5.03	8.42
AP00383	PHYT00004	Defensin (At-AFP1 (LCR67))	GAGKAVLGK	5.68	8.41
AP00612	PHYT00047	Defensin (MsDef1 (defensin 1.3))	ELENLAAMDLELQK	5.21	8.21
AP1220	PHYT00001	Defensin (Ah-AMP1)	GIKDWIK	5.88	7.71
AP00138	PHYT00258	MBP-1	RGCKRR	4.14	11.85

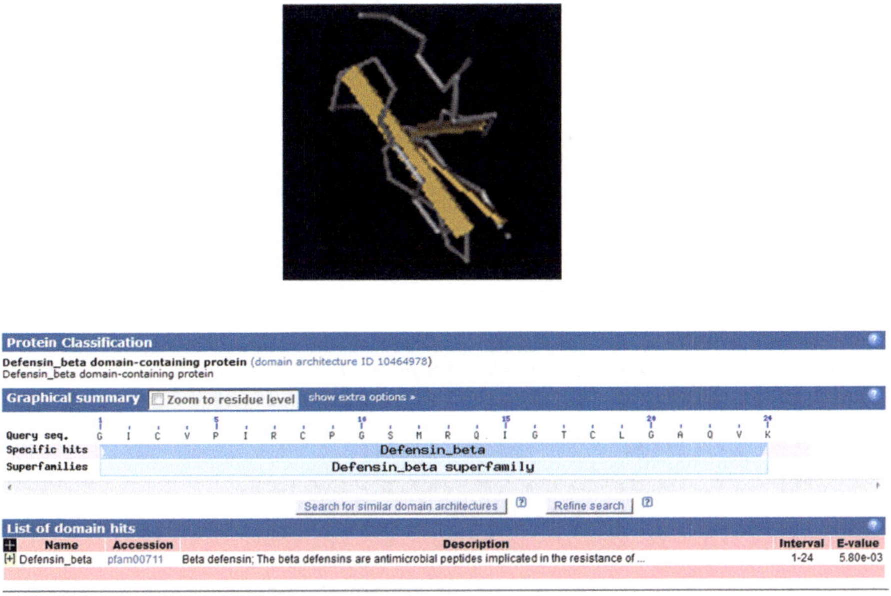

Fig. 15.1 Black pepper AMP 4 belonged to defensin beta superfamily

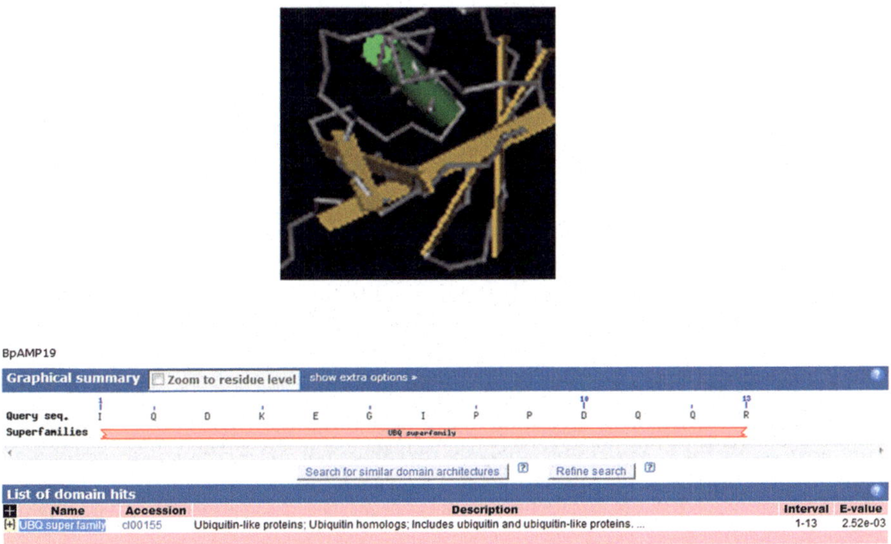

Fig. 15.2 Black pepper AMP 4 belonged to UBQ superfamily

Table 15.3 The abundance of cysteine-free AMPs in black pepper upon infection with *Phytophthora*

Black pepper AMP	Amino acid composition (%)	Peptide abundance in *Phytophthora*-infected leaf at 24 h (fold change)
BPAMP 1	Ala, Ser (7.14); ASP, Glu, Gly, Val, Lys, Leu (14.29)	High (1.17)
Bp AMP 3	Ala, Phe, Pro, Gln, Arg, Val (11.11); **Gly (22.22)**	High (5.88)
BpAMP6	Phe, Lys, Met, Ser (9.09); Ile, Leu (18.18); **Gly (27.27)**	Low (3.85)
BpAMP9	Ala, Lys (22.22); **Gly (33.33)**; Leu, Val (11.11)	High (3.53)
BpAMP10	Glu, Gly, Ile, Leu, Lys, Arg, Try (14.29)	High (2.04)
BpAMP11	Ala, Phe, Asn, Pro, Ser, Thr (11.11); **Lys (33.33)**	High (10.28)
BpAMP12	Ala (14.29); Asp, Lys, Met, Asn, Gln (7.14); **Glu(21.43);** Leu (**28.47**)	High (2.53)
BpAMP18	Gly, Lys, Arg (14.29); **Ser, Val (28.57)**	Low (2.01)
BpAMP19	Asp, Ile, Pro (15.38); **Gln(23.08)**; Glu, Gly, Lys, Arg (7.69)	Low (2.49)
BpAMP21	Asp, Gly, Trp (14.29); **Ile, Lys (28.57)**	Low (3.59)
BpAMP22	Ala, Gly, Lys, Thr (14.29); **Leu (42.86)**	High (1.74)
BpAMP23	**Ala, Gly, Thr (16.67)**; Asp, Ile, Leu, Gln, Ser, Val (8.33)	Low (10.59)
BpAMP24	Gly, His, Leu, Met, Asn, Gln, Arg, Val, Trp, Tyr (8.33)**; Lys (16.67)**	High (1.41)

showed the pathogen-specific expression dynamics mediated by the AtPep peptides (AtPep 1-6). The AtPep1 receptor was found to be a membrane-anchored leucine-rich repeat (LRR) receptor kinase named PEPR1 (Yamaguchi et al. 2006). The binding of AtPep1 to PEPR1 was shown to amplify the innate response to pathogens. The AtPep peptides are found to be involved in the regulation of SA and JA/Et signaling pathways. The overexpression of PROPEP genes resulted in constitutive expression of PR-1 and PDF1.2 in Arabidopsis leaves (Huffaker and Ryan 2007). Yamaguchi et al. (2010) identified the PERT2 receptor with 76% amino acid similarity to PEPR1. Molecular analysis revealed that the PEPR1 is a receptor for Pep1–6 and PEPR2 is a receptor for Pep1 and Pep2.

Following this, bioactive peptide ZmPep1 maize (*Zea mays*) gene, ZmPROPEP1. ZmPROPEP1 was identified as an ortholog of AtPROPEP1 gene. Expression studies revealed that ZmPROPEP1 gene is inducible by fungal infection and treatment with jasmonic acid or ZmPep1. ZmPep1 activates jasmonic acid and ethylene synthesis. The application of ZmPep1 induced the defense proteins endochitinase A, PR-4, Benzoxazineless1 (gene in biosynthesis of benzoxazinoid), and 2-hydroxy-4,

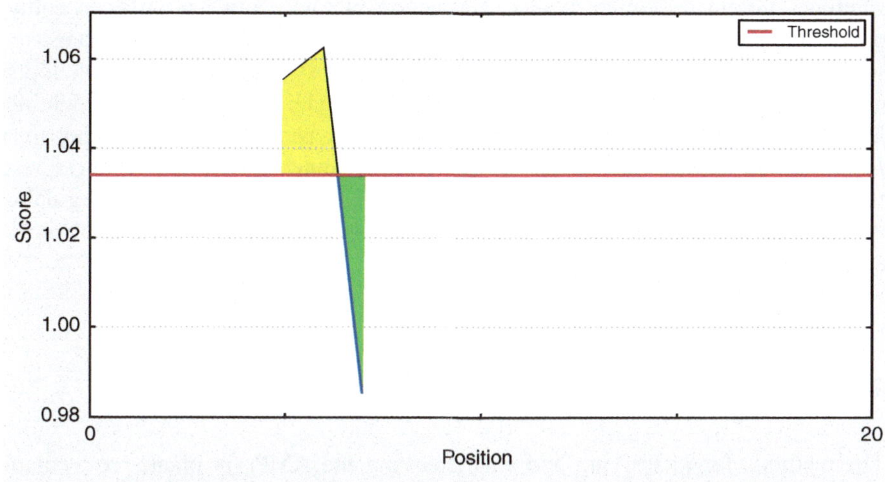

Average: 1.034 **Minimum:** 0.985 **Maximum:** 1.062

Predicted residue scores:

Position	Residue	Start	End	Peptide	Score
4	I	1	7	NQCINLE	1.033
5	N	2	8	QCINLEK	1.055
6	L	3	9	CINLEKA	1.062
7	E	4	10	INLEKAR	0.985

Fig. 15.3 Showing the Kolaskar and Tongaonkar antigenicity prediction of BpAMP 8 (NQCINLEKAR). The score denotes the antigenic nature of the peptide with the particular sequence

7-dimethoxy-1, 4-benzoxazin-3-one (glucoside in leaves). The pretreatment of maize with ZmPep1 was found to enhance resistance to pathogens, viz., *Cochliobolus heterostrophus* and *Colletotrichum graminicola* (Huffaker et al. 2011).

Utkina et al. (2013) identified Tk-AMP-X1 and Tk-AMP-X2, the AMPs from wheat *Triticum kiharae.* The sequences analysis showed that the peptides were of a-hairpinin structural family of plant peptides with a characteristic C1XXXC2-X (n)-C3XXXC4 motif. The infection of wheat by *Aspergillus niger, Bipolaris sorokiniana, and Fusarium oxysporum* upregulated the 4-cys gene expression.

The AMPS are found to have an action similar to the elicitors in plants. The peptide with 12 aa (GmSubPep) isolated from soybean leaves increased the pH in suspension-cultured cell media in 10 min time. This response for the exogenous addition of the AMP at low nanomolar concentration was very similar to the response exerted by other endogenous peptide elicitors and pathogen-derived elicitors (Pearce et al. 2010), and it induced the defense-related genes, viz., Cyp93A1, Chib-1b, PDR12, and achs. This unique cryptic peptide is located within the subtilase protein in legume plants. The action of this peptide in soybean culture provides information on the role of AMPs in the defense mechanism in plants.

The profiling of AMP signature under pathogen stress in plants was attempted using label-free quantitative proteomic strategy (Umadevi et al. 2018). The identified 24 AMPs were of cationic, anionic, cysteine-rich, and cysteine-free group upon *Phytophthora* infection in black pepper. The cysteine-rich AMP signatures were found to have 1–4 cysteine residues. The AMPs showed up- and downregulation during pathogen attack which showed a complex interaction and combinatorial strategy by the plant in avoiding the pathogen.

15.4 Conclusion and Future Prospects

The methods for identifying and characterizing the AMPs in plants are available nowadays. Research on plant AMPs would provide insights into the immunology and evolutionary significance of AMPs in plants, utilizing some of the AMPs as next-generation fungicide molecules, developing AMP-based markers and developing disease-resistant plants through gene editing approaches. With the availability of bioinformatics platform, the development of plant AMP databases is the most important area by which research on AMPs would speed up.

References

Asiegbu FO, Choi W, Li G, Nahalkova J, Dean RA (2003) Isolation of a novel antimicrobial peptide gene (SpAMP) homologue from *Pinus sylvestris* (scots pine) following infection with the root rot fungus *Heterobasidion annosum*. FEMS Microbiol Lett 228:27–31

Cammue BPA, De Bolle MFC, Terras FRG, Proost P, Van Damme J, Rees SB, Vanderleyden J, Broekaert WF (1992) Isolation and characterization of a novel class of plant antimicrobial peptides from *Mirabilis jalapa* L. seeds. J Biol Chem 267:2228–2233

Egorov TA, Odintsova TI, Vitaliy A, Pukhalsky VA, Grishin EV (2005) Diversity of wheat antimicrobial peptides. Peptides 26:2064–2073

Gillon AD, Saska I, Jennings CV, Guarino RF, Craik DJ, Anderson MA (2008) Biosynthesis of circular proteins in plants. Plant J 53:505–515

Huffaker A, Ryan CA (2007) Endogenous peptide defense signals in Arabidopsis differentially amplify signaling for the innate immune response. Proc Natl Acad Sci USA 104:10732–10736

Huffaker A, Pearce G, Ryan CA (2006) An endogenous peptide signal in Arabidopsis activates components of the innate immune response. Proc Natl Acad Sci USA 103:10098–10103

Huffaker A, Dafoe NJ, Schmelz EA (2011) ZmPep1, an Ortholog of Arabidopsis elicitor peptide 1, regulates maize innate immunity and enhances disease resistance. Plant Physiol 155:1325–1338

Ke T, Cao H, Huang J, Hu F, Huang J, Dong C, Ma X, Yu J, Mao H, Wang X, Niu Q, Hui F, Liu S (2015) EST-based in silico identification and in vitro test of antimicrobial peptides in *Brassica napus*. BMC Genomics 16:653

Kovtun A, Shelenkov A, Odintsova T (2018) The diversity of putative antimicrobial peptides revealed in wheat by high throughput next generation transcriptome sequencing. 43rd FEBS Congress, Prague

Mirouze M, Sels J, Richard O, Czernic P, Loubet S, Jacquier A, François IEJA, Cammue BPA, Lebrun M, Berthomieu P, Marque L (2006) A putative novel role for plant defensins: a defensin from the zinc hyper-accumulating plant, *Arabidopsis halleri*, confers zinc tolerance. Plant J 47:329–342

Nawrot R, Barylski J, Nowicki G, Broniarczyk J, Buchwald W, Gozdzicka-Jozefiak A (2014) Plant antimicrobial peptides. Folia Microbiol 59(3):181–196

Park CJ, Park CB, Hong SS, Lee HS, Lee SY, Kim SC (2000) Characterization and cDNA cloning of two glycine- and histidine-rich antimicrobial peptides from the roots of shepherd's purse, *Capsella bursa-pastoris*. Plant Mol Biol 44:187–197

Pearce G, Yamaguchi Y, Barona G, Clarence A (2010) Ryan A subtilisin-like protein from soybean contains an embedded, cryptic signal that activates defense related genes. PNAS 107 (33):14921–14925

Ramada MHS, Brand GD, Abrão FY, Oliveira M, Cardozo Filho JL, Galbieri R, Gramacho KP, Prates MV, Bloch C Jr (2017) Encrypted antimicrobial peptides from plant proteins. Sci Rep 7:13263

Salas CE, Badillo-Corona JA, Ramírez-Sotelo G, Oliver-Salvador C (2014) Biologically active and antimicrobial peptides from plants. Biomed Res Int 2015:11

Silva ON, Porto WF, Migliolo L, Mandal SM, Gomes DG, Holanda HH, Silva RS, Dias SC, Costa MP, Costa CR, Silva MR, Rezende TM, Franco OL (2012) Cn-AMP1: a new promiscuous peptide with potential for microbial infections treatment. Biopolymers 98(4):322–331

Slavokhotova AA, Shelenkov AA, Odintsova TI (2015) Prediction of *Leymus arenarius* (L.) antimicrobial peptides based on de novo transcriptome assembly. Plant Mol Biol 89:203–214

Umadevi P, Soumya M, George JK, Anandaraj M (2018) Proteomics assisted profiling of antimicrobial peptide signatures from black pepper (*Piper nigrum* L.). Physiol Mol Biol Plants 24 (3):379–387

Utkina LL, Yaroslav AA, Eugene AR, Tatyana VK, Anna AS, Peter BO, Vassilevski AA, Eugene VG, Tsezi AE, Tatyana IO (2013) Genes encoding 4-Cys antimicrobial peptides in wheat *Triticum kiharae* Dorof. et Migush.: multimodular structural organization, intraspecific variability, distribution and role in defence. FEBS J 280:3594–3608

Wang Q, Yanga S, Liua J, Terecskeib K, Ábrahámb E, Gombárc A, Domonkosc A, Szucsb A, Körmöczib P, Wangb T, Fodorc L, Maod L, Feid Z, Kondorosib E, Kalóc P, Keresztb A, Zhua H (2017) Host-secreted antimicrobial peptide enforces symbiotic selectivity in *Medicago truncatula*. PNAS 114(26):6854–6859

Yamaguchi Y, Pearce G, Ryan CA (2006) The cell surface leucine-rich repeat receptor for AtPep1, an endogenous peptide elicitor in Arabidopsis, is functional in transgenic tobacco cells. Proc Natl Acad Sci U S A 103:10104–10109

Yamaguchi Y, Huffaker A, Bryan AC, Tax FE, Ryan CA (2010) PEPR2 is a second receptor for the Pep1 and Pep2 peptides and contributes to defense responses in Arabidopsis. Plant Cell 22:508–522

Zipfel C (2009) Early molecular events in PAMP-triggered immunity. Curr Opin Plant Biol 12:414–420

Chapter 16
Microbe-Mediated Abiotic Stress Alleviation: Molecular and Biochemical Basis

Pandiyan Kuppusamy, Samadhan Yuvraj Bagul, Sudipta Das, and Hillol Chakdar

Abstract Abiotic stress is one of the major factors limiting the crop production globally. Plants experience diverse abiotic stresses including higher concentration of salt (salinity), temperature extremities, and water shortage (drought or dehydration). Such stressors impair the normal metabolic functioning of the plant leading to poor growth and development. A wide range of adaptations and mitigation strategies are required to efficiently manage the deleterious impacts of such stresses. Development of tolerant varieties, shifting the crop calendars, and resource management practices are some examples of such strategies. However, most of these technologies are cost-intensive and are beyond the reach of the small and marginal farmers. Microorganisms are naturally endowed with the ability to sustain extreme environmental conditions and also help other living beings in vicinity/association to cope with such stress to certain extent. Microorganisms through induction of systemic tolerance, modulation of plant defense mechanisms, and improvement of nutrition and growth can effectively alleviate or reduce the effect of stress. Hence, application of stress-alleviating microorganisms for crop production holds considerable potential to become a sustainable option to combat abiotic stresses.

16.1 Introduction

Global climate change perceived as heat waves and altered rainfalls leading to drought has been witnessed during the twenty-first century. A variety of stresses are encountered by our agro-ecosystems which are negatively influencing the crop productivity. The stresses are classified as biotic and abiotic stress which contribute 30 and 50%, respectively, for loss in agricultural productivity worldwide (Kumar and Verma 2018). According to a FAO report (2007), a significant portion of the total land

P. Kuppusamy · S. Y. Bagul · S. Das · H. Chakdar (✉)
ICAR-National Bureau of Agriculturally Important Microorganisms (NBAIM), Kushmaur, Maunath Bhanjan, Uttar Pradesh, India

A. Varma et al. (eds.), *Plant Biotic Interactions*,
https://doi.org/10.1007/978-3-030-26657-8_16

area in the globe is affected by various abiotic stresses with varying degrees (e.g., water deficit and low temperature have affected 64 and 57% of the global land area, respectively). Among various abiotic stresses, drought has the most serious impact on agricultural production. It causes diminished water potential and turgor pressure, closure of stomata, and membrane instability along with denaturation of protein. Stomatal closure during water stress reduces the photosynthesis due to lowered gas exchange and ultimately reduced yield (Yang et al. 2010). It has been reported that abiotic stresses are going to be one of the major limiting factors for the future of food security (Singh et al. 2011). Hence, sustainable management of abiotic stress becomes imperative. Developing stress-tolerant varieties is time-consuming, while development of transgenic crops with improved tolerance to abiotic stresses is costly with low rate of success. One of the possible alternatives to combat the effect of abiotic stress is application of plant growth-promoting bacteria (PGPB) and mycorrhizal fungi which help plants to survive under stress conditions. Due to their exposure to diverse stress conditions, microbes have developed appropriate responses and mechanisms to survive. These mechanisms range from modulation of gene expression to production of stress-adaptive proteins, compatible solutes, and sporulation. Heat-shock proteins (HSPs) identified from various organisms play a critical role in cell homeostasis. HSPs include chaperones and proteases that are presumably essential for maintaining the cell in abnormal conditions such as protein denaturation. Cold-shock proteins (CSPs) are synthesized when the cell is shifted to low temperature and decreased when the cell gets acclimatized to low temperature and growth resumes. The use of CSPs is in the limelight now due to its commercial and health implications. A microorganism during osmotic stress accumulates a number of organic solutes in the cytoplasm either through synthesis or uptake from external environment to maintain the intracellular osmotic balance (Galinski and Trüper 1994; Miller and Wood 1996). They are mainly composed of amino acids (e.g., proline), sugars/polyols (e.g., mannitol, trehalose), quaternary amines (e.g., glycine betaine), and derivatives of amino acids and esters (e.g., ectoine).

The stress-tolerant microbes can confer stress tolerance to plants also through a number of means (Finkel et al. 2017). They include production of key enzymes such as ACC (1-aminocyclopropane 1-carboxylate) deaminase, production of exopolysaccharides and rhizobitoxine, etc. (Vardharajula et al. 2011) that help plants to cope with adverse conditions. ACC deaminase and rhizobitoxine have been attributed to plant growth promotion through inhibition of ethylene production (Kumar et al. 2009). PGPB mitigate the impact of abiotic stress on plants through induced systemic tolerance (IST), which includes modulation of plant hormone levels by various means, production of antioxidants, etc. (Milosevic et al. 2012). Plant tolerance to abiotic stresses is associated with alleviation of reactive oxygen species (ROS). They include hydrogen peroxide (H_2O_2), singlet oxygen (O_2^-), and $HO^·$ and damage the cellular macromolecules ultimately leading to cell death in plants. The ROS scavenging enzymes include superoxide dismutases (SOD), catalases (CAT), ascorbate or thiol-dependent peroxidases (APX), glutathione reductases (GR), and monodehydroascorbate reductases (DHAR), and the scavengers ascorbate, tocopherol, and glutathione (Rouhier et al. 2008) help to protect the plants from ROS. PGPB play a vital role in inducing the synthesis of these ROS scavenging enzymes. Endophytic microbes can also contribute to abiotic stress tolerance of plants. Endophytes can induce stress-related genes as well as ROS scavengers to alleviate stress (Lata et al. 2018). These stress-tolerant PGPB not only help in combating the abiotic stresses but also enhance the crop productivity including rice, maize, and soybean (Glick 2014; Finkel et al. 2017).

Microorganisms, due to their multifaceted plant beneficial traits, are very promising for sustainable management of abiotic stress. In this chapter, potential application of microbes in managing major abiotic stresses like drought, salinity, and cold will be highlighted.

16.2 Microbe-Mediated Abiotic Stress Alleviation

Plant growth-promoting bacteria can induce systemic tolerance (IST) represented by a number of physicochemical changes in plant cell leading to better tolerance to abiotic stress (Yang et al. 2009). Many recent works show microbe-mediated abiotic stress tolerance such as salinity, drought, heat, chilling injury, and metal toxicity (Ait Barka et al. 2006; Palaniyandi et al. 2014; Gagné-Bourque et al. 2016; Tripathi et al. 2017; Ali et al. 2018; Duc et al. 2018; Yuan et al. 2018; Zaidi et al. 2018). Most of the microorganisms produce osmoprotectants under salt stress, exopolysaccharides under water stress, heat-shock proteins under heat stress, and cold-shock proteins under low temperatures (Sandhya et al. 2009b; Tripathi et al. 2017). Apart from this, microorganisms also produce plant growth-promoting substances as well as regulate stomatal movement as a defense strategy against a number of stresses (Gontia et al. 2011; Kim et al. 2014; Lim et al. 2015). Figure 16.1 summarizes some of the prominent mechanisms by which microbe can alleviate major abiotic stresses. Table 16.1 presents some recent reports of microbe-mediated abiotic stress alleviation.

16.2.1 Drought

As mentioned earlier, drought is one of the major limiting factors in crop growth and productivity in different parts of the world (Naveed et al. 2014). During drought, accumulation of ROS in plant cells results in damage of the cellular macromolecules (Sairam et al. 2005). Production of antioxidant enzymes such as SOD, CAT, and APX could prevent oxidative damage by eliminating ROS. Gusain et al. (2014) investigated inoculation of *Trichoderma harzianum* T35 in rice and concluded that the organism promoted the activity of these enzymes and enhanced drought tolerance in rice. Inoculation of *Pseudomonas aeruginosa* GGRJ32 in mung bean resulted in increased production of ROS scavenging enzymes and cellular osmolytes leading to alleviation of drought stress (Sarma and Saikia 2014). Secretion of plant hormones such as IAA, gibberellic acid (GA), and cytokinins has been reported to

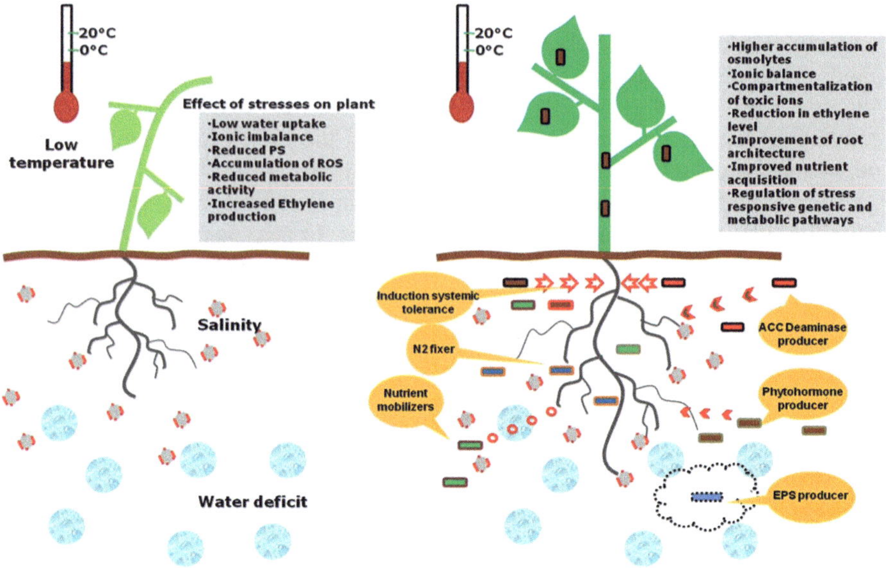

Fig. 16.1 Effect of abiotic stresses on plant and its alleviation by microorganisms

promote root development resulting in enhanced uptake of nutrients under stress condition (Gagné-Bourque et al. 2015). Cytokinin production by *Bacillus subtilis* conferred drought tolerance to *Platycladus orientalis* seedlings (Liu et al. 2013). Abscisic acid (ABA) also plays a key role in physiological processes in plants under drought stress (Cohen et al. 2008). Similarly, production of sugars like sucrose and fructans and amino acids such as asparagines, glutamic acid, and glutamine was increased in shoot and roots of *Bacillus subtilis* B26-colonized plant. The reports also showed an increase in GABA, a nonprotein amino acid in stressed plant after inoculation (Gagné-Bourque et al. 2016). Grover et al. (2014) investigated the role of *Bacillus* sp. under moisture stress when inoculated with sorghum and found increased sugars and proline content as compared to uninoculated plant. Ethylene plays a key role in plant homeostasis, thus resulting in underdevelopment of shoot and root growth. However, bacterial ACC deaminase confiscates and degrades the plant ACC to supply nitrogen and energy. In addition, it also reduces the detrimental effect of ethylene by removing plant ACC, thus enhancing plant growth and rectifying plant stress (Glick et al. 2007). Production of 1-aminocyclopropane-1-carboxylate (ACC) deaminase has been shown to be involved in stress alleviation by minimizing ethylene production (Azevedo Neto and Tabosa 2000; Mayak et al. 2004; Zahir et al. 2008). A study showed inoculation of *Bacillus licheniformis* K11 with pepper alleviated drought stress by ACC deaminase production and upregulation of other stress proteins (Lim and Kim 2013). *Piriformospora indica* colonized the roots of maize and showed increased antioxidant enzyme activity and proline content, whereas decrease in malondialdehyde (MDA) accumulation was reported which is responsible for membrane damage (Xu et al. 2017). Similarly, Sun et al. (2010)

Table 16.1 Examples of microbe-mediated alleviation of abiotic stresses

Stress	Microorganism	Crop	Mechanism	References
	Glomus intraradices	Lettuce	High antioxidant enzyme activity	Kohler et al. (2008)
	Pseudomonas putida GAP-P45	Sunflower	Exopolysaccharide production	Sandhya et al. (2009a)
	Pseudomonas putida MTCC5279	Chickpea	Production of ROS scavenging enzyme	Tiwari et al. (2016)
	Pseudomonas aeruginosa GGRJ21	Mung bean	Production of ROS scavenging enzyme and osmolytes	Sarma and Saikia (2014)
	Trichoderma harzianum	Tomato	Increased secondary metabolite and proline content	Alwhibi et al. (2017)
	Pseudomonas azotoformans ASS1	*Trifolium arvense*	Bioaccumulation, plant growth promotion	Ma et al. (2017)
	Paenibacillus polymyxa and *Rhizobium tropici*	Common bean	Phytohormone production	Figueiredo et al. (2008)
	Cladosporium cladosporioides and Tobamovirus	*Nicotiana benthamiana*	Elevated level of anti-oxidant enzyme, soluble protein, proline content	Dastogeer et al. (2017)
	Alcaligenes faecalis AF3	Maize	Exopolysacchharide production	Naseem and Bano (2014)
	Klebsiella variicola F2, *Pseudomonas fluorescens* YX2, *Raoultella planticola* YL2	Maize	Accumulation of gly-cine betaine	Gou et al. (2015)
	Phyllobacterium brassicacearum	*Arabidopsis thaliana*	Increased level of ABA	Bresson et al. (2013)
Salinity	*Azospirillum* sp.	Wheat	Production of phytohormone	Pereyra et al. (2012)
	Pseudomonas pseudoalcaligenes	Rice	Accumulation of gly-cine betaine	Jha et al. (2011)
	Trichoderma asperellum Q1	Cucumber	P solubilization and phytohormone production	Zhao and Zhang (2015)
	Pseudomonas putida R4	Cotton	IAA modulation	Egamberdieva et al. (2015)
	Bacillus sp., *Zhihengliuella halotolerans, Staphylo-coccus succinus*	Wheat	Plant growth promotion	Orhan (2016)
	Arbuscular mycorrhizal fungi (AMF)	Fenugreek	Increased levels of pro-line content and antioxidative enzymes	Rabab and Reda (2018)
	AMF	Tomato	High antioxidant enzyme activity	Ebrahim and Saleem (2017)

(continued)

Table 16.1 (continued)

Stress	Microorganism	Crop	Mechanism	References
	Burkholderia sp. MTCC 12259	Rice	ACC deaminase production	Sarkar et al. (2018a, b)
	Enterobacter sp.	Rice	ACC deaminase production	Sarkar et al. (2017)
	Curtobacterium albidum	Rice	EPS production, ACC deaminase, antioxidant enzyme activity	Vimal et al. (2018a, b)
	Arthrobacter protophormiae	Pea	ACC deaminase activity	Barnwal et al. (2014)
	Klebsiella sp. SBP8	Wheat	ACC deaminase activity	Singh et al. (2015)
	Bacillus amyloliquefaciens RWL1	Rice	ABA production	Shahzad et al. (2017a, b)
	Penicillium sp.	Sesame	High sugar concentrations, fatty acid content, reduced oxidative damage	Radhakrishnan and Lee (2015)
	Enterobacter cloacae HSNJ4	Canola	IAA, ACC deaminase activity	Li et al. (2017)
	Methylobacterium oryzae CBMB20 *and Glomus etunicatum*	Maize	Reduced proline content and Na uptake	Lee et al. (2015)
Cold stress	*Pseudomonas vancouverensis* OB155, *P. frederiksbergensis* OS261		Increased antioxidant enzyme, higher APX and glutathione reductase expression	Subramanian et al. (2016)
	Clavibacter sp.	*Chorispora bungeana* Enf12	Increased antioxidant enzyme, proline content	Ding et al. (2011)
	Pseudomonas sp.	Wheat	Reduced electrolyte leakage, phytohormone production	Mishra et al. (2011)
	Burkholderia phytofirmans PsJN	Grape	Modulation of carbohydrate metabolism	Fernandez et al. (2012)

recorded increased antioxidant enzyme and expression of drought-related genes when co-inoculated with Chinese cabbage. Inoculation with *Trichoderma harzianum* strain S2 with rainfed rice showed stress tolerance (Zaidi et al. 2018).

16.2.2 Salinity

Soils with electrical conductivity (EC) > 4 $dSm^{-}1$, NaCl concentration of approximately 40 mM at 25 °C, and 15% exchangeable sodium (Shrivastava and Kumar

2015) are generally considered as saline soils. In this situation most of the crops give lower yield. Estimates show that 20% of total cultivated and 33% of irrigated agricultural land throughout the world is salinity affected (Jamil et al. 2011). Soil with increased salt concentration becomes problematic to plants by reducing the ability of plants to absorb water, hindering the metabolic processes. It also affects the water uptake capacity in the root zone as a result of decrease in soil water potential. Prolonged salt stress causes ion toxicity due to intracellular accumulation of Na^+ and Cl^-. Under such circumstances, oxidative stress is induced by production of ROS which is harmful to viability of cells (Meena et al. 2017). Microbes could be useful in alleviating salt stress by formation of biofilm, phytohormone production, antioxidant production, osmolyte accumulation, and enhanced nutrient acquisition (Upadhyay et al. 2012; Barnwal et al. 2014; Khan et al. 2016; Petrova and Sauer 2016). Inoculation of arbuscular fungi and endophytic bacteria improves plant growth under salt stress (Shahzad et al. 2017a; Rabab and Reda 2018). Researchers also used halotolerant plant growth-promoting bacteria to alleviate salt stress; these microbes could produce IAA and ACC deaminase, solubilize phosphate, and fix nitrogen (Orhan 2016; Li et al. 2017; Sarkar et al. 2018a). Egamberdieva et al. (2015) studied the role of IAA-producing *Pseudomonas putida* R4 and *P. chlororaphis* R5 under salt stress in cotton (Egamberdieva et al. 2015). Similarly, Shahzad et al. (2017a, b) examined the role of endophyte *Bacillus amyloliquefaciens* RWL-1 in rice; the isolate was able to produce abscisic acid which conferred tolerance to rice (Shahzad et al. 2017b). Elevated activity of antioxidative enzymes like catalase, superoxide dismutase, peroxidase, and ascorbate peroxidase and K^+ uptake was recorded in paddy inoculated with *Curtobacterium albidum* SRV4 (Vimal et al. 2018a). Zhao and Zhang (2015) reported production of phytohormone by *Trichoderma asperellum* Q1 to promote cucumber growth under salt stress (Zhao et al. 2014). The isolate also showed ability to produce IAA, GA, and ABA. Under salt stress microbes also produce enhanced level of photosynthetic pigments; sugars such as sucrose, glucose, and fructose; fatty acids including palmitic, linolenic, and arachidic acids; and ion transporting K and Ca (Radhakrishnan and Lee 2015).

16.2.3 Cold Stress

Cold stress limits plants' growth directly through inhibition of metabolic reactions and indirectly through cold-induced osmotic (includes reduction in water uptake due to chilling and cellular dehydration due to ice-crystal formation), oxidative, and related stresses (Chinnusamy et al. 2007). Microorganisms through production of compatible solutes and ROS scavenging machineries can also alleviate the ill effects of low temperature. *Burkholderia phytofirmans* PsJN could confer chilling tolerance in grapewine by reducing toxic H_2O_2 production (Theocharis et al. 2012). Subramanian et al. (2016) showed endophytic bacterial inoculation led to chilling tolerance in tomato by producing higher amounts of antioxidant enzymes (Subramanian et al. 2016). A similar result was also reported by Ding et al. (2011)

in *Chorispora bungeana* plant at 0 °C, when inoculated with *Clavibacter* sp. (Ding et al. 2011). Arbuscular mycorrhizal fungi have also been reported to confer chilling tolerance to *Cyclamen persicum* plant. The plant showed enhanced activity of SOD and APX, ascorbic acid, and polyphenol content. Overall improved plant growth and biomass was observed (Matsubara et al. 2014). Increased water content, reduced electrolyte leakage, and Na^+/K^+ ratio were recorded in wheat inoculated with *Pseudomonas* sp. Significant increase in amino acids and proline content and improved cellular metabolites conferred cold stress at 8 °C in wheat seedlings (Mishra et al. 2011).

16.3 Biochemical and Molecular Basis of Microbe-Mediated Stress Alleviation

16.3.1 Mechanisms Involved in Drought Tolerance

The biochemical and molecular networking involved in drought stress alleviation by microorganisms are very complex and yet to be understood completely. Microbe-mediated drought alleviation can be through a coordinated or individual activation of osmolyte production, reduction in ethylene production, modulation of ABA biosynthesis and responsive cascades, exopolysaccharide production, induction of ROS scavenging system, etc. All these physiological changes are actually the reflection of a significant modulation of different stress-responsive genes (Table 16.2).

Microbes can modulate the expression of key genes associated with osmolyte biosynthesis. Δ1-Pyrroline-5-carboxylate synthase (P5CS) is one such enzyme involved with proline biosynthesis. Bacterial inoculation has been reported to upregulate the expression of *P5CS* which helped in free proline synthesis to maintain cellular osmotic balance (Vaishnav and Choudhary 2018). Improved drought tolerance of plants was elicited by *Pseudomonas putida* KT2440 expressing the *ots*AB genes for trehalose biosynthesis *in trans* (Vílchez et al. 2016). Transgenic plants carrying *mtlD* (encodes mannitol 1-phosphate dehydrogenase) convert mannitol 1-phosphate to mannitol via nonspecific phosphatases, thereby enhancing the plant tolerance to drought.

During biosynthesis of ethylene, 1-aminocyclopropane-1-carboxylate synthase (ACS) converts S-adenosylmethionine into 1-aminocyclopropane-1-carboxylate (ACC)—the immediate precursor of ethylene. ACC deaminase enzyme cleaves the ACC into ammonia and α-ketobutyrate (Honma and Smmomura 1978). Under stress conditions, ethylene regulates plant homeostasis which results in reduced plant growth and even death. By reducing the ACC levels in plants, ACC deaminase-producing microbes can reduce the plant ethylene levels which are deleterious to plants (Glick et al. 2007; Glick 2014). Stress induces ACC oxidase in plants resulting in first low ethylene level which in turn activates transcription of plant defensive genes. Bacterial ACC deaminase is activated by increased ACC level

Table 16.2 Microbe-mediated modulation of abiotic stress-related genes in plants

Stresses	Microbial strains	Plant host	Plant genes induced	Physiological changes in plants	References
Drought	*Piriformospora indica*	*Brassica campestris* ssp. *chinensis*	Upregulation of *DREB2A, CBL1, ANAC072,* and *RD29A*	Increased levels of superoxide dismutases, catalases, and peroxidases	Sun et al. (2010)
	Bacillus amyloliquefaciens 5113 *Azospirillum brasilense* NO40	*Triticum aestivum*	Upregulation of *APX1, SAMS1, and HSP17.8*	Increased levels of ascorbate peroxidase, S-adenosyl-methionine synthetase and heat-shock proteins	Kasim et al. (2016)
	Gluconacetobacter diazotrophicus	*Saccharum officinarum*	Upregulation of ERD15, DREB1A/CBF3, and DREB1B/CBF	Production of IAA and proline	Vargas et al. (2014)
	Sinorhizobium meliloti	*Medicago sativa*	Upregulation of *sod*	Increased levels of superoxide dismutase	Naya et al. (2007)
Salt	*Pseudomonas putida* UW4	*Solanum lycopersicum*	Upregulation of Toc GTPases	Chloroplast protein import apparatus	Yan et al. (2014)
	Bacillus amyloliquefaciens SN13	*Oryza sativa*	Upregulation of *SOS1, EREBP, SERK1, and NADP-Me2* and downregulation of *GIG* and *SAPK4*	Increased levels of ethylene-responsive element-binding proteins, somatic embryogenesis receptor-like kinase, and NADP-malic enzyme Decreased levels of glucose-insensitive growth and serine-threonine protein kinase	Nautiyal et al. (2013)
	B. subtilis GB03	*Puccinellia tenuiflora*	Upregulation of *PtHKT1* and *PTSOS1* Downregulation of *PtHKT2*	Decreased accumulation of Na^+ under high salinity	Niu et al. (2016)
	Enterobacter sp. EJ01	*Arabidopsis thaliana*	Upregulation of DREB2b, RD29A, RD29B, LEA, P5CS1and P5CS2, MPK3 and MPK6	Increased levels of DRE-binding proteins, relative to desiccation, late embryogenesis abundance, proline biosynthesis, and stress-inducible priming processes	Kim et al. (2014)
	Arthrobacter sp. and *Bacillus* sp.	*Capsicum annuum*	Downregulation of *CaACCO* and *CaLTP1*	Decreased levels of ACC oxidase and lipid transfer protein	Sziderics et al. (2007)

resulting in reduction of ethylene through degradation of ACC. As ACC oxidase has stronger affinity to ACC than ACC deaminase, plant ethylene levels depend on the ratio of ACC oxidase to ACC deaminase (Glick et al. 1998). Hence, for effective reduction of ethylene, activity of ACC deaminase should start before ACC oxidase is synthesized (Glick 2014). Inoculation of pepper plants with *Bacillus* sp. TW showed a tolerance to osmotic stress due to downregulation of *caACCO* (encoding ACC oxidase) and *caLTP1* (an abiotic stress-inducible gene encoding a lipid transfer protein). Saikia et al. (2018) also reported downregulation of ACO encoding gene in pulse crops inoculated with rhizobacteria under water stress (Saikia et al. 2018). ACC deaminase-mediated reduction in ethylene levels can downregulate *EREB* (a transcription factor which is a subcomponent of the ERF family and involved in ethylene signaling and the response pathway) which in turn reduces deleterious effects of ethylene in bacterial-inoculated plants under drought stress (Vaishnav and Choudhary 2018).

Microorganisms are known to produce auxin and its derivatives which can affect plant root architecture. At high levels, IAA stimulates lateral root and root hair formation (Gupta et al. 2015). The increased root projection area enhances water and nutrient uptake, thereby helping plants to cope with water stress (Vurukonda et al. 2016). Abscisic acid (ABA) is the major stress hormone playing an important role during drought. Under drought stress, the ABA level rises and results in physiological changes influencing plant growth (Farooq et al. 2009). Microorganisms can increase the level of ABA and improve the plant's tolerance towards drought. ABA has been reported to improve drought tolerance through regulation of transpiration, root hydraulic conductance, and upregulation of aquaporins (Porcel et al. 2006; Zhou et al. 2012). *Paenibacillus polymyxa* has been reported to improve the drought tolerance of *Arabidopsis* by activating late embryogenesis abundant (LEA) genes like *ERD15* and *RAB18* (Timmusk and Wagner 1999). The LEA genes are reported to be upregulated under water-deficit conditions and involved in the protection and repair of cellular macromolecules. Tiwari et al. (2016) showed that the expression of LEA genes increased along with increase in osmolyte production in rice (Tiwari et al. 2016). *Trichoderma harzianum* has been reported to upregulate dehydrin and other genes while alleviating stress in rice (Pandey et al. 2016; Meena et al. 2017). Likewise, the expression of LEA increased upon inoculation of *Brachypodium* with *B. subtilis* under drought stress as well as *P. putida* in chickpea (Gagné-Bourque et al. 2015; Tiwari et al. 2016).

Exopolysaccharides (EPS), produced by microorganisms, can bind soil particles and form micro- as well as macroaggregates which help plant roots to fit in the pores between the aggregates and thereby stabilize macroaggregates. Plants inoculated with EPS-producing bacteria showed increased tolerance to water stress due to improved soil structure (Sandhya et al. 2009b). Inoculation of *Pseudomonas fluorescens* DR7 with ACC deaminase- and EPS-producing activity along with efficient colonization of root adhering soil increased the soil moisture and root adhering soil/root tissue ratio in foxtail millet (Niu et al. 2016).

16.3.2 Mechanisms Involved in Salinity Tolerance

Microorganisms can produce a number of phytohormone and their derivatives which can evoke several physiological responses to the plants. Microorganisms are known to produce indole acetic acid (IAA), cytokinin, gibberellin, abscisic acid (ABA), etc. which are known to modulate root development and architecture along with signaling in response to salinity. Higher root proliferation due to microbially derived plant hormones also helps in acquisition of nutrients in a better way. Microbes which can mineralize or solubilize essential nutrients in the rhizosphere can also help to reduce the salinity stress through better acquisition of nutrients. Besides microorganisms can also modulate the phytohormone production in plants which in turn contributes to salinity tolerance. ABA acts as an important internal signaling molecule in plants under adverse environmental conditions like drought and salinity. It has been speculated that aquaporins and hydraulic conductance are affected by ABA and microorganisms can influence the levels of ABA (Aroca et al. 2007). Salinity stress tolerance may be mediated through ABA-independent pathways also. A detailed study on salinity tolerance in Arabidopsis through inoculation with *Enterobacter* sp. EJ01 revealed less induction of *RD29B* (a gene known to be regulated only by ABA-dependent pathway) and no induction in *AAO3* (a ABA biosynthetic gene) indicating independence from ABA (Kim et al. 2014). *AUX/IAA* genes present early auxin response genes coding transcriptional repressors of auxin-responsive genes (Li et al. 2015). Degradation of AUX/IAA proteins activates *ARF* transcription factors and consequently expression of auxin-responsive genes (Hagen and Guilfoyle 2002). Inoculation of microorganisms has been reported to influence the expression of *AUX/IAA* genes which in turn modulate auxin signaling pathway to protect plant from salinity and other stress conditions (Barnawal et al. 2017).

Salt Overly Sensitive (SOS) pathways also participate in mitigation of salinity stress in plants. WRKY transcription factors (WRKY TFs) are reported to have key roles in regulation of water stress through maintenance of osmotic balance, activation of ROS scavenging mechanism, and expression of different stress-responsive genes (Agarwal et al. 2011). Microorganisms are known to modulate the expression of *SOS* genes and WRKY TFs which can contribute to plant salinity stress response. Nidhi et al. (2016), while studying salinity stress alleviation in wheat using *Dietzia natronolimnaea*, hypothesized that *D. natronolimnaea* STR1 induced tolerance to salinity through both ABA-mediated and SOS-mediated pathways by increasing the expression of ABA-signaling cascade genes (*TaABARE* and *TaOPR1*) which in turn induced *TaMYB* and *TaWRKY* expression ultimately resulting in expression of a number of stress-related genes (Bharti et al. 2016). Modulation of *SOS1* and *SOS4* genes involved in SOS pathway and ion transporters TaNHX1, TaHAK, and TaHKT1 was also observed in *D. natronolimnaea* STR1-inoculated plants. Expression of level of different antioxidant enzyme coding genes and elevated proline concentration contributed to salinity tolerance in the inoculated wheat plants. Rice inoculated with *B. amyloliquefaciens* SN13 showed increased salinity tolerance and upregulation of *SOS1, EREBP, SERK1,* and *NADP-Me2* genes (Nautiyal et al. 2013).

Microbes are known to adjust the uptake of ions and nutrients by roots mostly via unknown mechanisms. Regulation of Na^+–K^+ influx-efflux through microbial inoculation can control the internal ionic balance required for a number of plant metabolic activities. Reduced efflux of potassium ions from roots and higher concentration in shoots and leaves can be helpful in salt stress adaptations in plants. It has been demonstrated that mycorrhiza, *Glomus intraradices,* can selectively take up elements such as K^+, Mg^+, and Ca^{2+} but exclude Na^+ uptake to keep the internal ratios of both Na^+:K^+ and Na^+:Ca^{2+} within the limit despite the magnitude of ion concentrations in the surrounding environment. This selective mechanism for ion uptake may partially alleviate the salinity stress in plants (Hammer et al. 2011). The volatile compounds released by PGPB downregulate the gene *hkt1* (high-affinity K^+ transporter 1 protein) expression in root while upregulating it in shoots, orchestrating lower Na^+ levels and recirculation of Na^+ in the plant under salinity stress (Zhang et al. 2008). The volatile metabolite, 2R, 3R-butanediol produced by *Pseudomonas chlororaphis* O6 induced tolerance in *Arabidopsis thaliana* to abiotic stresses. It also indicated that the induction requires salicylic acid (SA), jasmonic acid (JA), and ethylene pathways (Cho et al. 2008). Yun et al. (2018) reported that inoculation of *Piriformospora indica* improved salinity tolerance in maize through better stomatal operation due to higher rate of K delivery into the shoots by restricting K^+ efflux from roots and higher K^+ loading on shoots (Yun et al. 2018).

16.3.3 Cold Tolerance

Cold-adapted bacteria are known to produce cold-shock or cold acclimation proteins which help in sustenance under low temperatures. Like peptidoglycans and lipopeptides of bacteria, these cold-shock or acclimation proteins can also act as microbe-associated molecular patterns (MAMPs) which are recognized through their receptors on plant cell surface (Subramanian et al. 2015). Recognition of such MAMPs can trigger the cold adaptive mechanisms in plants. Bacteria are also known to induce expression of CRT repeat binding factors which are the transcriptional activators inducing cold-responsive (COR) gene expression under low temperature (Subramanian et al. 2015; Wang et al. 2016). In Arabidopsis, the major COR genes have been reported to encode highly hydrophilic and thermostable proteins which help in membrane stabilization and protein stability during freeze-induced dehydration (Thomashow 1999). Expression of key genes (*P5CS*-Δ1-pyrroline-5-carboxylate synthase) involved in proline biosynthesis in tomato has also been reported to significantly increase upon inoculation with a consortium of plant growth-promoting bacteria under chilling stress (Wang et al. 2016). Upregulation of proline biosynthesis can help to maintain the cellular osmotic balance which gets distorted under cold stress. Besides, bacterial inoculation has also been reported to compensate the reduction in expression of genes encoding osmotin, ascorbate peroxidase, monodehydroascorbic acid reductase, etc. under chilling stress (Wang et al. 2016).

16.4 Conclusion and Future Prospects

Microorganisms due to their immense metabolic diversity are endowed with unique adaptation mechanisms to endure multiple stresses of varying degrees. Being an integral part of the soil-plant ecosystem, microorganisms play a vital role in the maintenance of soil fertility and plant survival. Microbes in the vicinity of the plant in particular can play a significant role in the management of different biotic and abiotic stresses affecting plant productivity. Among such abiotic stresses, drought and salinity are the two major stresses which affect the crop production significantly. Agronomic management, changes in crop calendar, use of tolerant varieties, etc. are commonly practiced to manage such stresses despite the fact that none of these practices can effectively manage the problems in a sustainable way. Use of microorganisms in agriculture is long known and they can also be used for the management of abiotic stresses as evident from a large number of scientific studies throughout the world. Microbes offer eco-friendly, sustainable, and low-cost technologies which can be very effective. Despite many scientific studies, only very few technologies for abiotic stress management are available globally that too are restricted in local or regional level. The results obtained in the laboratory should effectively be translated into the farmers' field to make a socioeconomic impact. Rigorous selection and validation under stress conditions in the field is compulsory to develop such technologies. Exploration of extreme habitats and plants surviving under extreme environments can also help to identify novel microorganisms which can be useful for abiotic stress management. Transplantation of whole microbiome can also be another avenue to be looked upon as this can also be a viable strategy for the management of abiotic stresses in the near future.

Acknowledgments The authors acknowledge the infrastructural facility provided by ICAR-NBAIM, Mau, under the projects entitled "Deciphering molecular mechanism for eliciting drought tolerance in model plant by drought stress alleviating bacteria", "Molecular mining of AIMs for abiotic stress tolerance", and "Bioprospecting extremophilic cyanobacteria for plant growth promoting attributes and high value pigments".

References

Agarwal P, Reddy MP, Chikara J (2011) WRKY: its structure, evolutionary relationship, DNA-binding selectivity, role in stress tolerance and development of plants. Mol Biol Rep 38:3883–3896

Ait Barka E, Nowak J, Clément C (2006) Enhancement of chilling resistance of inoculated grapevine plantlets with a plant growth-promoting rhizobacterium, Burkholderia phytofirmans strain PsJN. Appl Environ Microbiol 72(11):7246–7252. https://doi.org/10.1128/AEM.01047-06

Ali AH, Abdelrahman M, Radwan U et al (2018) Effect of Thermomyces fungal endophyte isolated from extreme hot desert-adapted plant on heat stress tolerance of cucumber. Appl Soil Ecol 124:155–162. https://doi.org/10.1016/j.apsoil.2017.11.004

Aroca R, Porcel R, Ruiz-Lozano JM (2007) How does arbuscular mycorrhizal symbiosis regulate root hydraulic properties and plasma membrane aquaporins in Phaseolus vulgaris under drought, cold or salinity stresses? New Phytol 173(4):808–816. https://doi.org/10.1111/j.1469-8137.2006.01961.x

Barnawal D, Bharti N, Pandey SS et al (2017) Plant growth-promoting rhizobacteria enhance wheat salt and drought stress tolerance by altering endogenous phytohormone levels and TaCTR1/TaDREB2 expression. Physiol Plant 161:502–514

Barnwal P, Singh KK, Sharma A et al (2014) Biochemical, antioxidant and thermal properties of cryogenic and ambient ground turmeric powder. Int Agric Eng J 23(1):39–46

Bharti N, Pandey SS, Barnawal D et al (2016) Plant growth promoting rhizobacteria Dietzia natronolimnaea modulates the expression of stress responsive genes providing protection of wheat from salinity stress. Sci Rep 6:34768. https://doi.org/10.1038/srep34768

Bresson J, Varoquaux F, Bontpart T et al (2013) The PGPR strain Phyllobacterium brassicacearum STM196 induces a reproductive delay and physiological changes that result in improved drought tolerance in Arabidopsis. New Phytol 200(2):558–569. https://doi.org/10.1111/nph.12383

Chinnusamy V, Zhu J, Zhu JK (2007) Cold stress regulation of gene expression in plants. Trends Plant Sci 12(10):444–451

Cho SM, Kang BR, Han SH et al (2008) 2R,3R-Butanediol, a bacterial volatile produced by *Pseudomonas chlororaphis* O6, is involved in induction of systemic tolerance to drought in *Arabidopsis thaliana*. Mol Plant-Microbe Interact 21(8):1067–1075. https://doi.org/10.1094/MPMI-21-8-1067

Cohen AC, Bottini R, Piccoli PN (2008) Azospirillum brasilense Sp 245 produces ABA in chemically-defined culture medium and increases ABA content in arabidopsis plants. Plant Growth Regul 54:97–103

Dastogeer KMG, Li H, Sivasithamparam K et al (2017) Metabolic responses of endophytic Nicotiana benthamiana plants experiencing water stress. Environ Exp Bot 143:59–71. https://doi.org/10.1016/j.envexpbot.2017.08.008

de Azevedo Neto AD, Tabosa JN (2000) Salt stress in maize seedlings: part I growth analysis. Rev Bras Eng Agrícola e Ambient 4:159–164

Ding S, Huang CL, Sheng HM et al (2011) Effect of inoculation with the endophyte Clavibacter sp. strain Enf12 on chilling tolerance in Chorispora bungeana. Physiol Plant 141(2):141–151. https://doi.org/10.1111/j.1399-3054.2010.01428.x

Duc NH, Csintalan Z, Posta K (2018) Arbuscular mycorrhizal fungi mitigate negative effects of combined drought and heat stress on tomato plants. Plant Physiol Biochem 132:297–307. https://doi.org/10.1016/j.plaphy.2018.09.011

Ebrahim MKH, Saleem A-R (2017) Alleviating salt stress in tomato inoculated with mycorrhizae: photosynthetic performance and enzymatic antioxidants. J Taibah Univ Sci 11:850–860

Egamberdieva D, Jabborova D, Hashem A (2015) Pseudomonas induces salinity tolerance in cotton (Gossypium hirsutum) and resistance to Fusarium root rot through the modulation of indole-3-acetic acid. Saudi J Biol Sci 22(6):773–779. https://doi.org/10.1016/j.sjbs.2015.04.019

Farooq M, Wahid A, Kobayashi N et al (2009) Plant drought stress: effects, mechanisms and management. In: Sustainable agriculture. Springer, Dordrecht, pp 153–188

Fernandez O, Theocharis A, Bordiec S et al (2012) *Burkholderia phytofirmans* PsJN acclimates grapevine to cold by modulating carbohydrate metabolism. Mol Plant-Microbe Interact 25(4):496–504. https://doi.org/10.1094/MPMI-09-11-0245

Figueiredo MVB, Burity HA, Martínez CR, Chanway CP (2008) Alleviation of drought stress in the common bean (Phaseolus vulgaris L.) by co-inoculation with Paenibacillus polymyxa and rhizobium tropici. Appl Soil Ecol 40(1):182–188. https://doi.org/10.1016/j.apsoil.2008.04.005

Finkel OM, Castrillo G, Herrera Paredes S et al (2017) Understanding and exploiting plant beneficial microbes. Curr Opin Plant Biol 38:155–163

Gagné-Bourque F, Mayer BF, Charron JB et al (2015) Accelerated growth rate and increased drought stress resilience of the model grass brachypodium distachyon colonized by bacillus subtilis B26. PLoS One 10(6):e0130456. https://doi.org/10.1371/journal.pone.0130456

Gagné-Bourque F, Bertrand A, Claessens A et al (2016) Alleviation of drought stress and metabolic changes in Timothy (Phleum pratense L.) colonized with Bacillus subtilis B26. Front Plant Sci 7:584. https://doi.org/10.3389/fpls.2016.00584
Galinski EA, Trüper HG (1994) Microbial behaviour in salt-stressed ecosystems. FEMS Microbiol Rev 15(2–3):95–108. https://doi.org/10.1111/j.1574-6976.1994.tb00128.x
Glick BR (2014) Bacteria with ACC deaminase can promote plant growth and help to feed the world. Microbiol Res 169(1):30–39. https://doi.org/10.1016/j.micres.2013.09.009
Glick BR, Penrose DM, Li J (1998) A model for the lowering of plant ethylene concentrations by plant growth-promoting bacteria. J Theor Biol 190(1):63–68. https://doi.org/10.1006/jtbi.1997.0532
Glick BR, Cheng Z, Czarny J, Duan J (2007) Promotion of plant growth by ACC deaminase-producing soil bacteria. In: New perspectives and approaches in plant growth-promoting Rhizobacteria research, pp 329–339
Gontia I, Kavita K, Schmid M et al (2011) Brachybacterium saurashtrense sp. nov., a halotolerant root-associated bacterium with plant growth-promoting potential. Int J Syst Evol Microbiol 61 (Pt 12):2799–2804. https://doi.org/10.1099/ijs.0.023176-0
Gou W, Tian L, Ruan Z et al (2015) Accumulation of choline and glycinebetaine and drought stress tolerance induced in maize (Zea mays) by three plant growth promoting rhizobacteria (PGPR) strains. Pakistan J Bot 47(2):581–586
Grover M, Madhubala R, Ali SZ et al (2014) Influence of Bacillus spp. strains on seedling growth and physiological parameters of sorghum under moisture stress conditions. J Basic Microbiol 54 (9):951–961. https://doi.org/10.1002/jobm.201300250
Gupta G, Parihar SS, Ahirwar NK et al (2015) Plant growth promoting rhizobacteria (PGPR): current and future prospects for development of sustainable agriculture. J Microb Biochem Technol 7:96–102
Gusain YS, Singh US, Sharma AK (2014) Enhance activity of stress related enzymes in rice (Oryza sativa L.) induced by plant growth promoting fungi under drought stress. Afr J Agric Res 9:1430–1434
Hagen G, Guilfoyle T (2002) Auxin-responsive gene expression: genes, promoters and regulatory factors. Plant Mol Biol 49:373–385
Hammer EC, Nasr H, Pallon J et al (2011) Elemental composition of arbuscular mycorrhizal fungi at high salinity. Mycorrhiza 21(2):117–129. https://doi.org/10.1007/s00572-010-0316-4
Honma M, Smmomura T (1978) Metabolism of 1-aminocyclopropane-1-carboxylic acid. Agric Biol Chem 42:1825–1831. https://doi.org/10.1080/00021369.1978.10863261
Jamil A, Riaz S, Ashraf M, Foolad MR (2011) Gene expression profiling of plants under salt stress. Crit Rev Plant Sci 30:435–458
Jha Y, Subramanian RB, Patel S (2011) Combination of endophytic and rhizospheric plant growth promoting rhizobacteria in Oryza sativa shows higher accumulation of osmoprotectant against saline stress. Acta Physiol Plant 33(3):797–802. https://doi.org/10.1007/s11738-010-0604-9
Kasim WA, Gaafar RM, Abou-Ali RM et al (2016) Effect of biofilm forming plant growth promoting rhizobacteria on salinity tolerance in barley. Ann Agric Sci 61(2):217–227. https://doi.org/10.1016/j.aoas.2016.07.003
Khan A, Sirajuddin, Zhao XQ et al (2016) Bacillus pumilus enhances tolerance in rice (Oryza sativa L.) to combined stresses of NaCl and high boron due to limited uptake of Na+. Environ Exp Bot 124:120–129. https://doi.org/10.1016/j.envexpbot.2015.12.011
Kim K, Jang Y-J, Lee S-M et al (2014) Alleviation of salt stress by Enterobacter sp. EJ01 in tomato and Arabidopsis is accompanied by up-regulation of conserved salinity responsive factors in plants. Mol Cells 37:109
Kohler J, Hernández JA, Caravaca F, Roldán A (2008) Plant-growth-promoting rhizobacteria and arbuscular mycorrhizal fungi modify alleviation biochemical mechanisms in water-stressed plants. Funct Plant Biol 35:141–151
Kumar A, Verma JP (2018) Does plant—microbe interaction confer stress tolerance in plants: a review? Microbiol Res 207:41–52

Kumar KV, Srivastava S, Singh N, Behl HM (2009) Role of metal resistant plant growth promoting bacteria in ameliorating fly ash to the growth of Brassica juncea. J Hazard Mater 170(1):51–57. https://doi.org/10.1016/j.jhazmat.2009.04.132

Lata R, Chowdhury S, Gond SK, White JF (2018) Induction of abiotic stress tolerance in plants by endophytic microbes. Lett Appl Microbiol 66(4):268–276

Lee Y, Krishnamoorthy R, Selvakumar G et al (2015) Alleviation of salt stress in maize plant by co-inoculation of arbuscular mycorrhizal fungi and Methylobacterium oryzae CBMB20. J Korean Soc Appl Biol Chem 58(4):553–540. https://doi.org/10.1007/s13765-015-0072-4

Li Q, Yin M, Li Y et al (2015) Expression of Brassica napus TTG2, a regulator of trichome development, increases plant sensitivity to salt stress by suppressing the expression of auxin biosynthesis genes. J Exp Bot 66:5821–5836

Li H, Lei P, Pang X et al (2017) Enhanced tolerance to salt stress in canola (Brassica napus L.) seedlings inoculated with the halotolerant Enterobacter cloacae HSNJ4. Appl Soil Ecol 119:26–34

Lim JH, Kim SD (2013) Induction of drought stress resistance by multi-functional PGPR Bacillus licheniformis K11 in pepper. Plant Pathol J 29(2):201–208. https://doi.org/10.5423/PPJ.SI.02.2013.0021

Lim CW, Baek W, Jung J et al (2015) Function of ABA in stomatal defense against biotic and drought stresses. Int J Mol Sci 16(7):15251–15270

Liu F, Xing S, Ma H et al (2013) Cytokinin-producing, plant growth-promoting rhizobacteria that confer resistance to drought stress in Platycladus orientalis container seedlings. Appl Microbiol Biotechnol 97(20):9155–9164. https://doi.org/10.1007/s00253-013-5193-2

Ma Y, Rajkumar M, Moreno A et al (2017) Serpentine endophytic bacterium Pseudomonas azotoformans ASS1 accelerates phytoremediation of soil metals under drought stress. Chemosphere 185:75–85. https://doi.org/10.1016/j.chemosphere.2017.06.135

Matsubara Y, Ishioka C, Maya MA et al (2014) Bioregulation potential of arbuscular mycorrhizal fungi on heat stress and anthracnose tolerance in cyclamen. Acta Hortic 1037:813–818. https://doi.org/10.17660/ActaHortic.2014.1037.108

Mayak S, Tirosh T, Glick BR (2004) Plant growth-promoting bacteria confer resistance in tomato plants to salt stress. Plant Physiol Biochem 42:565–572

Meena KK, Sorty AM, Bitla UM et al (2017) Abiotic stress responses and microbe-mediated mitigation in plants: the Omics strategies. Front Plant Sci 8:172. https://doi.org/10.3389/fpls.2017.00172

Miller KJ, Wood JM (1996) Osmoadaptation by Rhizosphere bacteria. Annu Rev Microbiol 50:101–136. https://doi.org/10.1146/annurev.micro.50.1.101

Milosevic N, Marinkovic J, Tintor B (2012) Mitigating abiotic stress in crop plants by microorganisms. Zb Matice Srp za Prir Nauk 2012(123):17–26. https://doi.org/10.2298/ZMSPN1223017M

Mishra PK, Bisht SC, Ruwari P et al (2011) Alleviation of cold stress in inoculated wheat (Triticum aestivum L.) seedlings with psychrotolerant pseudomonads from NW Himalayas. Arch Microbiol 193(7):497–513. https://doi.org/10.1007/s00203-011-0693-x

Naseem H, Bano A (2014) Role of plant growth-promoting rhizobacteria and their exopolysaccharide in drought tolerance of maize. J Plant Interact 9:589–701. https://doi.org/10.1080/17429145.2014.902125

Nautiyal CS, Srivastava S, Chauhan PS et al (2013) Plant growth-promoting bacteria Bacillus amyloliquefaciens NBRISN13 modulates gene expression profile of leaf and rhizosphere community in rice during salt stress. Plant Physiol Biochem 66:1–9

Naveed M, Mitter B, Reichenauer TG et al (2014) Increased drought stress resilience of maize through endophytic colonization by Burkholderia phytofirmans PsJN and Enterobacter sp. FD17. Environ Exp Bot 97:30–39. https://doi.org/10.1016/j.envexpbot.2013.09.014

Naya L, Ladrera R, Ramos J et al (2007) The response of carbon metabolism and antioxidant Defenses of alfalfa nodules to drought stress and to the subsequent recovery of plants. Plant Physiol 144(2):1104–1114. https://doi.org/10.1104/pp.107.099648

Nidhi B, Pandey SS, Barnwal D, Patel VK, Kalra A (2016) Plant growth promoting rhizobacteria Dietzia natronolimnaea modulates the expression of stress responsive genes providing protection of wheat from salinity stress. Sci Rep 6:34768. https://doi.org/10.1038/srep34768

Niu SQ, Li HR, Paré PW et al (2016) Induced growth promotion and higher salt tolerance in the halophyte grass Puccinellia tenuiflora by beneficial rhizobacteria. Plant Soil 407(1–7):217–230. https://doi.org/10.1007/s11104-015-2767-z

Orhan F (2016) Alleviation of salt stress by halotolerant and halophilic plant growth-promoting bacteria in wheat (Triticum aestivum). Brazilian J Microbiol 47(3):621–627. https://doi.org/10.1016/j.bjm.2016.04.001

Palaniyandi SA, Damodharan K, Yang SH, Suh JW (2014) Streptomyces sp. strain PGPA39 alleviates salt stress and promotes growth of 'micro Tom'tomato plants. J Appl Microbiol 117:766–773

Pandey V, Ansari MW, Tula S et al (2016) Dose-dependent response of Trichoderma harzianum in improving drought tolerance in rice genotypes. Planta 243(5):1251–1264. https://doi.org/10.1007/s00425-016-2482-x

Pereyra MA, García P, Colabelli MN et al (2012) A better water status in wheat seedlings induced by Azospirillum under osmotic stress is related to morphological changes in xylem vessels of the coleoptile. Appl Soil Ecol 53:94–97. https://doi.org/10.1016/j.apsoil.2011.11.007

Petrova OE, Sauer K (2016) Escaping the biofilm in more than one way: desorption, detachment or dispersion. Curr Opin Microbiol 30:67–78

Porcel R, Aroca R, Azcón R, Ruiz-Lozano JM (2006) PIP aquaporin gene expression in arbuscular mycorrhizal Glycine max and Lactuca sativa plants in relation to drought stress tolerance. Plant Mol Biol 60(3):389–404. https://doi.org/10.1007/s11103-005-4210-y

Rabab AM, Reda EA (2018) Synergistic effect of arbuscular mycorrhizal fungi on growth and physiology of salt-stressed Trigonella foenum-graecum plants. Biocatal Agric Biotechnol 16 (6):538–544

Radhakrishnan R, Lee IJ (2015) Penicillium-sesame interactions: a remedy for mitigating high salinity stress effects on primary and defense metabolites in plants. Environ Exp Bot 116:47–60. https://doi.org/10.1016/j.envexpbot.2015.03.008

Rouhier N, Koh CS, Gelhaye E et al (2008) Redox based anti-oxidant systems in plants: biochemical and structural analyses. Biochim Biophys Acta Gen Subj 1780:1249–1260

Saikia J, Sarma RK, Dhandia R et al (2018) Alleviation of drought stress in pulse crops with ACC deaminase producing rhizobacteria isolated from acidic soil of Northeast India. Sci Rep 8 (1):3560. https://doi.org/10.1038/s41598-018-21921-w

Sairam RK, Srivastava GC, Agarwal S, Meena RC (2005) Differences in antioxidant activity in response to salinity stress in tolerant and susceptible wheat genotypes. Biol Plant 49:85. https://doi.org/10.1007/s10535-005-5091-2

Sandhya V, Ali S, Grover M et al (2009a) Pseudomonas sp. strain P45 protects sunflowers seedlings from drought stress through improved soil structure. J Oilseed Res 26:600–601

Sandhya V, Ali SKZ, Grover M et al (2009b) Alleviation of drought stress effects in sunflower seedlings by the exopolysaccharides producing Pseudomonas putida strain GAP-p45. Biol Fertil Soils 46(1):17–26. https://doi.org/10.1007/s00374-009-0401-z

Sarkar A, Ghosh PK, Pramanik K et al (2017) A halotolerant Enterobacter sp. displaying ACC deaminase activity promotes rice seedling growth under salt stress. Res Microbiol 169(1):20–32

Sarkar A, Pramanik K, Mitra S et al (2018a) Enhancement of growth and salt tolerance of rice seedlings by ACC deaminase-producing Burkholderia sp. MTCC 12259. J Plant Physiol 231:434–442

Sarkar J, Chakraborty B, Chakraborty U (2018b) Plant growth promoting Rhizobacteria protect wheat plants against temperature stress through antioxidant signalling and reducing chloroplast and membrane injury. J Plant Growth Regul 37:1396–1412

Sarma RK, Saikia R (2014) Alleviation of drought stress in mung bean by strain Pseudomonas aeruginosa GGRJ21. Plant Soil 377(1–2):111–126. https://doi.org/10.1007/s11104-013-1981-9

Shahzad R, Khan AL, Bilal S et al (2017a) Plant growth-promoting endophytic bacteria versus pathogenic infections: an example of *Bacillus amyloliquefaciens* RWL-1 and *Fusarium oxysporum* f. sp. *lycopersici* in tomato. PeerJ 5:e3107. https://doi.org/10.7717/peerj.3107

Shahzad R, Khan AL, Bilal S et al (2017b) Inoculation of abscisic acid-producing endophytic bacteria enhances salinity stress tolerance in Oryza sativa. Environ Exp Bot 136:68–77. https://doi.org/10.1016/j.envexpbot.2017.01.010

Shrivastava P, Kumar R (2015) Soil salinity: a serious environmental issue and plant growth promoting bacteria as one of the tools for its alleviation. Saudi J Biol Sci 22(2):123–131

Singh JS, Pandey VC, Singh DP (2011) Efficient soil microorganisms: a new dimension for sustainable agriculture and environmental development. Agric Ecosyst Environ 140(3–4):339–353

Singh RP, Jha P, Jha PN (2015) The plant-growth-promoting bacterium Klebsiella sp. SBP-8 confers induced systemic tolerance in wheat (Triticum aestivum) under salt stress. J Plant Physiol 184:57–67

Subramanian P, Mageswari A, Kim K et al (2015) Psychrotolerant endophytic *Pseudomonas* sp. strains OB155 and OS261 induced chilling resistance in tomato plants (*Solanum lycopersicum* mill.) by activation of their antioxidant capacity. Mol Plant-Microbe Interact 28(10):1073–1081. https://doi.org/10.1094/MPMI-01-15-0021-R

Subramanian P, Kim K, Krishnamoorthy R et al (2016) Cold stress tolerance in psychrotolerant soil bacteria and their conferred chilling resistance in tomato (Solanum lycopersicum mill.) under low temperatures. PLoS One 11(8):e0161592. https://doi.org/10.1371/journal.pone.0161592

Sun C, Johnson JM, Cai D et al (2010) Piriformospora indica confers drought tolerance in Chinese cabbage leaves by stimulating antioxidant enzymes, the expression of drought-related genes and the plastid-localized CAS protein. J Plant Physiol 167(12):1009–1017. https://doi.org/10.1016/j.jplph.2010.02.013

Sziderics AH, Rasche F, Trognitz F et al (2007) Bacterial endophytes contribute to abiotic stress adaptation in pepper plants (*Capsicum annuum* L.). Can J Microbiol 53(11):1195–1202. https://doi.org/10.1139/W07-082

Theocharis A, Bordiec S, Fernandez O et al (2012) *Burkholderia phytofirmans* PsJN primes *Vitis vinifera* L. and confers a better tolerance to low nonfreezing temperatures. Mol Plant-Microbe Interact 25(2):241–249. https://doi.org/10.1094/MPMI-05-11-0124

Thomashow MF (1999) Plant cold acclimation: freezing tolerance genes and regulatory mechanisms. Annu Rev Plant Physiol Plant Mol Biol 50:571–599. https://doi.org/10.1146/annurev.arplant.50.1.571

Timmusk S, Wagner EGH (1999) The plant-growth-promoting rhizobacterium Paenibacillus polymyxa induces changes in Arabidopsis thaliana gene expression: a possible connection between biotic and abiotic stress responses. Mol Plant-Microbe Interact 12:951–959

Tiwari S, Lata C, Chauhan PS, Nautiyal CS (2016) Pseudomonas putida attunes morphophysiological, biochemical and molecular responses in Cicer arietinum L. during drought stress and recovery. Plant Physiol Biochem 99:108–117. https://doi.org/10.1016/j.plaphy.2015.11.001

Tripathi P, Singh PC, Mishra A et al (2017) Arsenic tolerant Trichoderma sp. reduces arsenic induced stress in chickpea (Cicer arietinum). Environ Pollut 223:137–145. https://doi.org/10.1016/j.envpol.2016.12.073

Upadhyay SK, Singh JS, Saxena AK, Singh DP (2012) Impact of PGPR inoculation on growth and antioxidant status of wheat under saline conditions. Plant Biol 14:605–611

Vaishnav A, Choudhary DK (2018) Regulation of drought-responsive gene expression in Glycine max L. Merrill is mediated through Pseudomonas simiae strain AU. J Plant Growth Regul 38 (1):333–342

Vardharajula S, Ali SZ, Grover M et al (2011) Drought-tolerant plant growth promoting bacillus spp.: effect on growth, osmolytes, and antioxidant status of maize under drought stress. J Plant Interact 6(1):1–14. https://doi.org/10.1080/17429145.2010.535178

Vargas L, Brígida ABS, Mota Filho JP et al (2014) Drought tolerance conferred to sugarcane by association with gluconacetobacter diazotrophicus: a transcriptomic view of hormone pathways. PLoS One 9(12):e114744. https://doi.org/10.1371/journal.pone.0114744

Vílchez JI, García-Fontana C, Román-Naranjo D et al (2016) Plant drought tolerance enhancement by trehalose production of desiccation-tolerant microorganisms. Front Microbiol 7:1577. https://doi.org/10.3389/fmicb.2016.01577

Vimal SR, Gupta J, Singh JS (2018a) Effect of salt tolerant Bacillus sp. and Pseudomonas sp. on wheat (Triticum aestivum L.) growth under soil salinity: a comparative study. Microbiol Res (Pavia) 9:1

Vimal SR, Patel VK, Singh JS (2018b) Plant growth promoting Curtobacterium albidum strain SRV4: an agriculturally important microbe to alleviate salinity stress in paddy plants. Ecol Indic

Vurukonda SSKP, Vardharajula S, Shrivastava M, SkZ A (2016) Enhancement of drought stress tolerance in crops by plant growth promoting rhizobacteria. Microbiol Res 184:13–24

Wang C, Wang C, Gao YL et al (2016) A consortium of three plant growth-promoting Rhizobacterium strains acclimates Lycopersicon esculentum and confers a better tolerance to chilling stress. J Plant Growth Regul 35(1):54–64. https://doi.org/10.1007/s00344-015-9506-9

Xu L, Wang A, Wang J, Wei Q, Zhang W (2017) Piriformospora indica confers drought tolerance on Zea mays L. through enhancement of antioxidant activity and expression of drought-related genes. Crop J 5(3):251–258

Yan J, Smith MD, Glick BR, Liang Y (2014) Effects of ACC deaminase containing rhizobacteria on plant growth and expression of Toc GTPases in tomato (*Solanum lycopersicum*) under salt stress. Botany 92(11):775–781. https://doi.org/10.1139/cjb-2014-0038

Yang J, Kloepper JW, Ryu C-M (2009) Rhizosphere bacteria help plants tolerate abiotic stress. Trends Plant Sci 14:1–4

Yang S, Vanderbeld B, Wan J, Huang Y (2010) Narrowing down the targets: towards successful genetic engineering of drought-tolerant crops. Mol Plant 3(3):469–490

Yuan DP, Zhang C, Wang ZY et al (2018) RAVL1 activates brassinosteroids and ethylene signaling to modulate response to sheath blight disease in rice. Phytopathology 108(9):1104–1113

Yun P, Xu L, Wang S-S et al (2018) Piriformospora indica improves salinity stress tolerance in Zea mays L. plants by regulating Na+ and K+ loading in root and allocating K+ in shoot. Plant Growth Regul 86:323–331

Zahir ZA, Munir A, Asghar HN et al (2008) Effectiveness of rhizobacteria containing ACC deaminase for growth promotion of peas (Pisum sativum) under drought conditions. J Microbiol Biotechnol 18(5):958–963

Zaidi NW, Singh M, Kumar S et al (2018) Trichoderma harzianum improves the performance of stress-tolerant rice varieties in rainfed ecologies of Bihar, India. Field Crop Res 220(97):104. https://doi.org/10.1016/j.fcr.2017.05.003

Zhang H, Kim M-S, Sun Y et al (2008) Soil bacteria confer plant salt tolerance by tissue-specific regulation of the sodium transporter HKT1. Mol Plant-Microbe Interact 21:737–744

Zhao L, Wang F, Zhang Y, Zhang J (2014) Involvement of Trichoderma asperellum strain T6 in regulating iron acquisition in plants. J Basic Microbiol 54(Suppl 1):S115–S124. https://doi.org/10.1002/jobm.201400148

Zhao L, Zhang Y (2015) Effects of phosphate solubilization and phytohormone production of Trichoderma asperellum Q1 on promoting cucumber growth under salt stress. J Integr Agric 14 (8):1588–1597

Zhou S, Hu W, Deng X et al (2012) Overexpression of the wheat aquaporin gene, TaAQP7, enhances drought tolerance in transgenic tobacco. PLoS One 7:e52439

Chapter 17
Microbial Diversity in Soil: Biological Tools for Abiotic Stress Management in Plants

Neera Garg, Kiran Saroy, Amandeep Cheema, and Aditi Bisht

Abstract Soil is a dynamic and complex biological system comprising of various populations of microbes. These microorganisms enhance soil richness or fertility and water retention ability and have a major impact on the growth of the vegetation present on the ground. The range of diversity of microbes plays an essential role in improvement of soil quality that varies with depth, pH and horizons as well as soil health. Abiotic stresses comprising drought, salinity, high and low temperature and heavy metal (HM) toxicity are the principal restraining factors for microbial inhabitants and agricultural productivity. Several reports have indicated that inoculation of microbes in nutrient-deficient or stressed soil has significantly improved soil as well as plant health and established supportable way of agriculture. Microorganisms interact with plants and regulate systemic and local mechanisms to provide defence under adverse environmental conditions. Besides providing non-nutritional effects such as limiting soil erosion and detoxifying insecticides as well as pesticides, microbes also help in atmospheric nitrogen (N_2) fixation; activation of antioxidants, phytohormones and osmolytes; solubilization of mineral phosphates; decomposition of organic wastes; regulation of gene expression; improvement of nutrient cycling and plant-water relation; etc. and assist the plant to survive under unfavourable conditions. In addition, use of harmonious multiple microbial associations (e.g. fungal and bacterial) provides several benefits and is a promising approach towards this direction. This review deals with the potential of soil and its inhabitant microbes to nurture plants, plant-microbe interactions and their roles in abiotic stress management.

17.1 Introduction

Microbes make up diversity of life in the soil and have a vital role in environmental processes by influencing the earth's biogeochemical cycles, nutrient acquisition and pedogenesis (Schulz et al. 2013; Smith et al. 2015b). Being important living components of the soil, they become an integral part of crop production system as

---

N. Garg (✉) · K. Saroy · A. Cheema · A. Bisht
Department of Botany, Panjab University, Chandigarh, India

A. Varma et al. (eds.), *Plant Biotic Interactions*,
https://doi.org/10.1007/978-3-030-26657-8_17

soon as a seed germinates in the soil to start its life cycle (Meena et al. 2017). The soil near plant roots (rhizosphere) is an important habitat and ecosystem for these microbes (Ullah et al. 2015) where they live in harmony with each other. Bender et al. (2016) have identified approximately 200 m fungal hyphae and 1 bn bacterial cells in 1 g of soil. Various factors of soil, namely, physical (soil texture, moisture, aeration, etc.), chemical (charge on soil particles, pH) and biological (other living organisms like roots) as well as plant factors (e.g. species, age), influence the diversity of microbes (Grayston et al. 1998). Many of these factors interrelate with each other and affect the soil microbial community directly or indirectly, e.g. fungal and bacterial groups differ extremely because of diverse habitats and soil horizons (Buyer et al. 2010; Buscardo et al. 2018). Fungi usually predominate in pH < 7 (acidic soils), while bacteria become more prevalent in pH > 7 (alkaline soils). Fungi flourish and disperse more intensively in the soil, whereas bacteria have access to smaller pore spaces only (Ritz and Young 2004; Ellouze et al. 2014). Moreover, within similar climatic zone, a forest soil is usually found to have a greater range of microorganisms as compared to agricultural soil.

Soil microbes play a very important role in the maintenance and improvement of soil structure by binding organic substances and small soil particles to form aggregates (Bach et al. 2010). Soil microorganisms secrete gums, glycoproteins and polysaccharides which adhere soil minerals together, forming the basis for soil structure (Hassink 1994). Bacteria living in the rhizosphere can influence pH through the release of organic acids, thus making the nutrients more available to plants (Dimkpa et al. 2009). Moreover, microbes improve the proportion of carbon and nitrogen in the soil which helps in maintaining soil texture and health (Nielsen et al. 2002; Bhatia 2008). Presently, soil bacteria have also been used in bioremediation of organic pollutants in soils by the process of mineralization (Garbisu et al. 2017).

One of the key challenges in present agriculture system is to achieve improved plant growth and yield even under unfavourable environmental circumstances. Various anthropogenic as well as natural activities have led to soil degradation due to several abiotic stresses which has an undesirable impact on the development and growth of plants (Gururani et al. 2015; Meena et al. 2017). These stresses involve salinity (due to improper irrigation practices), temperature (change in environment), heavy metals (by mining, volcanic eruption), flood/drought (excess/less rainfall), etc. Abiotic stresses in increasing order of their affected areas worldwide are salinity (6%), flood, anoxia (13%) and drought (64%) (Mittler 2006; Cramer et al. 2011). Plants show symptoms like wilting, chlorosis, necrosis, reduced biomass, reduction in photosynthetic and respiration rates, generation of reactive oxygen species, etc. under these stresses. Most of the microbes have the ability to survive under stressed soil in an inactive state and get activated when favourable conditions are maintained. In addition, microbes have developed several molecular and physiological mechanisms to strive under stressed environments. Such mechanisms include modification in the cell wall (bacteria), making of cell surface epitope to defend the cell from host immune reactions (in fungi) (Ayalaja and Pedro 2012; Latge 2017), altering metabolic responses and changing gene expression through mutations against various stresses (Mendes et al. 2013; Hartmann et al. 2017).

In the last few decades, role of microbes in amelioration of diverse abiotic stresses in plants has been the area of great concern (Mittler 2006; de Zelicourt et al. 2013; Ahanger et al. 2017). The microbes involved in stress managements are some endophytes (bacteria and fungi) with special emphasis on plant growth-promoting rhizobacteria (PGPRs) (e.g. *Azospirillum*, *Enterobacter*, *Pseudomonas*, *Rhizobium*, *Bacillus*, cyanobacteria, etc.) and fungi (i.e. saprophytes and mutualists arbuscular mycorrhizal fungi (AMF)) (Barea et al. 2005; Hassen et al. 2016). These microorganisms have attained diverse mechanisms to strive against these stresses. In soil, PGPRs enhance the iron (Fe) accessibility by producing Fe-chelating agents (siderophores) and releasing organic acids (malate, citrate, etc.) which decline pH and promote plant development by regulating nitrogen fixation and plant hormones, secreting enzymes, enhancing the antioxidant system, emitting volatile organic compounds (VOCs), etc. (Singh and Jha 2016; Cohen et al. 2017; İpek and Eşitken 2017). Moreover, under adverse conditions, few bacteria have sigma factors to overcome negative effects by upregulating gene expression (Taniguchi and Wendisch 2015). Besides bacteria, AMF modify rhizospheric environment by amassing glomalin-related soil protein (GRSP) and increasing absorptive surface area (by forming arbuscules and vesicles in root cortical cells) and water holding capacity (Yin et al. 2016). Fungal hyphae also increase supply of nutrients (by upregulating P transporters) and change transcript levels of genes as well as activity of antioxidant enzymes involved in stress responses (Manchanda and Garg 2011; Talaat and Shawky 2011; Maya and Matsubara 2013).

The above-mentioned benefits of microbes are the results of exchange of signals during plant-microbe interactions which play a pivotal role in sustainable ecosystem. These interactions take place in the rhizospheric environment by diverse mechanisms, such as plant or microbial signalling, volatile production and quorum sensing (QS). Plant roots send signals to microbes by producing exudates such as amino acids (AAs), sugars, several dicarboxylic acids (fumarate, succinate, malate) and organic compounds which act as chemoattractants for microorganisms and influence their assemblage in the rhizosphere (Garg and Geetanjali 2007; Rosier et al. 2016). In bacteria, fungi and plants, volatile metabolites (alkyl sulphides, terpenes, indole, etc.) manipulate physiological processes and also play an essential role in long-distance microorganism interactions by distributing easily through air- and gas-filled pores in the soil (Audrain et al. 2015; Schmidt et al. 2015). Quorum sensing (QS) is facilitated by tiny diffusive signal molecules (autoinducers) regulating gene expression activities of population in response to their mass (Hooshangi and Bentley 2008; Garg et al. 2014a). This review gives a summary of recent knowledge on microbial diversity in soil, their interactions with plants and their roles in imparting abiotic stress tolerance.

17.1.1 Soil Quality, Habitat, Microbial Diversity

Soil contains organic matter, minerals, water, air and living entities that cover the necessary components and amenities to sustain life. In present time, with increase in the demand for crop productivity, use of chemical products has also increased which has negatively affected the soil quality (SQ). The concept of SQ comprises of its capacity for carbon sequestration, water retention, crop productivity and other functions (Warkentin 1995; Bünemann et al. 2018). Soil organic matter (SOM) or soil organic carbon (SOC) and soil pH are commonly known as key chemical parameters of soil quality which influence water relations, soil porosity and gas exchange process. Moreover, SOM has a great impact on several fundamental, biological and chemical practices (e.g. adsorption, chemical form, precipitation) which play a crucial role in nutrient release and accessibility (Gale et al. 1991; Nambiar 1997). Soil habitat is defined as the whole living organisms inhabiting in the soil which contains microbes, animals, plants and their abiotic atmosphere (Garbeva et al. 2004). In soil, microorganisms (bacteria, fungi, nematodes, protozoa, etc.) exist in a three-dimensional (3D) physical atmosphere that vary with time and space and balance the flows of energy, nutrients, water, etc. (Young and Ritz 2000). Most abundant microbes in soil are bacteria, trailed by the actinomycetes, fungi, algae and protozoa (Davinic et al. 2012; Meliani et al. 2012). Different niches and alterations in organic matter and environmental conditions (such as oxygen concentration) within/between soil aggregates result into microbial diversity (Sexstone et al. 1985), e.g. large macroaggregates (>2 mm) involve more filamentous fungi, which add to macroaggregate formation and stabilization, whereas microaggregates confined more bacteria which help in maintaining soil health and quality (Rillig et al. 2015; Mummey and Stahl 2004; Davinic et al. 2012; Luo et al. 2016). Microbes are necessary for the various soil functions in normal as well as managed agricultural system because of their participation in processes, namely, toxin removal, decomposition of organic matter, biogeochemical cycle (C, N, P and S) and soil structure formation (Das and Varma 2010). Although several reports illustrate the relationships between abundance of microbial variations in soil, plant growth and ecosystem sustainability, PGPRs and AMFs have recently emerged as major area of research. PGPRs are involved in diverse useful events within the soil like disintegration of crop residues, mineralization and synthesis of SOM, immobilization of mineral nutrients, nitrification and nitrogen fixation, phosphate solubilization, plant hormone production (biostimulants), etc. which helps in crop production and protection. Additionally, PGPRs produce auxins, which help in promoting the expansion of mycorrhizal mycelium (Fernández-Bidondo et al. 2011). This mycorrhizal mycelium colonizes the plant roots, causing changes in root exudates and selectively affecting (negative or positive) the communities of other microbes present in the rhizosphere (Cornejo et al. 2017). For example, *Lolium multiflorum* inoculated with *Glomus versiforme* have shown more copiousness of nitrobacters, denitrifying, ammonificant and phosphorus bacteria, which increased alkaline phosphatase and urease activity in soil (Ye et al. 2015). Moreover, AMF produced protein, i.e. glomalin (GRSP—glomalin-related soil protein), which plays an essential role in the soil aggregation and affects carbon (C) dynamics in agroecosystems.

17.2 Plant-Microbe Interactions

The space nearby the roots is microbially dynamic (Hiltner 1904) and is an essential niche for plant-microbe interactions in terms of nutrient transformations (biogeochemical cycles) and uptake, exudation, etc. (White 2003). Major beneficial microbial diversities which interact with plants are PGPRs (*Alcaligenes*, *Beijerinckia*, *Burkholderia*, *Enterobacter*, *Flavobacterium*, *Serratia*, *Pseudomonas*, *Azotobacter*, *Azospirillum*, *Rhizobium*, *Pantoea*, *Bacillus*, *Bradyrhizobium*, cyanobacteria, etc.) and AMF (*Glomus* spp., *Archaeospora*, *Acaulospora*, etc.) (Sturz et al. 2000; Sudhakar et al. 2000; Wang et al. 2004; Karlidag et al. 2007; Ilyas and Bano 2012; Evelin et al. 2009). These microbes interact with plant via various processes such as plant-microbial signalling, production of volatile compounds (VCs) and quorum sensing (QS) which are discussed as follows:

17.2.1 Plant-Microbial Signalling

Higher plants have developed familiar, multifaceted, subtle and relatively persistent relationships with a group of microorganisms, known as the phytomicrobiome (Smith et al. 2015a, b). Microbial assemblages in the rhizosphere are influenced by various plant signals, and both can use numerous molecular signals to intercommunicate with each other. It is well-defined for the legume-rhizobial symbiosis (N_2-fixation) and realistically revealed for mycorrhizal colonizations (Smith et al. 2015a, b; Rosier et al. 2016). Strigolactones and flavonoids act as essential plant signals that are professed by the AMF and rhizobium to elicit symbiotic responses with their host plants (Holmer et al. 2017).

Symbiosis between rhizobium and legumes initiates with an intricate signal exchange process (Hirsch and Fujishige 2012). Firstly, the root of leguminous plants secretes exudate compounds, namely, flavonoids, amino acids (AAs) and phenolic compounds (flavone, flavanones, isoflavones, etc.) (Cooper 2007; Badri et al. 2009; De-la-Peña et al. 2010). In bacteria these compounds behave as nodulation gene inducers by beginning several chemotactic reactions (Hirsch and Fujishige 2012; Ryu et al. 2012). After the activation of nodulation genes, the rhizobia release Nod factors (lipo-chitooligosaccharides) associated with rhizobial root infection which is specific to symbiotic association, and these factors also induce cellular alterations (Oldroyd and Downie 2004; Jones et al. 2007). Perception of Nod factors by plants incites physiological and molecular responses, such as amassing of early nodulin transcripts (i.e. genes induced during nitrogen-fixing symbiosis in plants), root hair curling and the development of the infection thread, where bacteria can multiply and move towards the cells of the root cortex (Gage 2004). In these cortical cells, symbiosomes (organelle-like structures), formed by bacteria, are converted into bacteroids and ready to fix atmospheric N_2 by the activity of the nitrogenase enzyme complex (Clúa et al. 2018). Root exudates (such as flavonoids) also play vital role in initiation of AMF interaction with the host plant via signal exchange mechanisms. Numbers of flavonoid have stimulatory effect on hyphal growth as well as branching, and these effects depend on the chemical composition of the complexes (Giovannetti et al. 1994; Scervino et al. 2006). Moreover, strigolactones (SLs) present in the exudates of plants are also an essential signalling compounds for the formation of the AMF symbiosis (Akiyama and Hayashi 2006; Gomez-Roldan et al. 2007). SLs act as primary signals for AMF hyphal branching and also play a vital role for guiding progress of AMF in the direction of roots (Sbrana and Giovannetti 2005; Steinkellner et al. 2007). The SL signalling apparatus is made up of (At)D14/DAD2/RMS3-α/β-fold hydrolase, MAX2/RMS4/D3-F-box leucine-rich protein and the repressor protein-D53 and belongs to SMAX1-like (SMXL) family proteins (Saeed et al. 2017). MAX2, leucine-rich F-box proteins, has been revealed a part of the SKP1-CUL1-F-box protein (SCF)-type ubiquitin ligase complex, which can ubiquitinate target proteins and lead to proteasomal disintegration (Stirnberg et al. 2007; Arite et al. 2009; Xie et al. 2013). Present systematic assumption of SL perception suggests that D14 acts in a signalling pathway imitating one of gibberellins (GA) and containing MAX2-F-box protein (Seto and Yamaguchi 2014; Saeed et al. 2017). D14 is a sole receptor and an active hydrolase unique enzyme unlike other plant hormones when GID1 doesn't transform its ligand. D14-mediated SL perception depends on three catalytic triads made up of AAs-Asp, His and Ser for hydrolysis and binding (Hamiaux et al. 2012). MAX2/D3 behaves in the SCF complex as identification subunit for SL-loaded D14 and downstream repressors. D14 with SLs transforms them, chiefly via hydrophobic ligand-binding pocket and specific nucleophilic residues in the amino acid triad, which strike carbonyl of butenolide of D ring and detached it from ABC complex (Scaffidi et al. 2012). SL-loaded D14 then employs to SCF^{MAX2}, which via the help of proteasome further leads to disintegration of intent proteins (e.g. D53) (Zhao et al. 2015). Additionally, D14 itself is also a target for late proteasomal disintegration, which is implicit to strengthen SL desensitization (Chevalier et al. 2014). However, certain orthologs of D53 were investigated in *Arabidopsis* (suppressor of MAX-2-LIKE 6-8 SMXL6-8) which intricates in suppression branching of stem and another action controlled via SLs (Soundappan et al. 2015). D53 is thought to interrelate through transcriptional corepressors named topless-related (TPR) proteins which ultimately form D53-TPR complex and supress the SL transcriptional action (Jiang et al. 2013; Smith and Li 2014).

17.2.2 Volatile Compounds (VCs)

Volatile compounds are chemicals ssecreted by all microbes as a part of their metabolism which facilitate signalling across all kingdoms (bacteria, fungi, plants, etc.) and play an important role in diverse microbial interactions as well as manipulate various physiological processes (Thorn et al. 2011; Schmidt et al. 2015). Varieties of VCs formed by bacteria include fatty acids (FAs) and their by-products, N_2-containing compounds, volatile S-compounds and aromatic complexes (Schulz and Dickschat

2007). Bacterial interactions with surrounding microorganisms via production of VCs enable bacteria to respond and adapt to environmental changes. In intercellular contacts, bacteria also secrete wide varieties of volatile secondary metabolites (Audrain et al. 2015). *Sorghum bicolor* inoculated with either one of the AMF (*G. mosseae* or *G. intraradices*) produced more alkenes, ethers, alcohols and acids than non-mycorrhizal plants, indicating that AMF have the capacity to alter the profile of VCs emitted by roots, thus helping the plants to acclimatize and modify soil environments (Sun and Tang 2013). Moreover, Cleason (2006) reported that wide range of VCs, aldehydes, amines and lightweight organic acids were emitted by a mixture of five fungi (*Aspergillus versicolor*, *Ulocladium botrytis*, *Penicillium chrysogenum*, *Wallemia sebi* and *Fusarium culmorum*) and bacterium (*Streptomyces albidoflavus*). Ketones and alcohols were the major VCs, whereas the bacterial culture mounting on TGEA emanated methylamine, diethylamine, ethylamine and ammonia.

17.2.3 Quorum Sensing (QS)

Quorum sensing or cell-cell communication is mediated by autoinducers, which can control the gene expression of bacterial communities in response to their firmness in the surroundings (Hooshangi and Bentley 2008). In response to autoinducers such as N-acyl-homoserine lactone (AHL) signals, regulation of bacterial gene expression varies with the oxidation state, length and degree of saturation of their acyl side chains, etc. (Miller and Bassler 2001; Fuqua et al. 2001; Badri et al. 2009). Two major proteins are associated with QS system in Gram-negative bacteria: transcriptional regulator R (LuxR) and autoinducer synthase I (LuxI). AHL ligates specifically to the transcriptional regulator LuxR, used for interspecies as well as intraspecies communication. This type of communication can also occur between different organisms (bacteria-bacteria, bacteria-fungi and bacteria-plants) and can perform diverse roles: virulence, fluorescence emission, competence, biofilm formation sporulation, antibiotic production, etc. (Rinaudi and González 2009). Additionally, numerous plant-associated bacteria need QS for colonization of the plant-associated surroundings and control various phenotypes including hydrolytic enzymes secretion, bacteria-legume symbiosis, secondary metabolites production, etc. (von Bodman et al. 2003).

One such mechanism of QS involves plant-associated bacteria, i.e. symbiotic rhizobium with legume which involves exchange of signals between rhizobium and plant by the help of AHLs which regulate expression and repression of the symbiotic genes during nodulation (Daniels et al. 2002; Khan 2005). In *R. leguminosarum*, the secretion of seven AHLs is managed by four LuxI homologs (i.e. RhiI, CinI, RaiI and TraI) (Lithgow et al. 2000). QS is involved in rhizosphere growth by influencing nodulation via regulating RhiR-mediated rhiABC operon. Mutation of rhiA or rhiR leads to reduction in nodulation potential of the mutant strains which were faulty in the production of the lipochitin Nod signal, whereas RhiI mutants had positive role in nodule formation. Daniels et al. (2002) reported that in *R. etli*, the presence of CIN system is necessary for proper nodule growth, and CinI is expressed during the formation of infection thread and formation of bacteroid colonies. In contrast, CinI mutants result in the decline in N_2 fixation efficacy, typical bacteroid morphology and growth. Some QS autoinducers (*N*-3-oxo-tetradecanoyl-L-homoserine lactone) can provoke resistance in *Arabidopsis* plants to bacteria *P. syringae* and *Golovinomyces orontii* and in *Hordeum vulgare* to *Blumeria graminis* f. sp. hordei (Schikora et al. 2011). Autoinducers play vital role in cell density-controlled processes as well as in disease suppression. Moreover, various plants have developed mechanisms to perceive nanomolar (nM) concentrations of bacterial QS compounds formed by symbiotic and pathogenic partners (Lehmann et al. 2014). Plants can sense these QS compounds by using them as signals to upregulate various responses and also stimulate their own 'signal-mimic' complexes to actively regulate bacterial processes (Lehmann et al. 2014). Halogenated furanones (structural homologous of AHL) are QS 'mimic' compounds manufactured by the different plants as well as some algae (Bauer and Mathesius 2004) which especially obstruct AHL-regulated activities in numerous bacteria. In bacteria, furanones bind to AHL receptor proteins and stimulate the proteolytic degradation of mentioned receptors (Manefield et al. 1999, 2002). QS behaves as antagonist or agonist of bacterial AHL systems which are hampered by plant-produced complexes (Bauer and Mathesius 2004).

17.3 Effect of Abiotic Stresses on Soil and Plants

Extreme natural calamities including flooding, drought, temperature change, etc. can alter soil properties which are the key factors influencing carbon (C) and nutrient cycling in soils (Liu et al. 2017). Abiotic factors such as soil physico-chemical characteristics (e.g. soil moisture and aeration, etc.) and climatic factors (e.g. temperature and precipitation) are usually recognized as fundamental interpreters of SOM mineralization. Moreover, fluctuations in soil moisture can modify the temperature responsiveness of SOM mineralization (Craine and Gelderman 2011). High soil moistness limit O_2 accessibility and, thus, subsidize to a decline in redox circumstances, which leads to a decrease in SOM mineralization and directly influences microbial metabolic processes (Liptzin et al. 2011; Qiu et al. 2018). Under various water stresses (flood and drought), changes in soil moisture affect microbial richness, structure, C and N transformation and metabolic processes (Schimel et al. 2007; Rousk et al. 2013; Brunner et al. 2015; Liu et al. 2017). Additionally, temperature is another abiotic factor determining microbial and biochemical soil activity (Järvan et al. 2014).

An exponential correlation has been detected between temperature and SOM mineralization (Suseela et al. 2013; Razavi et al. 2016; Friedlingstein et al. 2006; Davidson and Janssens 2006). The temperature of soil relies on the climatic zone, physico-chemical properties and natural geography (Singh et al. 2010; Kodaira 2014; Borowik and Wyszkowska 2016). Alteration in soil temperature influences the interspheric activities of gas exchange between the atmosphere, soil and biological processes (root growth, the availability of nutrients, seed germination, etc.)

(Probert 2000; Lehnert 2013; Onwuka and Mang 2018). Mineralization of diverse organic materials, rate of organic matter decomposition, water holding capacity, transport and accessibility of nutrients (N, P, K, S, etc.) and water uptake to the plants are also affected (Davidson and Janssens 2006; Onwuka and Mang 2018). Moreover, contamination of soils by HMs (Cu, Ni, Cd, Zn, Cr, Pb, etc.) is one of the major problems, which leads to negative influence on soil characteristics, plant growth, yield and microbial community (Xie et al. 2016; Chu 2018). The toxicity of HMs for the plant life and soil properties depends on the temperature, pH, hydrous metal oxides, inorganic ions, etc. (Giller et al. 1998; Friedlová 2010). HMs indirectly influence microbial community which produces various enzymes (e.g. nitrate reductase, urease and amidase) (Huang et al. 2009). During moisture limitation in soil, microbes can survive by amassing solutes (amino acids (AAs)) to decrease their water potential and avoid dehydration and death; however, the cost of accumulating AAs is dynamically expensive (Harris 1981; Schimel et al. 2007).

Besides soil, plants also face some abiotic environmental stresses, e.g. salinity, drought, flooding, high and low temperature, mineral imbalance, HM toxicity, UV-B radiations, etc. (Kumar et al. 2015; Raineri et al. 2015; Assumpção et al. 2018). The negative impacts of drought and salinity on plant growth and metabolism are related with nutritional imbalance, low osmotic potential (water stress), effect of specific ions (salt stress) or associations of all these factors (Parvaiz and Satyawati 2008; Balliu et al. 2015) which lead to the decline in leaf water potential, closing of stomata and decrease of cell enlargement and growth (Hu and Schmidhalter 2005; Jaleel et al. 2007; Golldack et al. 2014). In plants, acute water stress results into the arrest of photosynthesis, translocation, respiration, ion uptake, nutrient availability, etc. which can cause disruption in metabolic processes and finally lead to the death (Jaleel et al. 2008; Farooq et al. 2008). On the other hand, flooding negatively affects plant biodiversity by imposing a selection pressure on plants since excess water in the rhizosphere can divest plant roots of O_2 (anoxia), CO_2 and light (Silvertown et al. 1999; Normile 2008). Heat stress causes accretion of secondary metabolites, superfluous generation of reactive oxygen species (ROS), loss of crop productivity, altering in metabolism, etc. (Mittler and Blumwald 2010; Mittler et al. 2012; Hasanuzzaman et al. 2013), whereas during low temperature (chilling), the seedling stage of plants is most sensitive, and chilling symptoms are tissue breakdown, discoloration, surface lesion, desiccation, enhanced senescence, reduce life span, etc. (Sharma et al. 2005). Chilling stress also influences the membrane fluidity due to fatty acid (FA) unsaturation and alteration in proportions of proteins to lipids in the cell membrane (Wang et al. 2006).

HMs lead to decrease in photosynthetic rates, leaf relative water contents due to decline in number and size of xylem vessels, transpiration rate, plant growth reduction, etc. (Sarwar et al. 2015). These HMs assimilate in various eatable plant parts and thus enter into human beings through the food chain (Sarwar et al. 2017). Moreover, UV-B radiations have also been reported to cause negative effects on all living beings including plant and microbes by reducing their growth and photosynthetic rates and causing oxidative stress, DNA damage, etc. (Frohnmeyer and Staiger 2003; Yao and Liu 2007; Jansen and Bornman 2012; Assumpção et al. 2018).

Although plants have developed protective mechanisms like developing cuticle, sunken stomata and producing antioxidants, osmolytes, etc., these mechanisms become insufficient and need a pushing force to overcome the effects of abiotic stresses. Therefore, microorganisms present in soil provide various strategies to counteract the unfavourable environment.

17.4 Microbe-Mediated Mitigation of Abiotic Stresses for Sustainable Agriculture

Microbes, as discussed earlier, are crucial for soil and plant health. In addition to this fact, they have the advantage over plants due to their existence beneath the soil surface which protect them from direct exposure of heat and water. There are numerous microbes in the soil which present as individual units and thus are more capable to counteract abiotic stresses. Microbial interactions with plants are significant to the survival and adaptation of both the symbiont partners during abiotic-stressed conditions. Endophytes such as bacteria and fungi (PGPR and AMF) impart tolerance to the host plant by up-regulating various mechanisms such as soil enrichment and nutrient acquisition, releasing organic acid into the soil, regulation of hormones, production of antioxidants, polyamine balance, activation of heat-shock transcription factors, etc. which are elucidated as follows:

17.4.1 Soil Enrichment and Nutrient Acquisition

Soil microbes enhance the nutrient composition of soil during stressed conditions by various physiological and metabolic processes like glomalin production, nitrogen fixation, siderophore production and exopolysaccharide secretion, increase the efficiency of nutrient acquisition by sensing strigolactones, etc. (Beneduzi et al. 2012; Sharma et al. 2013; Chowdhury et al. 2015; Kuan et al. 2016).

Arbuscular mycorrhizal fungi release glomalin-related soil proteins (GRSP) into the soil to upgrade soil texture (by regulating water-plant-soil relations) which are vital for ecosystem functioning and ecological restoration (Wu and Zou 2017). Glomalin behaves as a hydrophobic glue that allows the soil fungus to deal with gas-water interfaces during aerial tissue growth of the plant and reduce macroaggregate interruption (Miller and Jastrow 2000; Singh et al. 2010). During drought stress, AMF (*G. diaphanum*, *G. versiforme* and *G. mosseae*) symbiosis with *Poncirus trifoliata* improved soil moisture retention capability and plant growth via the production of glomalins (Wu et al. 2008). The glomalin also sequesters, chelates and stabilizes HMs, thus reducing their availability and toxicity to soil microbes and plants in contaminated sites (Gonzalez-Chavez et al. 2004; Vodnik et al. 2008). The production and deposition of GRSP in soils are essential mechanisms for the maintenance of soils (Cornejo et al. 2008). AMF mycelia have a beneficial effect on plant nutrient transport and accessibility, root development and activities as well as soil characteristics (Neumann and George 2010). AM hyphae are able to access smaller soil pores (due to narrow diameter) outside the root depletion zone in soil for providing water and nutrients to the plants (Hohnjec et al. 2005). For example, under salt stress wheat plants inoculated with *Glomus* spp. have been shown to improve nutrient level by enhancing P, N, K^+, Zn, Fe and Cu acquisition and declining sodium ion uptake (Talaat and Shawky 2013). Under water deficit stress, *Pistacia* sp. inoculated with *G. intraradices* and *G. mosseae* exhibited noteworthy upsurges in nutrient (K, P, Mn and Zn) levels (Bagheri et al. 2012). Establishment of association with *G. mosseae* has been reported to reduce AsV (arsenate) uptake and increase plant growth by enhancing the levels of N, K^+ and P in *Pisum sativum* (Garg and Singla 2012). Garg and Kashyap (2017) and Garg and Singh (2018) reported more efficient uptake of nutrients in *Cajanus cajan* inoculated with *R. irregularis* under As and Zn/Cd stresses, respectively. Additionally, in sunflowers, *G. mosseae* improved tolerance of the plant towards drought stress and seed yield (oil) by enhancing P concentration. AMF have also been shown to increase crop yield and uptake of gradually diffusing ions, e.g. phosphate (PO_4^{3-}) and immovable nutrients (Cu and Zn) (Liu et al. 2002; Smith et al. 2010). Moreover, in Cu-contaminated soil, AMF was able to continue an effective symbiosis with jack bean (*Canavalia ensiformis*) and can increase nutritional and physiological alterations. Some essential nutrients have common transporters as of toxic elements, e.g. arsenate and phosphate, to which the plants cannot distinguish. AMF colonization, in these cases, protects the plant by down-regulating the high-affinity PO_4^{3-}/AsV transport system and increasing the efflux of AsIII (arsenite) from mycorrhizal-infected roots (Chen et al. 2007; Garg and Singla 2012). Therefore, these fungi constitute an integral and important component of agricultural systems and become increasingly urgent to be used as biofertilizers, dropping the usage of agrochemicals (Harrier and Watson 2004).

Host plants release strigolactones (SLs) in soil to attract the microbes for the establishment of symbiotic relationships. Strigolactones (SLs) help in imparting abiotic stress tolerance both directly (as a hormone) and indirectly (establishing AMF symbiosis) (Pandey et al. 2016). SLs have an adaptive response to phosphate (PO_4^{3-}) deficiency because higher SL amount in roots and root exudates under low PO_4^{3-} conditions has been conferred to increase nodulation potential and AMF colonization, which make plants more efficient for nutrient acquisition (Yoneyama et al. 2007; Xie et al. 2010). Under drought conditions, SLs get highly accumulated in the roots of lettuce and tomato plants that in turn influence AMF symbiosis which helps to cope up with abiotic stresses (Ruiz-Lozano et al. 2016). Arite et al. (2012) investigated positive impact on root elongation of rice plants, when inoculated with GR24 (synthetic SL analogue) under phosphate deficiency. Aroca et al. (2013) reported that under salinity conditions, *G. intraradices*-colonized lettuce plants (*Lactuca sativa*) prompted SL secretion which triggered AMF growth and thus helped to overcome various stresses (Sun et al. 2014).

Plant growth-promoting rhizobacteria help in fixing atmospheric N_2 by forming symbiotic relationship with host plants. In addition to N_2-fixing rhizobium bacteria, other soil PGPRs including species of *Azospirillum*, *Klebsiella pneumoniae*, *Pantoea agglomerans*, *Azoarcus*, *Bradyrhizobium*, *Enterobacter*, *Pseudomonas*, *Beijerinckia*, etc. have been reported to convert atmospheric N_2 to available forms by the help of enzyme nitrogenase and make it accessible to the plants (Riggs et al. 2001; Bhattacharyya and Jha 2012). During metal-stressed and poor N condition, however, synthesis of nitrogenase in legumes gets reduced, and these PGPRs have the ability to provide abundant nitrogen to associated plants by synthesizing nitrogenase enzyme. From the stem part of *Populus trichocarpa* (poplar) and *Salix sitchensis* (willow), Doty et al. (2009) have detected diazotrophic endophytic bacteria (*Acinetobacter*, *Sphingomonas*, *Rahnella* and *Burkholderia*) which have the capability to synthesize nitrogenase enzyme and fix atmospheric N_2 during its deficiency. Moreover, inoculation with *Enterobacter* sp. SBP-6 containing nitrogen fixation ability imparted salt tolerance to wheat plants and significantly improved plant biomass, K^+/Na^+ ratio, K^+ uptake and chlorophyll content (Singh and Jha 2016). Garg and Singla (2004) reported lesser decline in chlorophyll content, nodulation, and Rubisco activity and increase in PEPCase activity in chickpea genotypes when inoculated with *Mesorhizobium ciceri* strain F:75 under salt stress. Similarly, under N_2-deficient conditions, kallar grass roots inoculated with *Azoarcus* BH72 (PGPR) showed increase in the dry weight (Sarkar and Reinhold-Hurek 2014).

Siderophores are low molecular weight (LMW) Fe-chelating complexes produced by PGPRs which bind Fe^{3+} and help in Fe uptake in plants. On the basis of their Fe-associated functional groups, types of ligands and structural characters, bacterial siderophores have been categorized into four classes (pyoverdines, carboxylate, phenol catecholates and hydroxamates) (Crowley 2006). Phenol catechols include linear catechols (secreted by *Agrobacterium*), pyochelin (*Pseudomonas aeruginosa*) and enterobactins (*Escherichia coli*); hydroxamates include arthrobactin (*Arthrobacter* spp.), schizokinen (*Bacillus megaterium*), pseudobactin (*Pseudomonas* sp. St B10) and ferrioxamine (*Arthrobacter* spp., *Streptomyces* spp., *Actinomyces* spp. and *Nocardia* spp.); and carboxylate rhizobactin (*R. meliloti*) and pyoverdines include pyoverdine (*Pseudomonas* sp.). Potato plant inoculated with *P. fluorescens* secretes a siderophore (pseudobactin pigment) which showed increase in the plant growth under Fe deficiency (Cobessi et al. 2005). Similarly, under salinity and cold stress, wheat plants inoculated with *Streptomyces* strain and cold-tolerant bacterium *Pseudomonas lurida* M2RH3 (MTCC 9245) having PGP activity of releasing siderophores significantly improved the germination (by solubilizing phosphate) and nutrient contents (N, P, Mn and Fe) in shoot, shoot length and dry weight (Selvakumar et al. 2011; Sadeghi et al. 2012). Microbial-secreted siderophores have also been stated to ameliorate HM-induced oxidative damage in plants by decreasing the formation of free radicals (Dimkpa et al. 2009). Pinter et al. (2017) reported that *Micrococcus luteus*, *P. fluorescens* and *B. licheniformis* imparted As tolerance in grapevine and increased its biomass by siderophore production, atmospheric N_2 fixation and phosphate solubilization processes. These studies concluded that application of atmospheric N_2 fixing bacteria and siderophore-producing bacteria acts as biofertilizers by providing Fe to the plants (Fernández et al. 2005; Radzki et al. 2013). AMF have also been reported to enhance siderophore production and proton gradient by increasing Fe availability in soil and hormone production, which triggers the development of plant (Jin et al. 2013). Fungal siderophores is mainly of three types, (1) rhodotorulic acids (di- or tri-hydroxamates secreted by *Aspergillus* and *Penicillium*), (2) ferrichrome and (3) fusarinines. Siderophore production by mycorrhizal (*G. etunicatum* and *G. intraradices*) sorghum roots has been reported under micronutrient-deficient condition (Aliasgharzad et al. 2009). Moreover, Sigel and Sigel (1998) reported expression of siderophore coding genes in microorganisms in response to Fe deficiency. Ectomycorrhizal fungi *Cenococcum geophilum* (*Gloniaceae*), *Hebeloma crustuliniforme* (*Hymenogastraceae*) and member of the family Ericaceae produce the ferricrocin or fusigen as the primary siderophores (Haselwandter 2008). *Zea mays* (maize) inoculated with *G. irregularis* have shown biofortification of grains even in the presence of low micronutrients due to capability of AMF to produce siderophores (Subramanian et al. 2013). Siderophore production by mycorrhizal fungi regulates iron phosphate solubilization and, under phosphate-limiting environments, enhances its availability to plants and microbes (Jayachandran et al. 1989).

Plant growth-promoting rhizobacteria secrete exopolysaccharides (EPS), i.e. high molecular weight polysaccharides, which invite the micro-colonies to form a matrix which provide protection against environmental variations, e.g. nutrient and water retaining and epiphytic establishment (Balsanelli et al. 2014; Ilangumaran and Smith 2017). Moreover, the soil rich in polysaccharides offers a cementing action to enhance stability for soil accretion which increases biomass and decomposition of organic material by PGPRs (Flouri et al. 1990). They also play necessary role in mature biofilm formation and nodular function during legume-rhizobia association (Stoodley et al. 2002; Skorupska et al. 2006). Grover et al. (2011) reported that wheat seedling inoculated with *Pantoea agglomerans* under salinity stress produced EPS which restricted Na^+ uptake and stimulated plant growth. Sunflower plants showed enhanced drought resistance by producing EPS when inoculated with *Pseudomonas putida* strain P45. Under drought stress, wheat (nonlegume) plant showed beneficial interaction with EPS secreting PGPRs *R. phaseoli* (MR-2), *R. leguminosarum* (LR-30) and *Mesorhizobium ciceri* (CR-30 and CR-39) (Vurukonda et al. 2016). Improvement in N_2 nutrition, dry matter and water uptake of plants was reported in inoculating sunflower with EPS producing rhizobial strain YAS34 under reduced water system (Alami et al. 2000). Kaci et al. (2005) observed that rhizobia (KYGT207) separated from rhizospheric dry soils of wheat seedlings enhanced the soil aggregation by EPS production and thus increased plant growth. Under high salinity, chickpea (*Cicer arietinum* var. CM-98) plants inoculated with EPS manufacturing salt-tolerant bacteria *Planococcus rifietoensis* RT4 and *Halomonas variabilis* HT1 showed increased growth and soil accretion within roots (Qurashi and Sabri 2012). Yang et al. (2016) observed that under saline irrigation seeds of *Chenopodium quinoa* inoculated with bacteria *Bacillus* sp. MN54 and *Enterobacter* sp. MN17, it showed enhancement in plant-water relation conditions by production of EPS. Moreover, PGPRs increased soil permeability, water holding and gas exchange capability, decreased bulk density and this property can be utilized to fulfil the future need of heavy crop yield and productivity (Gauri et al. 2012). Therefore, applications of EPS have continued to gain interest and found a good commercial market in food and health care industry.

17.4.2 Releasing Organic Acids into the Soil

Organic acids (gluconic, oxalic, acetic and malic acids, acid phosphatase) released by microbes act as natural chelating agents, precipitate HMs outside the mycelium, sequester soluble metal ions and lower soil pH, thus enhancing the bioavailability of Cd, Pb and Zn (Turnau and Kottke 2005; Ullah et al. 2015; Gube 2016; Seneviratne et al. 2017). Sheng et al. (2011) reported higher biomass and accumulation of organic compounds and free AAs in leaves of maize plants inoculated with *Glomus mosseae* under salt stress. In addition, AM symbiosis in maize lifted the concentrations of total soluble sugar, reducing sugars, organic acids, malic acid, fumaric acid, oxalic acid, citric acid, etc. in leaves and dropped the concentrations of succinic acid, proline, total free AAs and formic acid (Raaijmakers and Weller 1998; Mendes et al. 2011). Soil microorganisms decompose organic matter with the help of enzyme acid phosphatase yielding mineral nutrients (N, P, S and trace elements), carbon dioxide, water, simple inorganic ions such as $NH4^+$, NO_3^- and SO_4^{2-} and also AAs (cysteine (cys) and glycerine (gly)) into the stressed soil from where they can be easily uptake by the plants (Stotzky 1997; Sharma and Dietz 2006). *Pisum sativum* inoculated with cold-tolerant PGPRs (*R. leguminosarum*-PR1 and *Pseudomonas* spp.) showed acclimation to cold and enhanced nutrient uptake, Fe acquisition and plant growth (Mishra et al. 2011). In addition, PGPR increased the extraction efficiency of plants via enhancing the HM ion mobility by releasing chelating agents, phosphate solubilization, acidification and redox changes (Ullah et al. 2015; Mishra et al. 2017). From metal-contaminated soil, Cd-tolerant *Bradyrhizobium* sp. extracted from *Glycine max* (soya bean) nodules displayed numerous PGP characters (e.g. siderophore and ACC deaminase production, synthesis of IAA), which, when inoculated to *Lolium multiflorum*, increased shoot dry biomass and Cd accumulation in roots (Guo and Chi 2014). Therefore, PGPRs and AMF can be used as biofertilizers due to their ability of solubilizing the naturally available P, in disparity to P fertilizers which causes surface and groundwater pollution, soil fertility depletion, accretion of toxic elements (As and Se) and waterway eutrophication (Alori et al. 2017).

17.4.3 Hormonal Regulation

Phytohormones [auxins (indole acetic acids (IAA)), cytokinins, gibberellins, ethylenes and abscisic acid (ABA)] formed by microbes are effective in promoting plant growth under stressed environments (Egamberdieva et al. 2017). During abiotic stress, IAA regulate the root and vascular tissue differentiation, shoot growth and cell division (Goswami et al. 2016). Sadeghi et al. (2012) investigated that, under salt stress, a *Streptomyces* strain amended the growth of wheat plant by producing IAA. IAA have positive impact on root growth by minimizing the level of ethylene (C_2H_4) directly or indirectly (Lugtenberg and Kamilova 2009). In roots, disintegration of the ethylene precursor-1-aminocyclopropane-1-carboxylate (ACC) into 2-oxobutanoate ($C_4H_5O_3^-$) and ammonia (NH_3) by bacterial ACC deaminase alleviates the C_2H_4 repression, drops concentration of C_2H_4 and indirectly improves plant biomass and growth (Glick 2014). Improvement in plant biomass was observed when the various crop plants were inoculated with ACC deaminase producing bacteria such as *Klebsiella* spp. SBP-8 under salt stress (Singh et al. 2015), *Achromobacter piechaudii* ARV8 under drought stress (Mayak et al. 2004), *Pseudomonas aeruginosa* 2CpS1 under temperature stress (Meena et al. 2015) and *Variovorax paradoxus* 5C-2-HM stress (Belimov et al. 2009). Arkhipova et al. (2007) reported that lettuce seedlings inoculated with cytokinin-producing *Bacillus subtilis* under water-deficit circumstances increased accumulation of shoot biomass and shortened the roots. Under salt stress (*Cucumis sativus*), plants inoculated with *Burkholderia cepacia* SE4, *Promicromonospora* sp. SE188 and *Acinetobacter calcoaceticus* SE370 had notably higher biomass by decline in electrolyte leakage (EL) and increase in water potential. Moreover, the inoculated cucumber plants showed upregulation of gibberellin (GA4), salicylic acid (SA) and down-regulation of ABA contents (Kang et al. 2014; Ilangumaran and Smith 2017). Under water deficiency, symbiosis with AMF has been reported to impart tolerance through closure of stomatal either by active or passive, i.e. via ABA-mediated or hydraulic-mediated mechanisms, respectively. In grapevine, stomata closure has been demonstrated to be driven by either or both mechanisms (Estrada-Luna and Davies 2003). Asensio et al. (2012) found increase in ABA content and in water-stressed non-mycorrhizal *Solanum lycopersicum* as compared to mycorrhizal (*Funneliformis mosseae* and *Rhizophagus irregularis* plants). Moreover, in *Solanum lycopersicum*, ABA upregulated AMF colonization (Herrera-Medina et al. 2007; Charpentier et al. 2014). In salinity stress, lettuce plants were detected to possess lower ABA levels when inoculated with *G. intraradices* (Jahromi et al. 2008).

17.4.4 Antioxidant Production

Abiotic stresses induce free radicals, which affects antioxidant defence and ROS resulting into oxidative stress in plants (Vurukonda et al. 2016). Some rhizobacteria can produce antioxidants which counterbalance the noxious effects of ROS in the plant cells and biomolecules under diverse abiotic stresses (Grover et al. 2011). Bacterial-inoculated potato plant grown under different stress (polyethylene glycol (PEG), NaCl and $ZnCl_2$) conditions resulted in elevated mRNA expression of ROS-scavenging enzymes (SOD and APX) (Gururani et al. 2013a). Similarly, the mRNA expression levels of genes encoding for other antioxidative enzymes, namely, CAT, DHAR and GR, also improved in the PGPR (*Bacillus* spp.)-treated potato plants (Gururani et al. 2013b). Mycorrhizal symbiosis enhanced ROS scavenging activity via increasing the concentration of non-enzymatic (ascorbate, glutathione, etc.) and enzymatic (POX, SOD, GR and CAT) antioxidant activity under salt stress. Under various abiotic stresses, wheat plant inoculated with Cu-resistant *Pseudomonas* sp. DGS6; lettuce plants with *Pseudomonas mendocina* and *G. mosseae* or *G. intraradices*; *Digitaria eriantha* with *R. irregularis*; pistachio (*Pistacia vera*) with *G. etunicatum*, pomegranate plants with *Rhizophagus intraradices, Cicer arietinum with Funneliformis mosseae, Cajanus cajan with G. mosseae*; and wheat plants with *Glomus* spp. had a positive effect on the upregulation of antioxidative defence responses, i.e. improved enzymatic (SOD, CAT, APX and GPX activities) and non-enzymatic (AsA/DHA, GSH/GSSG, ascorbic acid, polyphenol and flavonoids) activities (Kohler et al. 2008; Abbaspour et al. 2012; Garg and Aggarwal 2012; Garg and Kaur 2013; Bompadre et al. 2014; Talaat and Shawky 2014; Garg and Singla 2016; Pedranzani et al. 2016).

17.4.5 Polyamine (PA) Balance

Polyamines (PAs) are aliphatic low-molecular-weight polycations that are cosmopolitan in all living organisms (bacteria, plants and animals) (Hussain et al. 2011). There are three PAs, diamine putrescine (Put), triamine spermidine (Spd) and tetramine spermine (Spm) in plants (Liu et al. 2015). Evidences have illustrated that PAs play an important role in regulating various physiological processes (cell growth and development) as well as imparting stress tolerance (Gill and Tuteja 2010). PGPRs and AMF have an important role in alleviation of abiotic stress in plant by altering endogenous PAs levels. Zhou et al. (2016) reported that in *Arabidopsis* under various osmotic stresses, Spd from *Bacillus megaterium* BOFC15 activated PA-mediated signalling pathways which significantly improved cellular PA accumulation (increase free Spd and Spm levels) and contributed towards the stress tolerance. Under salt stress, *Triticum aestivum* L. plant inoculated with *Glomus* spp. demonstrated change in PA content as well as lesser activities of diamine oxidase (DAO) and PA oxidase (PAO) (Talaat and Shawky 2013). In *Medicago sativa*, *G. fasciculatum* enhanced free PAs and soluble nitrogenous compound levels under water stress conditions (Rapparini and Peñuelas 2014). Spm and Spd contents in the leaves of control or salt-treated grape rootstock 'Dogridge' have considerably improved by AMF inoculation. Similarly, *Citrus tangerine* seedlings inoculated with *G. mosseae* enhanced both leaf and root ornithine decarboxylase (ODC) and arginine decarboxylase (ADC) activity by the synthesis of Put (Wu et al. 2012). Cicatelli et al. (2010) investigated that inoculation by either *G. intraradices* or *G. mosseae* modified PA (PaSPDS1, PaSPDS2, PaMT1, PaMT2, PaMT3 and PaADC) gene expression and growth in white poplar grown in HM-contaminated (Cu and Zn) soil. In addition, inoculated plants of *Populus alba* clone AL_{35} have been shown to possess improved capability for stabilization of HM soils, together with enhanced growth due to upregulation of numerous transcriptional stress genes and the defensive role of PAs. Therefore, PA balance is a need of plant during abiotic stresses, which can be fulfilled by the rhizosphere microbes having the ability to regulate it.

17.4.6 Photosynthetic Capacity

Photosynthetic capacity (PC) is defined as the maximum rate at which plant leaves are capable to fix accessible carbon (C) during the process of photosynthesis and normally calculated as the amount of carbon dioxide (CO_2) that is fixed as μ mol m^{-2} s^{-1}. Due to various abiotic stresses, when this capacity of plants gets reduced, microbes help in rejuvenating the photosystems and diverse enzymes. *Populus cathayana* and grapevine plants inoculated with *R. irregularis* and PGPBs (*Acinetobacter* and *Pseudomonas*) displayed enhanced plant water status, photosynthetic capability as well as biomass production under salt and drought stress respectively (Ibaraki 2008; Rolli et al. 2015; Vurukonda et al. 2016). Similarly, under drought and water stress, *Triticum aestivum* seedlings and basil plants primed with *Bacillus thuringiensis* AZP2 and *Pseudomonades* sp. resulted in increased photosynthetic pigment concentration, antioxidant activity, plant biomass and manyfold higher survival rate (Heidari and Golpayegani 2012; Timmusk et al. 2014; Vurukonda et al. 2016). *G. irregulare* ameliorated Cd stress in *Medicago truncatula* by supporting photosynthesis, enhancing aminolevulinic acid synthesis, facilitating the electron transport and protochlorophyllide photoreduction and increasing the density of photosynthetic units and photosynthesis-related proteins (Aloui et al. 2011). AMF symbiosis (*F. mosseae* and *R. intraradices*) decreased Pb accumulation in roots and leaves and promoted photosynthetic rate as well as biomass in *Robinia pseudoacacia* L. with higher gas exchange ability (Yang et al. 2015). *Cajanus cajan* inoculated with *G. mosseae* (Cd stress) and *Cicer arietinum* with *Funneliformis mosseae* (HM and salt stress) exhibited significant increase in chlorophyll contents (Garg and Chandel 2012; Garg et al. 2014a, b).

17.4.7 Production of Volatile Organic Compounds (VOCs)

Diverse PGPRs strains have capacity to release VOCs that function in imparting abiotic stress tolerance and increase plant biomass, disease resistance, etc., e.g. 2,3-butanediol ($C_4H_{10}O_2$) (a VOC) released by *Pseudomonas chlororaphis* improved salicylic acid (SA) concentration and thus induced the closure of stomata and provided drought resistant (Cho et al. 2008). PGPRs mediate methylation of metals (As, Hg, Pb, Se, Sn and Tn) resulting in formation of methylated volatile metal compounds that can effortlessly escape the soil horizon (Bolan et al. 2014). AMF, however, do not directly release volatile compounds but provide better nutrient accessibility and therefore lead to increase in primary productivity that offers more resources to the plant for the biosynthesis of defensive metabolites (VOCs) (Fontana et al. 2009). Asensio et al. (2012) reported that under drought stress, *Solanum lycopersicum* inoculated with AMF (*Glomus* spp.) showed enhancement in production of VOCs (isoprenoids) and provide tolerance.

17.4.8 Osmoprotection

Osmoprotectants (compatible solutes) are LMW molecules or compounds which are highly soluble, electrically neutral and non-toxic at molar concentrations and have the ability to stabilize membranes and proteins and also lessen the osmotic potential of membranes to check dehydration inside the cell (Ahn et al. 2011; Wani et al. 2013). Diverse microorganisms have been reported to alleviate the undesirable effect of abiotic stresses by enhancing osmoprotectant accumulation; e.g. both Gram-positive (*Bacillus* spp.) and Gram-negative (*Azospirillum* and *Pseudomonas* spp.) bacteria stimulated water stress resistance in basil plants by motivating the production of soluble carbohydrates and proline especially leaf and root tissues. In addition, under drought stress, inoculated basil *Ocimum* sp. had also more chlorophyll content, confirming the optimistic effect of bacteria (Heidari et al. 2011). Similarly, *Macadamia tetraphylla* inoculated with AMF (*Glomus* spp.) have shown decrease in osmotic potential during water stresses due to the accumulation of sugars and proline (Yooyongwech et al. 2013). Abdel Latef and Chaoxing (2011) reported that under cold stress, mycorrhizal inoculation (*G. mosseae*) in tomato (*Lycopersicon esculentum* cv Zhongzha105) plant displayed enhanced accumulation of osmoprotectants, i.e. proline, soluble sugar and soluble protein, and imparted tolerance. During low temperature stress, cucumber plant inoculated with AMF (*F. mosseae*) showed enhancement in accretion of lignin, phenolics and flavonoids accompanied by significant decrease in the H_2O_2 accumulation (Chen et al. 2013). Under water deficiency in soil, symbiotic association between higher plants and AMF enhanced water uptake and can also upregulate stomatal aperture. In water-deficit *Poncirus trifoliata* plants, AMF upgraded water regulation by stimulating the accumulation of osmolytes or modulating hormonal signalling (Fan and Liu 2012). *Arabidopsis* plant under osmotic stress conditions when inoculated by bacterium *B. subtilis* GB03 accumulated increased levels of choline ($C_5H_{14}NO$) and glycine betaine (GB), which act as significant osmoprotectants towards dehydration tolerance (Zhang et al. 2010). Under HM stress (Cd), AMF (*F. mosseae*)-inoculated pigeon pea plant showed high accumulation of osmolytes (proline, glycine betaine) and provide better tolerance (Kaur and Garg 2017). Garg and Baher (2013) reported that *Cicer arietinum* plant inoculated with *G. mosseae* under salinity stress has boosted level of proline biosynthesis which helps in stress tolerance.

17.4.9 Adsorption, Exclusion and Immobilization

Soil microbes have the ability to alleviate the stress caused by HMs on plant growth by soil bioremediation (bio-augmentation). It involves absorption of metals from polluted areas (rhizofiltration) and immobilization of pollutants by microbes (bio-degradation) in soil (Chaudhry et al. 1998; Khan 2005). Microorganisms have more surface area to volume ratio that proposes a huge contact area which assists the interaction with metal ions in the nearby environment. High adsorption capacity was detected for thorium (Th) by Gram-positive bacteria (*Micrococcus luteus*, *Arthrobacter nicotianae*, etc.) and for uranium (U) by microorganisms (*A. nicotianae*, *S. levoris* and *S. flavoviridis*) (Nakajima and Tsuruta 2004). Microbes use trace elements or HMs as terminal electron acceptors from which they obtain the required energy for detoxification of metals via non-enzymatic and enzymatic activities (Dixit et al. 2015). Bala and Joshi (2016) isolated bacterial strains from HM-contaminated sites and found their biosorption capacity in the order: *Acinetobacter calcoaceticus* > *Enterobacter* spp. > *Staphylococcus gallinarum* > *Pantoea agglomerans* towards Zn, Ni, Cr and Pb. Mullen et al. (1989) observed capability of *P. Aeruginosa*, *E. coli*, *Bacillus cereus* and *B. subtilis* in binding Cu^{2+}, Ag^{+}, Cd^{2+} and La^{3+}. AM fungi are capable to detoxify metal ions by absorbing them into their mycelium and spores through intracellular and extracellular precipitation, active accumulation and valency transformation; hence, they are prospective biocatalysts for the remediation of HMs. Janoušková et al. (2006) established that AMF (*G. intraradices*) are capable to reduce the deleterious effects of cadmium (Cd) on lettuce plant development and growth by phytostabilization. Joner et al. (2000) observed the sorption capacity of *Glomus mosseae* for Cd and found that it can adsorb up to 0.5 mg Cd/mg dry matter. Rhizospheric fungi and bacteria produce metallothioneins (MT) and phytochelatins (PC) to combat HMs stress which result in their accumulation in microbial cells (Miransari 2011). PGPBs that have the capability of being metal-resistant can enhance the plant growth and development by obstructing the transport of superfluous metal ions and thus prevent its bio-amplification in the ecosystem. Clover plant inoculated with a Cd-resistant PGPR (*Brevibacillus* sp.) in Cd-contaminated soil resulted in little Cd uptake in plants, which enhances the plant growth. PGPRs can accumulate Cd ions in their cells and can reduce their bioavailability as well as its uptake by rhizobia and plants (Vivas et al. 2005).

17.4.10 Gene Expression Regulation

Various PGPRs and AMFs are found to impart tolerance in host plant by regulating the expression of stress-responsive genes, which strengthen plant inherent tolerance machinery and make it capable of withstanding the unfavourable conditions. In wheat plants, halotolerant bacterium *Dietzia natronolimnaea* STR1 induced salt tolerance via modulating ABA-signalling cascade, which upregulated TaABARE (ABA-responsive gene), *TaST* (salinity stress-induced gene) and TaOPR1 (12-oxophytodienoate reductase 1) and then ultimately lead to *TaWRKY* and *TaMYB* gene stimulation, trailed by expression of stress response genes (Bharti et al. 2016). Transgenic *A. thaliana* with proBA genes obtained from *B. subtilis* conferred plant tolerance to salinity by improving proline (pro) synthesis (Chen et al. 2007). Under salt stress, *Phaseolus vulgaris* inoculated with *Rhizobium tropici* and *Paenibacillus polymyxa* upregulated trehalose 6-phosphate gene and improved nodulation potential, N content and plant growth (Figueiredo et al. 2008). The expressional sequences of genes (SOS1, HKT1, NHX2 and KT1) involved in ion balancing were transformed under stress conditions, and instant molecular changes were persuaded by PsJN (Pinedo et al. 2015). *Puccinellia tenuiflora*, i.e. a halophytic grass under high salt concentration, when inoculated with *B. subtilis* GB03, displayed less Na^+ accretion and confirmed upregulation of genes (*PtSOS1* and *PtHKT1*) in roots (Niu et al. 2016; Ilangumaran and Smith 2017). Different species of AMF have also been found to impart tolerance to host plant by upregulating the expression of its stress-responsive genes, whose functions help in sequestering the excess ions or toxic metals in vacuoles or their efflux out of the cell. Porcel et al. (2016) reported that, in aerial tissue of plant (*Oryza sativa*), AM symbiosis through expression of numerous genes (OsSOS1, OsNHX3, OsHKT2;1 and OsHKT1;5) played important role in Na^+ unloading from the xylem, their extrusion from cytoplasm, sequestration into the vacuole and recirculation of Na^+ from photosynthetic tissues to roots encoding plant transporters intricate in ion homeostasis. Under water-deficit stress, AMF inoculations have been found to alleviate the oxidative stress and can detoxify cytotoxic compounds in *Poncirus trifoliata* by enhancing the mRNA levels of four stress-responsive genes, *MIOX1* (myo-inositol oxygenase), *GLX1* (glyoxalase), *CSD1* (copper/zinc SOD) and TTC5 (transparent testa 5) (Fan and Liu 2012). *G. intraradices* inoculated *Zea mays* plants have been shown to increase water uptake under water scarcity by inducing the expression of various stress-related genes including two aquaporin genes (*GintAQPF1* and *GintAQPF2*) (Li et al. 2013). Aroca et al. (2007) found that AM symbiosis enhanced *Phaseolus vulgaris* tolerance to various stresses (cold, drought and salt stress) by regulating root hydraulic properties via phosphatidylinositol 4, 5-bisphosphate (PIP2) protein levels and phosphorylation state regulation. To overcome heat stress, organism expresses chaperons or heat-shock proteins (HSPs) to accomplish biological processes. HSPs are intricated in diverse regular biological functions such as transcription, translation and posttranslational modifications, protein folding, aggregation and disaggregation of proteins (Scharf et al. 2012; Tiwari et al. 2015; Lata et al. 2018). In sorghum seedlings, thermotolerant PGPR *Pseudomonas* sp. AKM-P6 helped to resist heat stress by inducing HSPs in leaves (Ali et al. 2009). In *A. thaliana* plants, *Paraphaeosphaeria quadriseptata* enhanced plant heat stress tolerance by initiating the expression of HSP70 and HSP101 proteins (McLellan et al. 2007). Expression of 14-3-3 HSPs has been found to be upregulated in *Daucus carota* under drought stress colonized by *G. intraradices* (Porcel et al. 2007). Moreover, under drought stress the gene for HSPs, a luminal binding protein in the endoplasmic reticulum

(ER), was upregulated in *G. intraradices* (Hammer and Rillig 2011). These reports suggest that AMF-induced modification in stress-tolerant gene has multiple effects on plant stress tolerance, growth, development and yield. As we have discussed, the microbes (AMF and PGPRs) play very important roles in the growth and development of plants under stressful environments. Recently, researchers have experimented the cumulative beneficial effects of combined inoculation of these microbes on plants in order to alleviate high degrees of stresses (Fig. 17.1).

17.5 Co-inoculation of PGPR and Mycorrhizal Fungi

In the soil, plant and microorganisms (AMF and PGPRs) live symbiotically and have a beneficial effect on plant growth by providing nutrients and in getting favourable niche from plants for their proliferation (Abd_Allah and Egamberdieva 2016). Marschner and Timonen (2005) reported that PGPRs play an essential role in the formation of AM symbiosis with host plant. A plant belonging to family Leguminosae or Fabaceae has frequently been shown to entail high levels of PO_4^{3-} for effective growth, nodulation and N_2 fixation (Gibson 1976). Primary effect of mycorrhizal colonization on nodulation is phosphate transport, and secondary effect includes supplying of plant hormones, trace elements and photosynthates which play a vital role in atmospheric N_2 fixation. AMF colonization has also been found to enhance the level of phytoalexins (isoflavonoid substances) in legume roots which are known to induce expression of Nod gene. From all these, it is clear that AMF have a significant role in the expression of nodulation gene in rhizobia (Suresh and Bagyaraj 2002; Desai et al. 2016). Lecomte et al. (2011) reported that bacteria may be found to adhere with the AMF hyphae as well as inserted within the AM spore walls and get profit by nourishing on hyphal exudates and/or use the mycelium as a vehicle for colonization of the root rhizosphere (Ramasamy et al. 2011). Therefore, the harmony within these soil microbes can be utilized for replenishing the plant growth and soil health. Under salt stress, positive interactions between AMF (*Funneliformis mosseae*, *Rhizophagus intraradices*, *Claroideoglomus etunicatum*) and endophytic bacteria (*B. subtilis*) enhanced nutrient acquisition, plant growth, nodulation and stress tolerance in *Acacia gerrardii* (Hashem et al. 2016). Khan and Zaidi (2007) reported that wheat plants inoculated with *Bacillus* sp., *Azotobacter chroococcum* and *G. fasciculatum* enhanced grain yield (twofold increase) as well as N and P content. *Acacia saligna* plant, when co-inoculated with *Acaulospora*, *Gigaspora*, *Glomus* sp. and *Sinorhizobium terangae*, showed increase in salinity tolerance (Soliman et al. 2012). Co-inoculation of *Lactuca sativa* (lettuce plant) with PGPB (*Pseudomonas mendocina*) alone or in combination with AMF (*G. intraradices* or *G. mosseae*) provided drought resistance by accretion of proline in leaves and triggering various enzyme (catalase, phosphatase and nitrate reductase) activities in plant roots (Kohler et al. 2008). *Zea mays* L. co-inoculated with AMF (*Glomus intraradices*), PGPB (*Azospirillum lipoferum*, *Azotobacter chroococcum*) and P-solubilizing bacteria

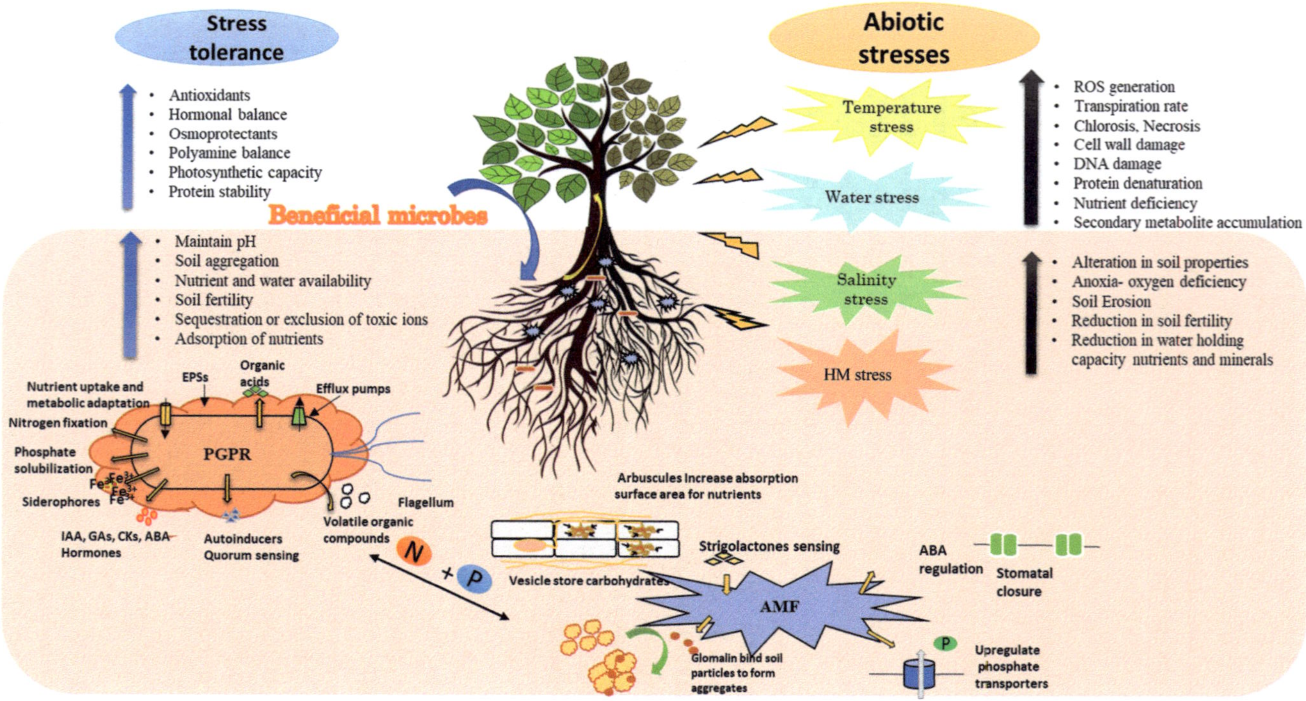

Fig. 17.1 Schematic representation of microbe-mediated abiotic stress management. Beneficial microbes such as PGPRs and AMFs impart stress tolerance in plants by establishing symbiotic relationships. PGPRs help by fixing nitrogen; producing siderophores, hormones, antioxidants, volatile organic compounds; mineralization and solubilization; etc., whereas AMFs add to the stability by secreting glomalin, regulating hormone balance, upregulating phosphate transporters, etc.

(*Bacillus megaterium*) enhanced yield and biological properties of soil (Jafari and Đurić 2012). Garg and Manchanda (2008) investigated a correlation between higher osmolyte accumulation, enhanced antioxidant enzyme activities and improved functional efficiency of nodules in pigeon pea plants when *Rhizobium* was co-inoculated with *G. mosseae* under salt-stressed conditions. The role of an efficient AM (*G. mosseae*) and *Sinorhizobium fredii* AR-4 symbiosis capable of attenuating salt stress has been observed which improved *Cajanus cajan* growth (Garg and Chandel 2012). Similarly, co-inoculating HM-stressed pigeon pea with *R. irregularis* and *Sinorhizobium fredii* AR-4 resulted in improved metal tolerance, nutrient acquisition and productivity (Garg and Kashyap 2017; Garg and Singh 2018). Moreover, chickpea (*Cicer arietinum*) inoculated with *G. mosseae* and *Mesorhizobium ciceri* PF:75 under salinity stress showed enhanced antioxidant defence system and provided tolerance (Garg and Bhandari 2016). AMF and PGPB co-inoculation considerably enhanced salinity tolerance of grape rootstock and by improvements in K^+: Na^+ ratio, plant-water balance, accumulation of PAs (spermine and spermidine), etc. (Upreti et al. 2016). *Cajanus cajan* inoculated with mixture of AMF (*F. mosseae*, *R. irregularis*) and PGPR (*Sinorhizobium fredii* AR-4) under salinity stress improved plant biomass and reduced oxidative damage by enhancing antioxidant enzyme activity (Pandey and Garg 2017). These studies indicate that PGPR and AMF are important biofertilizers that could be used in an environmentally friendly manner as the interactions between these microbes and plants result in improved abiotic stress tolerance, nutrient availability, biocontrol and phytoremediation for sustainable agriculture.

17.6 Conclusion

Nowadays, in the environment, abiotic stresses are one of the major obstructions for agricultural productivity in the world. Plant-associated microbes can play an essential role in imparting abiotic stress tolerance by refining soil texture and physico-chemical properties of soil, particularly aggregate formation. Rhizosphere microorganisms, particularly beneficial bacteria and fungi, can control abiotic stresses and are considered as eco-friendly strategies to improve crop yield. They can trigger various mechanisms for plant growth like supplying nutrients and hormones, acting as biocontrol agents, osmotic response and induction of novel genes. At present, very few strains of *Rhizobium*, *Azotobacter* and *Azospirillum* and few strains of AMF, i.e. *Glomus* sp. and *Acaulospora*, are being commercialized, and there is a need to identify superior and more effective strains by genetic manipulation techniques. Application of synergistic combinations of microbes includes PGPRs, saprophytic fungi and other helper microbes which will help farmers to develop sustainable agricultural system. The production of stress-tolerant crop varieties by using plant breeding and genetic engineering techniques is necessary, but both are long and costly, while microbial inoculations are cost-effective and eco-friendly alternatives to alleviate abiotic stress-induced responses in plants. To conclude, application of various microorganisms, isolated from different environmental conditions, is necessary for the alleviation of abiotic stresses and development of sustainable agroecosystem.

Acknowledgement The authors are grateful to Department of Biotechnology (DBT), Ministry of Science and Technology, Govt. of India and University Grants Commission (UGC) for financial assistance in carrying out related research.

References

Abbaspour H, Saeidi-Sar S, Afshari H, Abdel-Wahhab MA (2012) Tolerance of mycorrhiza infected pistachio (*Pistacia vera* L.) seedling to drought stress under glasshouse conditions. J Plant Physiol 169:704–709

Abd_Allah EF, Egamberdieva D (2016) Arbuscular mycorrhizal fungi enhance basil tolerance to salt stress through improved physiological and nutritional status. Pak J Bot 48:37–45

Abdel Latef A, Chaoxing H (2011) Arbuscular mycorrhizal influence on growth, photosynthetic pigments, osmotic adjustment and oxidative stress in tomato plants subjected to low temperature stress. Acta Physiol Plant 33:1217–1225

Ahanger MA, Akram NA, Ashraf M, Alyemeni MN, Wijaya L, Ahmad P (2017) Plant responses to environmental stresses—from gene to biotechnology. AoB Plants 9(4):plx025. https://doi.org/10.1093/aobpla/plx025

Ahn C, Park U, Park PB (2011) Increased salt and drought tolerance by D-ononitol production in transgenic *Arabidopsis thaliana*. Biochem Biophys Res Commun 415:669–674

Akiyama K, Hayashi H (2006) Strigolactones: chemical signals in fungal symbionts and parasitic weeds in plant roots. Ann Bot 97:925–931

Alami Y, Achouak W, Marol C, Heulin T (2000) Rhizosphere soil aggregation and plant growth promotion of sunflowers by an exopolysaccharide-producing *Rhizobium* sp. Strain isolated from sunflower roots. Appl Environ Microbiol 66:3393–3398

Ali SZ, Sandhya V, Grover M, Kishore N, Rao LV, Venkateswarlu B (2009) *Pseudomonas* sp. strain AKM-P6 enhances tolerance of *sorghum* seedlings to elevated temperatures. Biol Fertil Soils 46:45–55

Aliasgharzad N, Shirmohamadi E, Oustan S (2009) Siderophore production by mycorrhizal sorghum roots under micronutrient deficient condition. Soil Environ 28:119–123

Alori ET, Glick BR, Babalola OO (2017) Microbial phosphorus solubilization and its potential for use in sustainable agriculture. Front Microbiol 8:971. https://doi.org/10.3389/fmicb.2017.00971

Aloui A, Recorbet G, Robert F, Schoefs B, Bertrand M, Henry C, Gianinazzi-Pearson V, Dumas-Gaudot E, Aschi-Smiti S (2011) Arbuscular mycorrhizal symbiosis elicits shoot proteome changes that are modified during cadmium stress alleviation in *Medicago truncatula*. BMC Plant Biol 11:75. https://doi.org/10.1186/1471-2229-11-75

Arite T, Umehara M, Ishikawa S, Hanada A, Maekawa M, Yamaguchi S, Kyozuka J (2009) d14, a strigolactone-insensitive mutant of rice, shows an accelerated outgrowth of tillers. Plant Cell Physiol 50:1416–1424

Arite T, Kameoka H, Kyozuka J (2012) Strigolactone positively controls crown root elongation in rice. J Plant Growth Regul 31:165–172

Arkhipova TN, Prinsen E, Veselov SU, Martinenko EV, Melentiev AI, Kudoyarova GR (2007) Cytokinin producing bacteria enhance plant growth in drying soil. Plant Soil 292:305–315

Aroca R, Porcel R, Ruiz-Lozano JM (2007) How does arbuscular mycorrhizal symbiosis regulate root hydraulic properties and plasma membrane aquaporins in *Phaseolus vulgaris* under drought, cold or salinity stresses? New Phytol 173:808–816

Aroca R, Ruiz-Lozano JM, Zamarreño AM, Paz JA, García-Mina JM, Pozo MJ, López-Ráez JA (2013) Arbuscular mycorrhizal symbiosis influences strigolactone production under salinity and alleviates salt stress in lettuce plants. J Plant Physiol 170:47–55

Asensio D, Rapparini F, Peñuelas J (2012) AM fungi root colonization increases the production of essential isoprenoids vs. nonessential isoprenoids especially under drought stress conditions or after jasmonic acid application. Phytochemistry 77:149–161

Assumpção CF, Hermes VS, Pagno C, Castagna A, Mannucci A, Sgherri C, Pinzino C, Ranieri A, Hickmann Flôres S, de Oliveira RA (2018) Phenolic enrichment in apple skin following post-harvest fruit UV-B treatment. Postharvest Biol Technol 138:37–45

Audrain B, Farag MA, Ryu CM, Ghigo JM (2015) Role of bacterial volatile compounds in bacterial biology. FEMS Microbiol Rev 39:222–233

Ayalaja CF, Pedro A (2012) Stress response in microbiology. In: Reguena JM (ed) Caister Academic, Madrid

Bach EM, Baer SG, Meyer CK, Six J (2010) Soil texture affects soil microbial and structural recovery during grassland restoration. Soil Biol Biochem 42:2182–2191

Badri DV, Weir TL, van der Lelie D, Vivanco JM (2009) Rhizosphere chemical dialogues: plant–microbe interactions. Curr Opin Biotechnol 20:642–650

Bagheri V, Shamshiri MH, Shirani H, Roosta H (2012) Nutrient uptake and distribution in mycorrhizal pistachio seedlings under drought stress. JAST 14:1591–1604

Bala A, Joshi N (2016) Role of bacteria in bio sorption of heavy metals. Int J Adv Res 4:416–424

Balliu A, Sallaku G, Rewald B (2015) AMF inoculation enhances growth and improves the nutrient uptake rates of transplanted, salt-stressed tomato seedlings. Sustainability 7:15967–15981

Balsanelli E, de Baura VA, Pedrosa FD, de Souza EM, Monteiro RA (2014) Exopolysaccharide biosynthesis enables mature biofilm formation on abiotic surfaces by *Herbaspirillum seropedicae*. PLoS One 9:e110392. https://doi.org/10.1371/journal.pone.0110392

Barea JM, Pozo MJ, Azcon R, Azcon-Aguilar C (2005) Microbial co-operation in the rhizosphere. J Exp Bot 56:1761–1778

Bauer WD, Mathesius U (2004) Plant responses to bacterial quorum sensing signals. Curr Opin Plant Biol 7:429–433

Belimov AA, Dodd IC, Hontzeas N, Theobald JC, Safronova VI, Davies WJ (2009) Rhizosphere bacteria containing 1-aminocyclopropane-1-carboxylate deaminase increase yield of plants grown in drying soil via both local and systemic hormone signalling. New Phytol 181:413–423

Bender SF, Wagg C, van der Heijden MG (2016) An underground revolution: biodiversity and soil ecological engineering for agricultural sustainability. Trends Ecol Evolut 31:440–452

Beneduzi A, Ambrosini A, Passaglia LM (2012) Plant growth-promoting rhizobacteria (PGPR): their potential as antagonists and biocontrol agents. Genet Mol Biol 35:1044–1051

Bharti N, Pandey SS, Barnawal D, Patel VK, Kalra A (2016) Plant growth promoting rhizobacteria *Dietzia natronolimnaea* modulates the expression of stress responsive genes providing protection of wheat from salinity stress. Sci Rep 6:34768. https://doi.org/10.1038/srep34768

Bhatia CR (2008) Role of microbial diversity for soil, health and plant nutrition. In: Nautiyal CS, Dion P (eds) Molecular mechanisms of plant and microbe coexistence. Springer, Berlin, pp 53–74

Bhattacharyya PN, Jha DK (2012) Plant growth-promoting rhizobacteria (PGPR): emergence in agriculture. World J Microbiol Biotechnol 28:1327–1350

Bolan N, Kunhikrishnan A, Thangarajan R, Kumpiene J, Park J, Makino T, Kirkham MB, Scheckel K (2014) Remediation of heavy metal (loid)s contaminated soils—to mobilize or to immobilize. J Hazard Mater 266:141–166

Bompadre MJ, Silvani VA, Bidondo LF, Ríos de Molina MDC, Colombo RP, Pardo AG, Godeas AM (2014) Arbuscular mycorrhizal fungi alleviate oxidative stress in pomegranate plants growing under different irrigation conditions. Botany 92:187–193

Borowik A, Wyszkowska J (2016) Impact of temperature on the biological properties of soil. Int Agrophys 30:1–8

Brunner I, Herzog C, Dawes MA, Arend M, Sperisen C (2015) How tree roots respond to drought. Front of Plant Sci 6:547

Bünemann EK, Bongiorno G, Bai Z, Creamer RE, De Deyn G, de Goede R, Pulleman M et al (2018) Soil quality—a critical review. Soil Biol Biochem 120:105–125

Buscardo E, Geml J, Schmidt SK, Silva AL, Ramos RT, Barbosa SM, Andrade SS, Dalla Costa R, Souza AP, Freitas H, Cunha HB (2018) Of mammals and bacteria in a rainforest: temporal dynamics of soil bacteria in response to simulated N pulse from mammalian urine. Funct Ecol 32:773–784

Buyer JS, Teasdale JR, Roberts DP, Zasada IA, Maul JE (2010) Factors affecting soil microbial community structure in tomato cropping systems. Soil Biol Biochem 42:831–841

Charpentier M, Sun J, Wen J, Mysore KS, Oldroyd GE (2014) Abscisic acid promotion of arbuscular mycorrhizal colonization requires a component of the PROTEIN PHOSPHATASE 2A complex. Plant Physiol 166:2077–2090

Chen M, Wei H, Cao J, Liu R, Wang Y, Zheng C (2007) Expression of *Bacillus subtilis* proBA genes and reduction of feedback inhibition of proline synthesis increases proline production and confers osmotolerance in transgenic *Arabidopsis*. J Biochem Mol Biol 40:396–403

Chen S, Jin W, Liu A, Zhang S, Liu D, Wang F, Lin X, He C (2013) Arbuscular mycorrhizal fungi (AMF) increase growth and secondary metabolism in cucumber subjected to low temperature stress. Sci Hortic 160:222–229

Chevalier F, Nieminen K, Sánchez-Ferrero JC, Rodríguez ML, Chagoyen M, Hardtke CS, Cubas P (2014) Strigolactone promotes degradation of DWARF14, an α/β hydrolase essential for strigolactone signaling in *Arabidopsis*. Plant Cell 26:1134–1150

Cho SM, Kang BR, Han SH, Anderson AJ, Park JY, Lee YH, Cho BH, Yang KY, Ryu CM, Kim YC (2008) 2R, 3R-butanediol, a bacterial volatile produced by *Pseudomonas chlororaphis* O6, is involved in induction of systemic tolerance to drought in *Arabidopsis thaliana*. MPMI 21:1067–1075

Chowdhury SP, Hartmann A, Gao X, Borriss R (2015) Biocontrol mechanism by root-associated *Bacillus amyloliquefaciens* FZB42—a review. Front Microbiol 6:780. https://doi.org/10.3389/fmicb.2015.00780

Chaudhry TM, Hayes WJ, Khan AG, Khoo CS (1998) Phytoremediation-focusing on accumulator plants that remediate metal-contaminated soils. Aust J Ecotoxicol 4:37–51

Chu D (2018) Effects of heavy metals on soil microbial community. In: IOP conference series: Environ Earth Sci 113(1):012009. IOP

Cicatelli A, Lingua G, Todeschini V, Biondi S, Torrigiani P, Castiglione S (2010) Arbuscular mycorrhizal fungi restore normal growth in a white poplar clone grown on heavy metal-contaminated soil, and this is associated with upregulation of foliar metallothionein and polyamine biosynthetic gene expression. Ann Bot 106:791–802

Cleason A (2006) Volatile organic compounds from microorganisms. Doctoral dissertation, Ph.D. thesis, Umeå University, Umeå

Clúa J, Roda C, Zanetti ME, Blanco FA (2018) Compatibility between legumes and rhizobia for the establishment of a successful nitrogen-fixing symbiosis. Genes 9:125

Cobessi D, Celia H, Folschweiller N, Schalk IJ, Abdallah MA, Pattus F (2005) The crystal structure of the pyoverdine outer membrane receptor FpvA from *Pseudomonas aeruginosa* at 3.6 Å resolution. J Mol Biol 347:121–134

Cohen AJ, Brauer M, Burnett R, Anderson HR, Frostad J, Estep K, Balakrishnan K, Brunekreef B, Dandona L, Dandona R, Feigin V (2017) Estimates and 25-year trends of the global burden of disease attributable to ambient air pollution: an analysis of data from the Global Burden of Diseases Study 2015. Lancet 389:1907–1918

Cooper JE (2007) Early interactions between legumes and rhizobia: disclosing complexity in a molecular dialogue. J Appl Microbiol 103:1355–1365. https://doi.org/10.1111/j.1365-2672.2007.03366.x

Cornejo P, Meier S, Borie G, Rillig MC, Borie F (2008) Glomalin-related soil protein in a Mediterranean ecosystem affected by a copper smelter and its contribution to Cu and Zn sequestration. Sci Total Environ 406:154–160

Cornejo P, Seguel A, Aguilera P, Meier S, Larsen J, Borie F (2017) Arbuscular mycorrhizal fungi improve tolerance of agricultural plants to cope abiotic stress conditions. In: Singh DP, Singh HB, Prabha R (eds) Plant-microbe interactions in agro-ecological perspectives. Springer, Singapore, pp 55–80

Craine JM, Gelderman TM (2011) Soil moisture controls on temperature sensitivity of soil organic carbon decomposition for a mesic grassland. Soil Biol Biochem 43:455–457

Cramer GR, Urano K, Delrot S, Pezzotti M, Shinozaki K (2011) Effects of abiotic stress on plants: a systems biology perspective. BMC Plant Biol 11:163. https://doi.org/10.1186/1471-2229-11-163

Crowley DE (2006) Microbial siderophores in the plant rhizospheric. In: Barton LL, Abadía J (eds) Iron nutrition in plants and rhizospheric microorganisms. Springer, Dordrecht, pp 169–198

Daniels R, De Vos DE, Desair J, Raedschelders G, Luyten E, Rosemeyer V, Verreth C, Schoeters E, Vanderleyden J, Michiels J (2002) The cin quorum sensing locus of *Rhizobium etli* CNPAF512 affects growth and symbiotic nitrogen fixation. J Biol Chem 277:462–468

Das SK, Varma A (2010) Role of enzymes in maintaining soil health. In: Shukla G, Varma A (eds) Soil enzymology. Springer, Berlin, pp 25–42

Davidson EA, Janssens IA (2006) Temperature sensitivity of soil carbon decomposition and feedbacks to climate change. Nature 440:165–173

Davinic M, Fultz LM, Acosta-Martinez V, Calderon FJ, Cox SB, Dowd SE, Allen VG, Zak JC, Moore-Kucera J (2012) Pyrosequencing and mid-infrared spectroscopy reveal distinct aggregate stratification of soil bacterial communities and organic matter composition. Soil Biol Biochem 46:63–72

De-la-Peña C, Badri DV, Lei Z, Watson BS, Brandão MM, Silva-Filho MC, Sumner LW, Vivanco JM (2010) Root secretion of defense-related proteins is development-dependent and correlated with flowering time. J Biol Chem 285:30654–30665

Desai S, Kumar GP, Amalraj LD, Bagyaraj DJ, Ashwin R (2016) Exploiting PGPR and AMF biodiversity for plant health management. In: Singh DP, Singh HB, Prabha R (eds) Microbial inoculants in sustainable agricultural productivity. Springer, New Delhi, pp 145–160

de Zelicourt A, Al-Yousif M, Hirt H (2013) Rhizosphere microbes as essential partners for plant stress tolerance. Mol Plant 6:242–245

Dimkpa C, Weinand T, Asch F (2009) Plant–rhizobacteria interactions alleviate abiotic stress conditions. Plant Cell Environ 32:1682–1694

Dixit R, Malaviya D, Pandiyan K, Singh UB, Sahu A, Shukla R, Singh BP, Rai JP, Sharma PK, Lade H, Paul D (2015) Bioremediation of heavy metals from soil and aquatic environment: an overview of principles and criteria of fundamental processes. Sustainability 7:2189–2212

Doty SL, Oakley B, Xin G, Kang JW, Singleton G, Khan Z, Vajzovic A, Staley JT (2009) Diazotrophic endophytes of native black cottonwood and willow. Symbiosis 47:23–33

Egamberdieva D, Wirth SJ, Alqarawi AA, Abd_Allah EF, Hashem A (2017) Phytohormones and beneficial microbes: essential components for plants to balance stress and fitness. Front Microbiol 8:2104. https://doi.org/10.3389/fmicb.2017.02104

Ellouze W, Esmaeili Taheri A, Bainard LD, Yang C, Bazghaleh N, Navarro-Borrell A, Hanson K, Hamel C (2014) Soil fungal resources in annual cropping systems and their potential for management. BioMed Res Int. https://doi.org/10.1155/2014/531824

Estrada-Luna AA, Davies FT (2003) Arbuscular mycorrhizal fungi influence water relations, gas exchange, abscissic acid and growth of micropropagated Chile ancho pepper (*Capsicum annuum*) plantlets during acclimatization and post-acclimatization. J Plant Physiol 160:1073–1083

Evelin H, Kapoor R, Giri B (2009) Arbuscular mycorrhizal fungi in alleviation of salt stress: a review. Ann Bot 104:1263–1280

Fan QJ, Liu JH (2012) Nitric oxide is involved in dehydration/drought tolerance in *Poncirus trifoliata* seedlings through regulation of antioxidant systems and stomatal response. Plant Cell Rep 31:145–154

Farooq M, Basra SMA, Wahid A, Cheema ZA, Cheema MA, Khaliq A (2008) Physiological role of exogenously applied glycinebetaine in improving drought tolerance of fine grain aromatic rice (*Oryza sativa* L.). J Agron Crop Sci 194:325–333

Fernández V, Ebert G, Winkelmann G (2005) The use of microbial siderophores for foliar iron application studies. Plant Soil 72:245–252

Fernández-Bidondo L, Silvani V, Colombo R, Pérgola M, Bompadre J, Godeas A (2011) Pre-symbiotic and symbiotic interactions between *Glomus intraradices* and two *Paenibacillus* species isolated from AM propagules. In vitro and in vivo assays with soybean (AG043RG) as plant host. Soil Biol Biochem 43:1866–1872

Figueiredo MVB, Burity HA, Martinez CR, Chanway CP (2008) Alleviation of drought stress in the common bean (*Phaseolus vulgaris* L.) by co-inoculation with *Paenibacillus polymyxa* and *Rhizobium tropici*. Appl Soil Ecol 40:182–188

Flouri F, Chatjipavdlidis C, Balis C (1990) Effect of olive oil mills liquid wastes on soil fertility. In: Reunion Znt, sobre: Tratamiento de alpechines. Cordoba, Spain

Fontana A, Reichelt M, Hempel S, Gershenzon J, Unsicker SB (2009) The effects of arbuscular mycorrhizal fungi on direct and indirect defense metabolites of *Plantago lanceolata* L. J Chem Ecol 35:833–843

Friedlingstein P, Cox P, Betts R, Bopp L, von Bloh W, Brovkin V, Cadule P, Doney S, Eby M, Fung I, Bala G (2006) Climate–carbon cycle feedback analysis: results from the C4MIP model intercomparison. J Clim 19:3337–3353

Friedlová M (2010) The influence of heavy metals on soil biological and chemical properties. Soil Water Res 5:21–27

Frohnmeyer H, Staiger D (2003) Ultraviolet-B radiation-mediated responses in plants. Balancing damage and protection. Plant Physiol 133:1420–1428

Fuqua C, Parsek MR, Greenberg EP (2001) Regulation of gene expression by cell-to-cell communication: acyl-homoserine lactone quorum sensing. Annu Rev Genet 35:439–468

Gage DJ (2004) Infection and invasion of roots by symbiotic, nitrogen-fixing rhizobia during nodulation of temperate legumes. Microbiol Mol Biol Rev 68:280–300

Gale MR, Grigal DF, Harding RB (1991) Soil productivity index: predictions of site quality for white spruce plantations. Soil Sci Soc Am J 55:1701–1708

Garbeva PV, Van Veen JA, Van Elsas JD (2004) Microbial diversity in soil: selection of microbial populations by plant and soil type and implications for disease suppressiveness. Annu Rev Phytopathol 42:243–270

Garbisu C, Garaiyurrebaso O, Epelde L, Grohmann E, Alkorta I (2017) Plasmid-mediated bioaugmentation for the bioremediation of contaminated soils. Front microbiol 8:1966. https://doi.org/10.3389/fmicb.2017.01966

Garg N, Aggarwal N (2012) Effect of mycorrhizal inoculations on heavy metal uptake and stress alleviation of *Cajanus cajan* (L.) Millsp. genotypes grown in cadmium and lead contaminated soils. Plant Growth Regul 66:9–26

Garg N, Baher N (2013) Role of arbuscular mycorrhizal symbiosis in proline biosynthesis and metabolism of *Cicer arietinum* L.(chickpea) genotypes under salt stress. J Plant Growth Regul 32:767–778

Garg N, Bhandari P (2016) Interactive effects of silicon and arbuscular mycorrhiza in modulating ascorbate-glutathione cycle and antioxidant scavenging capacity in differentially salt-tolerant *Cicer arietinum* L. genotypes subjected to long-term salinity. Protoplasma 253:1325–1345

Garg N, Chandel S (2012) Role of arbuscular mycorrhizal (AM) fungi on growth, cadmium uptake, osmolyte, and phytochelatin synthesis in *Cajanus cajan* (L.) Millsp. under NaCl and Cd stresses. J Plant Growth Regul 31:292–308

Garg N, Geetanjali N (2007) Symbiotic nitrogen fixation in legume nodules: process and signaling. A review. Agron Sustain Dev 27:59–68

Garg N, Kashyap L (2017) Silicon and *Rhizophagus irregularis*: potential candidates for ameliorating negative impacts of arsenate and arsenite stress on growth, nutrient acquisition and productivity in *Cajanus cajan* (L.) Millsp. genotypes. Environ Sci Pollut Res Int 24:18520–18535

Garg N, Kaur H (2013) Impact of cadmium-zinc interactions on metal uptake, translocation and yield in pigeonpea genotypes colonized by arbuscular mycorrhizal fungi. J Plant Nutr 36:67–90
Garg N, Manchanda G (2008) Effect of arbuscular mycorrhizal inoculation on salt-induced nodule senescence in Cajanus cajan (pigeonpea). J Plant Growth Regul 27:115. https://doi.org/10.1007/s00344-007-9038-z
Garg N, Singh S (2018) Arbuscular mycorrhiza *Rhizophagus irregularis* and silicon modulate growth, proline biosynthesis and yield in *Cajanus cajan* L. Millsp. (pigeonpea) genotypes under cadmium and zinc stress. J Plant Growth Regul 37:46–63
Garg N, Singla R (2004) Growth, photosynthesis, nodule nitrogen and carbon fixation in the chickpea cultivars under salt stress. BJPP 16:137–146
Garg N, Singla P (2012) The role of *Glomus mosseae* on key physiological and biochemical parameters of pea plants grown in arsenic contaminated soil. Sci Hortic 143:92–101
Garg N, Singla P (2016) Stimulation of nitrogen fixation and trehalose biosynthesis by naringenin (Nar) and arbuscular mycorrhiza (AM) in chickpea under salinity stress. Plant Growth Regul 80:5–22
Garg N, Manchanda G, Kumar A (2014a) Bacterial quorum sensing: circuits and applications. Antonie Van Leeuwenhoek 105:289–305
Garg N, Manchanda G, Singla P (2014b) Analysis of emergence stage facilitates the evaluation of chickpea (*Cicer arietinum* L.) genotypes for salinity tolerance imparted by mycorrhizal colonization. Acta Physiol Plant 36:2651–2669
Gauri SS, Mandal SM, Pati BR (2012) Impact of *Azotobacte*r exopolysaccharides on sustainable agriculture. Appl Microbiol Biotechnol 95:331–338
Gibson AH (1976) Recovery and compensation by nodulated legumes to environmental stress. In: Nutman PS (ed) Symbiotic nitrogen fixation. Cambridge University Press, Cambridge, pp 385–404
Gill SS, Tuteja N (2010) Polyamines and abiotic stress tolerance in plants. Plant Signal Behav 5:26–33
Giller KE, Witter E, McGrath SP (1998) Toxicity of heavy metals to microorganisms and microbial processes in agricultural soils. Soil Biol Biochem 30:1389–1414
Giovannetti M, Sbrana C, Logi C (1994) Early processes involved in host recognition by arbuscular mycorrhizal fungi. New Phytol 127:703–709
Glick BR (2014) Bacteria with ACC deaminase can promote plant growth and help to feed the world. Microbiol Res 169:30–39
Golldack D, Li C, Mohan H, Probst N (2014) Tolerance to drought and salt stress in plants: unraveling the signaling networks. Front Plant Sci 5:151. https://doi.org/10.3389/fpls.2014.00151
Gomez-Roldan V, Roux C, Girard D, Bécard G, Puech V (2007) Strigolactones: promising plant signals. Plant Signal Behav 2:163–164
Gonzalez-Chavez MC, Carrillo-Gonzalez R, Wright SF, Nichols KA (2004) The role of glomalin, a protein produced by arbuscular mycorrhizal fungi, in sequestering potentially toxic elements. Environ Pollut 130:317–323
Goswami D, Thakker JN, Dhandhukia PC (2016) Portraying mechanics of plant growth promoting rhizobacteria (PGPR): a review. Cogent Food Agric 2:1127500. https://doi.org/10.1080/23311932.2015.1127500
Grayston SJ, Wang SQ, Campbell CD, Edwards AC (1998) Selective influence of plant species on microbial diversity in the rhizosphere. Soil Biol Biochem 30:369–378
Grover M, Ali Sk Z, Sandhya V, Venkateswarlu B (2011) Role of microorganisms in adaptation of agricultural crops to abiotic stresses. World J Microbiol Biotechnol 27:1231–1240
Gube M (2016) Fungal molecular response to heavy metal stress. In: Hoffmeister D (ed) Biochemistry and molecular biology. Springer, Cham, pp 47–68
Guo J, Chi J (2014) Effect of Cd-tolerant plant growth-promoting rhizobium on plant growth and Cd uptake by *Lolium multiflorum* Lam. and *Glycine max* (L.) Merr. in Cd-contaminated soil. Plant Soil 375:205–214

Gururani MA, Upadhyaya CP, Strasser RJ, Yu JW, Park SW (2013a) Evaluation of abiotic stress tolerance in transgenic potato plants with reduced expression of PSII manganese stabilizing protein. Plant Sci 198:7–16

Gururani MA, Upadhyaya CP, Upadhyaya CP, Baskar B, Venkatesh J, Nookaraju A, Park SW (2013b) Plant growth promoting rhizobacteria enhance abiotic stress tolerance in *Solanum tuberosum* through inducing changes in the expression of ROS scavenging enzymes and improved photosynthetic performance. J Plant Growth Regul 32:245–258

Gururani MA, Venkatesh J, Tran LSP (2015) Regulation of photosynthesis during abiotic stress-induced photoinhibition. Mol Plant 8:1304–1320

Hamiaux C, Drummond RSM, Janssen BJ, Ledger SE, Cooney JM, Newcomb RD, Snowden KC (2012) DAD2 is an α/β hydrolase likely to be involved in the perception of the plant branching hormone, strigolactone. Curr Biol 22:2032–2036

Hammer EC, Rillig MC (2011) The influence of different stresses on glomalin levels in an arbuscular mycorrhizal fungus-salinity increases glomalin content. PLoS One 6:e28426. https://doi.org/10.1371/journal.pone.0028426

Harrier LA, Watson CA (2004) The potential role of arbuscular mycorrhizal (AM) fungi in the bioprotection of plants against soil-borne pathogens in organic and/or other sustainable farming systems. Pest Manag Sci 60:149–157

Harris RF (1981) Effect of water potential on microbial growth and activity. In: Parr JF, Gardner WR, Elliott LF (eds) Water potential relations in soil microbiology. Soil Science Society of America, Madison, WI, pp 23–96

Hartmann FE, Vallet AS, McDonald BA, Croll D (2017) A fungal wheat pathogen evolved host specialization by extensive chromosomal rearrangements. ISME J 11:1189–1204. https://doi.org/10.1038/15mej.2016.196

Hasanuzzaman M, Nahar K, Alam MM, Roychowdhury R, Fujita M (2013) Physiological, biochemical, and molecular mechanisms of heat stress tolerance in plants. Int J Mol Sci 14:9643–9684

Haselwandter K (2008) Structure and function of siderophores produced by mycorrhizal fungi. Miner Mag 72:61–64

Hashem A, Abd_Allah EF, Alqarawi AA, Al-Huqail AA, Wirth S, Egamberdieva D (2016) The interaction between arbuscular mycorrhizal fungi and endophytic bacteria enhances plant growth of *Acacia gerrardii* under salt stress. Front Microbiol 19(7):1089

Hassen AI, Bopape FL, Sanger LK (2016) Microbial inoculants as agents of growth promotion and abiotic stress tolerance in plants. In: Singh DP, Singh HB, Prabha R (eds) Microbial inoculants in sustainable agricultural productivity. Springer, New Delhi, pp 23–36

Hassink J (1994) Effect of soil texture on the size of the microbial biomass and on the amount of C and N mineralized per unit of microbial biomass in Dutch grassland soils. Soil Biol Biochem 26:1573–1581

Heidari M, Golpayegani A (2012) Effects of water stress and inoculation with plant growth promoting rhizobacteria (PGPR) on antioxidant status and photosynthetic pigments in basil (*Ocimum basilicum* L.). J Saudi Soc Agric 11:57–61

Heidari M, Mousavinik SM, Golpayegani A (2011) Plant growth promoting rhizobacteria (PGPR) effect on physiological parameters and mineral uptake in basil (*Ociumum basilicm* L.) under water stress. ARPN J Agric Biol Sci 6:6–11

Herrera-Medina MJ, Steinkellner S, Vierheilig H, Ocampo Bote JA, García Garrido JM (2007) Abscisic acid determines arbuscule development and functionality in the tomato arbuscular mycorrhiza. New Phytol 175:554–564

Hiltner L (1904) Uber neuere Erfahrungen und Probleme auf dem Gebiete der Bodenbakteriologie unter bessonderer Berücksichtigung der Gründung und Brache. Arb Dtsch Landwirtsch Ges Berl 98:59–78

Hirsch AM, Fujishige NA (2012) Molecular signals and receptors: communication between nitrogen-fixing bacteria and their plant hosts. In: Witzany G, Baluška F (eds) Biocommunication of plants. Springer, Berlin, pp 255–280

Hohnjec N, Vieweg MF, Pühler A, Becker A, Küster H (2005) Overlaps in the transcriptional profiles of *Medicago truncatula* roots inoculated with two different Glomus fungi provide insights into the genetic program activated during arbuscular mycorrhiza. Plant Physiol 137:1283–1301

Holmer R, Rutten L, Kohlen W, van Velzen R, Geurts R (2017) Commonalities in symbiotic plant-microbe signalling. Adv Bot Res 82:187–221

Hooshangi S, Bentley WE (2008) From unicellular properties to multicellular behavior: bacteria quorum sensing circuitry and applications. Curr Opin Biotechnol 19:550–555

Hu Y, Schmidhalter U (2005) Drought and salinity: a comparison of their effects on mineral nutrition of plants. J Soil Sci Plant Nutr 168:541–549

Huang SH, Bing P, Yang ZH, Chai LY, Zhou LC (2009) Chromium accumulation, microorganism population and enzyme activities in soils around chromium-containing slag heap of steel alloy factory. T Nonferr Metal Soc 19:241–248

Hussain SS, Ali M, Ahmad M, Siddique KH (2011) Polyamines: natural and engineered abiotic and biotic stress tolerance in plants. Biotechnol Adv 29:300–311

Ibaraki Y (2008) Evaluation of photosynthetic capacity in micropropagated plants by image analysis. In: Dutta Gupta S, Ibaraki Y (eds) Plan tissue culture engineering. Springer, Dordrecht, pp 15–29

Ilangumaran G, Smith DL (2017) Plant growth promoting rhizobacteria in amelioration of salinity stress: a systems biology perspective. Front Plant Sci 8:1768. https://doi.org/10.3389/fpls.2017.01768

Ilyas N, Bano A (2012) Potential use of soil microbial community in agriculture. In: Bacteria in agrobiology: plant probiotics, 1st edn. Springer, Berlin, pp 45–64

İpek M, Eşitken A (2017) The actions of PGPR on micronutrient availability in soil and plant under calcareous soil conditions: an evaluation over Fe nutrition. In: Singh DP, Singh HB, Prabha R (eds) Plant-microbe interactions in agro-ecological perspectives. Springer, Singapore, pp 81–100

Jafari TH, Đurić S (2012) Inoculation of maize hybrids: the effect on grain yield and number of microorganisms in soil. In: International conference on bioscience: biotechnology and biodiversity, pp 401–415

Jahromi F, Aroca R, Porcel R, Ruiz-Lozano JM (2008) Influence of salinity on the *in vitro* development of *Glomus intraradices* and on the *in vivo* physiological and molecular responses of mycorrhizal lettuce plants. Microb Ecol 55:45–53

Jaleel CA, Manivannan P, Sankar B, Kishorekumar A, Gopi R, Somasundaram R, Panneerselvam R (2007) Induction of drought stress tolerance by ketoconazole in *Catharanthus roseus* is mediated by enhanced antioxidant potentials and secondary metabolite accumulation. Colloids Surf B Biointerfaces 60:201–206

Jaleel CA, Gopi R, Panneerselvam R (2008) Growth and photosynthetic pigments responses of two varieties of *Catharanthus roseus* to triadimefon treatment. C R Biol 331:272–277

Janoušková M, Pavlíková D, Vosátka M (2006) Potential contribution of arbuscular mycorrhiza to cadmium immobilisation in soil. Chemosphere 65:1959–1965

Jansen MAK, Bornman JF (2012) UV-B radiation: from generic stressor to specific regulator. Physiol Plant 145:501–504

Järvan M, Edesi L, Adamson A, Võsa T (2014) Soil microbial communities and dehydrogenase activity depending on farming systems. Plant Soil Environ 60:459–463

Jayachandran K, Hetrick BAD, Schwab AP (1989) Mycorrhizal mediation of phosphorus availability: synthetic iron chelate effects on phosphorus solubilization. Soil Sci Soc Am J 53:1701–1706

Jiang L, Liu X, Xiong G, Liu H, Chen F, Wang L, Meng X, Liu G, Yu H, Yuan Y, Yi W (2013) DWARF 53 acts as a repressor of strigolactone signalling in rice. Nature 504:401–405

Jin CW, Ye YQ, Zheng SJ (2013) An underground tale: contribution of microbial activity to plant iron acquisition via ecological processes. Ann Bot 113:7–18

Joner EJ, Briones R, Leyval C (2000) Metal-binding capacity of arbuscular mycorrhizal mycelium. Plant Soil 226:227–234

Jones KM, Kobayashi H, Davies BW, Taga ME, Walker GC (2007) How rhizobial symbionts invade plants: the *Sinorhizobium-Medicago* model. Nat Rev Microbiol 5:619–633

Kaci Y, Heyraud A, Barakat M, Heulin T (2005) Isolation and identification of an EPS-producing *Rhizobium* strain from arid soil (Algeria): characterization of its EPS and the effect of inoculation on wheat rhizosphere soil structure. Res Microbiol 156:522–531

Kang SM, Khan AL, Waqas M, You YH, Kim JH, Kim JG, Hamayun M, Lee IJ (2014) Plant growth-promoting rhizobacteria reduce adverse effects of salinity and osmotic stress by regulating phytohormones and antioxidants in *Cucumis sativus*. J Plant Interact 9:673–682

Karlidag H, Esitken A, Turan M, Sahin F (2007) Effects of root inoculation of plant growth promoting rhizobacteria (PGPR) on yield, growth and nutrient element contents of leaves of apple. Sci Hortic 114:16–20

Kaur H, Garg N (2017) Recent perspectives on cross talk between cadmium, zinc, and arbuscular mycorrhizal fungi in plants. J Plant Growth Regul:1–14

Khan AG (2005) Role of soil microbes in the rhizospheres of plants growing on trace metal contaminated soils in phytoremediation. J Trace Elem Med Biol 18:355–364

Khan MS, Zaidi A (2007) Synergistic effects of the inoculation with plant growth-promoting rhizobacteria and an arbuscular mycorrhizal fungus on the performance of wheat. Turk J Agric For 31:355–362

Kodaira H (2014) Responding to climate change and expectations for research. Paddy Water Environ 12:211–212

Kohler J, Hernández JA, Caravaca F, Roldán A (2008) Plant-growth-promoting rhizobacteria and arbuscular mycorrhizal fungi modify alleviation biochemical mechanisms in water-stressed plants. Funct Plant Biol 35:141–151

Kuan KB, Othman R, Rahim KA, Shamsuddin ZH (2016) Plant growth-promoting rhizobacteria inoculation to enhance vegetative growth, nitrogen fixation and nitrogen remobilisation of maize under greenhouse conditions. PLoS One 11:e0152478. https://doi.org/10.1371/journal.pone.0152478

Kumar A, Dames JF, Gupta A, Sharma S, Gilbert JA, Ahmad P (2015) Current developments in arbuscular mycorrhizal fungi research and its role in salinity stress alleviation: a biotechnological perspective. Crit Rev Biotechnol 35:461–474

Lata R, Chowdhury S, Gond SK, White JF Jr (2018) Induction of abiotic stress tolerance in plants by endophytic microbes. Lett Appl Microbiol 66:268–276

Latge JP (2017) Immune evasion: face changing in the fungal opera. Nat Microbiol 2:16266–16269

Lecomte J, St-Arnaud M, Hijri M (2011) Isolation and identification of soil bacteria growing at the expense of arbuscular mycorrhizal fungi. FEMS Microb Lett 317:43–51

Lehmann KDS, Goldman BW, Dworkin I, Bryson DM, Wagner AP (2014) From cues to signals: evolution of interspecific communication via aposematism and mimicry in a predator–prey system. PLoS One 9:e91783. https://doi.org/10.1371/journal.pone.0091783

Lehnert M (2013) The soil temperature regime in the urban and suburban landscapes of olomoric. Czech Repub MGR 21:27–36

Li T, Hu YJ, Hao ZP, Li H, Chen BD (2013) Aquaporin genes GintAQPF1 and GintAQPF2 from Glomus intraradices contribute to plant drought tolerance. Plant Signal Behav 8:e24030. https://doi.org/10.4161/psb.24030

Liptzin D, Silver WL, Detto M (2011) Temporal dynamics in soil oxygen and greenhouse gases in two humid tropical forests. Ecosystems 14:171–182

Lithgow JK, Wilkinson A, Hardman A, Rodelas B, Wisniewski-Dye F, Williams P, Downie JA (2000) The regulatory locus *cinRI* in *Rhizobium leguminosarum* controls a network of quorum-sensing loci. Mol Microbiol 37:81–97

Liu A, Hamel C, Elmi A, Costa C, Ma B, Smith DL (2002) Concentrations of K, Ca, and Mg in maize colonized by arbuscular mycorrhizal fungi under field conditions. Can J Soil Sci 82:271–278

Liu D, An Z, Mao Z, Ma L, Lu Z (2015) Enhanced heavy metal tolerance and accumulation by transgenic sugar beets expressing *Streptococcus thermophilus* StGCS-GS in the presence of Cd, Zn and Cu alone or in combination. PLoS One 10:e0128824. https://doi.org/10.1371/journal.pone.0128824

Liu YR, Delgado-Baquerizo M, Trivedi P, He JZ, Wang JT, Singh BK (2017) Identity of biocrust species and microbial communities drive the response of soil multifunctionality to simulated global change. Soil Biol Biochem 107:208–217

Lugtenberg BJ, Kamilova F (2009) Plant-growth-promoting rhizobacteria. Annu Rev Microbiol 63:541–556

Luo X, Fu X, Yang Y, Cai P, Peng S, Chen W, Huang Q (2016) Microbial communities play important roles in modulating paddy soil fertility. Sci Rep 6:20326. https://doi.org/10.1038/srep20326

Manchanda G, Garg N (2011) Alleviation of salt-induced ionic, osmotic and oxidative stresses in *Cajanus cajan* nodules by AM inoculation. Plant Biosyst 145:88–97

Manefield M, de Nys R, Kumar N, Read R, Givskov M, Steinberg P, Kjelleberg S (1999) Evidence that halogenated furanones from *Delisea pulchra* inhibit acylated homoserine lactone (AHL)-mediated gene expression by displacing the AHL signal from its receptor protein. Microbiology 145:283–291

Manefield M, Rasmussen TB, Henzter M, Andersen JB, Steinberg P, Kjelleberg S, Givskov M (2002) Halogenated furanones inhibit quorum sensing through accelerated LuxR turnover. Microbiology 148:1119–1127

Marschner P, Timonen S (2005) Interactions between plant species and mycorrhizal colonization on the bacterial community composition in the rhizosphere. Appl Soil Ecol 28:23–36

Maya MA, Matsubara Y (2013) Influence of arbuscular mycorrhiza on the growth and antioxidative activity in cyclamen under heat stress. Mycorrhiza 23:381–390

Mayak S, Tirosh T, Glick BR (2004) Plant growth-promoting bacteria that confer resistance to water stress in tomatoes and peppers. Plant Sci 166(2):525–530

McLellan CA, Turbyville TJ, Wijeratne EK, Kerschen A, Vierling E, Queitsch C, Whitesell L, Gunatilaka AL (2007) A rhizosphere fungus enhances *Arabidopsis* thermotolerance through production of an HSP90 inhibitor. Plant Physiol 145:174–182

Meena H, Ahmed MA, Prakash P (2015) Amelioration of heat stress in wheat, Triticum aestivum by PGPR (*Pseudomonas aeruginosa* strain 2CpS1). Biosci Biotechno Res 8(2):171–174

Meena KK, Sorty AM, Bitla UM, Choudhary K, Gupta P, Pareek A, Singh DP, Prabha R, Sahu PK, Gupta VK, Singh HB (2017) Abiotic stress responses and microbe-mediated mitigation in plants: the omics strategies. Front Plant Sci 8:172. https://doi.org/10.3389/fpls.2017.00172

Meliani A, Bensoltane A, Mederbel K (2012) Microbial diversity and abundance in soil: related to plant and soil type. Am J Plant Nutr Fertil Technol 2:10–18

Mendes R, Kruijt M, De Bruijn I, Dekkers E, Van der Voort M, Schneider JHM, Piceno YM, DeSantis TZ, Andersen GL, Bakker PAHM, Raaijmakers JM (2011) Deciphering the rhizosphere microbiome for disease-suppressive bacteria. Science 332:1097–1100

Mendes R, Garbeva P, Raaijmakers JM (2013) The rhizosphere microbiome: significance of plant beneficial, plant pathogenic, and human pathogenic microorganisms. FEMS Microbiol Rev 37:634–663

Miller MB, Bassler BL (2001) Quorum sensing in bacteria. Annu Rev Microbiol 55:165–199

Miller RM, Jastrow JD (2000) Mycorrhizal fungi influence soil structure. In: Kapulnik Y, Douds DD (eds) Arbuscular mycorrhizas: physiology and function. Kluwer Academic, Dordrecht, pp 3–18

Miransari M (2011) Hyperaccumulators, arbuscular mycorrhizal fungi and stress of heavy metals. Biotechnol Adv 29:645–653

Mishra PK, Bisht SC, Ruwari P, Joshi GK, Singh G, Bisht JK, Bhatt JC (2011) Bioassociative effect of cold tolerant Pseudomonas spp. and Rhizobium leguminosarum-PR1 on iron acquisition, nutrient uptake and growth of lentil (*Lens culinaris* L.). Eur J Soil Biol 47:35–43

Mishra J, Singh R, Arora NK (2017) Alleviation of heavy metal stress in plants and remediation of soil by rhizosphere microorganisms. Front Microbiol 8:1706. https://doi.org/10.3389/fmicb.2017.01706

Mittler R (2006) Abiotic stress, the field environment and stress combination. Trends Plant Sci 11:15–19

Mittler R, Blumwald E (2010) Genetic engineering for modern agriculture: challenges and perspectives. Ann Rev Plant Biol 61:443–462

Mittler R, Finka A, Goloubinoff P (2012) How do plants feel the heat? Trends Biochem Sci 37:118–125

Mullen MD, Wolf DC, Ferris FG, Beveridge TJ, Flemming CA, Bailey GW (1989) Bacterial sorption of heavy metals. Appl Environ Microbiol 55:3143–3149

Mummey DL, Stahl PD (2004) Analysis of soil whole-and inner-microaggregate bacterial communities. Microb Ecol 48:41–50

Nakajima A, Tsuruta T (2004) Competitive biosorption of thorium and uranium by *Micrococcus luteus*. J Radioanal Nucl Chem 260:13–18

Nambiar EKS (1997) Sustained productivity of forests as a continuing challenge to soil science. Soil Sci Soc Am J 60:1629–1642

Neumann E, George E (2010) Nutrient uptake: the arbuscular mycorrhiza fungal symbiosis as a plant nutrient acquisition strategy. In: Kapulnik Y, Douds DD (eds) Arbuscular mycorrhizas: physiology and function. Springer, Dordrecht, pp 137–167

Nielsen MN, Winding A, Binnerup S (2002) Microorganisms as indicators of soil health. National Environmental Research Institute, Denmark

Niu SQ, Li HR, Paré PW, Aziz M, Wang SM, Shi H, Li J, Han QQ, Guo SQ, Li J, Guo Q (2016) Induced growth promotion and higher salt tolerance in the halophyte grass *Puccinellia tenuiflora* by beneficial rhizobacteria. Plant Soil 407:217–230

Normile D (2008) Agricultural research. Reinventing rice to feed the world. Science 321:330–333

Oldroyd GED, Downie JA (2004) Calcium, kinases and nodulation signalling in legumes. Nat Rev Mol Cell Biol 5:566–576

Onwuka B, Mang B (2018) Effects of soil temperature on some soil properties and plant growth. Adv Plants Agric Res 8(1):00288. https://doi.org/10.15406/apar.2018.08.00288

Pandey R, Garg N (2017) High effectiveness of *Rhizophagus irregularis* is linked to superior modulation of antioxidant defence mechanisms in *Cajanus cajan* (L.) Millsp. genotypes grown under salinity stress. Mycorrhiza 27:669–682

Pandey A, Sharma M, Pandey GK (2016) Emerging roles of strigolactones in plant responses to stress and development. Front Plant Sci 7:434

Parvaiz A, Satyawati S (2008) Salt stress and phyto-biochemical responses of plants—a review. Plant Soil Environ 54:89–99

Pedranzani H, Rodríguez-Rivera M, Gutiérrez M, Porcel R, Hause B, Ruiz-Lozano JM (2016) Arbuscular mycorrhizal symbiosis regulates physiology and performance of *Digitaria eriantha* plants subjected to abiotic stresses by modulating antioxidant and jasmonate levels. Mycorrhiza 26(2):141–152

Pinedo I, Ledger T, Greve M, Poupin MJ (2015) *Burkholderia phytofirmans* PsJN induces long-term metabolic and transcriptional changes involved in Arabidopsis thaliana salt tolerance. Front Plant Sci 6:466. https://doi.org/10.3389/fpls.2015.00466

Pinter IF, Salomon MV, Berli F, Bottini R, Piccoli P (2017) Characterization of the As (III) tolerance conferred by plant growth promoting rhizobacteria to in vitro-grown grapevine. Appl Soil Ecol 109:60–68

Porcel R, Aroca R, Cano C, Bago A, Ruiz-Lozano JM (2007) A gene from the arbuscular mycorrhizal fungus *Glomus intraradices* encoding a binding protein is up-regulated by drought stress in some mycorrhizal plants. Environ Exp Bot 60:251–256

Porcel R, Aroca R, Azcon R, Ruiz-Lozano JM (2016) Regulation of cation transporter genes by the arbuscular mycorrhizal symbiosis in rice plants subjected to salinity suggests improved salt tolerance due to reduced Na^+ root-to-shoot distribution. Mycorrhiza 26:673–684

Probert RJ (2000) The role of temperature in the regulation of seed dormancy and germination. In: Fenner M (ed) Seeds: the ecology of regeneration in plant communities. CABI, Wallingford, pp 261–292

Qiu H, Ge T, Liu J, Chen X, Hu Y, Wu J, Su Y, Kuzyakov Y (2018) Effects of biotic and abiotic factors on soil organic matter mineralization: experiments and structural modeling analysis. Eur J Soil Biol 84:27–34

Qurashi AW, Sabri AN (2012) Bacterial exopolysaccharide and biofilm formation stimulate chickpea growth and soil aggregation under salt stress. Braz J Microbiol 43:1183–1191

Raaijmakers JM, Weller DM (1998) Natural plant protection by 2, 4-diacetylphloroglucinol-producing *Pseudomonas* spp. in take-all decline soils. MPMI 11:144–152

Radzki W, Mañero FG, Algar E, García JL, García-Villaraco A, Solano BR (2013) Bacterial siderophores efficiently provide iron to iron-starved tomato plants in hydroponics culture. Antonie Leeuwenhoek 104:321–330

Raineri J, Ribichich KF, Chan RL (2015) The sunflower transcription factor HaWRKY76 confers drought and flood tolerance to *Arabidopsis thaliana* plants without yield penalty. Plant Cell Rep 34:2065–2080

Ramasamy K, Joe MM, Kim KY, Lee SM, Shagol C, Rangasamy A, Chung JB, Islam MR, Sa TM (2011) Synergistic effects of arbuscular mycorrhizal fungi and plant growth promoting rhizobacteria for sustainable agricultural production. Korean J Soil Sci Fertil 44:637–649

Rapparini F, Peñuelas J (2014) Mycorrhizal fungi to alleviate drought stress on plant growth. In: Miransari M (ed) Use of microbes for the alleviation of soil stresses. New York, Springer, pp 21–42

Razavi BS, Blagodatskaya E, Kuzyakov Y (2016) Temperature selects for static soil enzyme systems to maintain high catalytic efficiency. Soil Biol Biochem 97:15–22

Riggs PJ, Chelius MK, Iniguez AL, Kaeppler SM, Triplett EW (2001) Enhanced maize productivity by inoculation with diazotrophic bacteria. Aust J Plant Physiol 28:829–836

Rillig MC, Aguilar-Trigueros CA, Bergmann J, Verbruggen E, Veresoglou SD, Lehmann A (2015) Plant root and mycorrhizal fungal traits for understanding soil aggregation. New Phytol 205:1385–1388

Rinaudi LV, González JE (2009) The low-molecular weight fraction of exopolysaccharide II from *Sinorhizobium meliloti* is a crucial determinant of biofilm formation. J Bacteriol 191:7216–7224

Ritz K, Young IM (2004) Interactions between soil structure and fungi. Mycologist 18:52–59

Rolli E, Marasco R, Vigani G, Ettoumi B, Mapelli F, Deangelis ML, Gandolfi C, Casati E, Previtali F, Gerbino R, Pierotti Cei F (2015) Improved plant resistance to drought is promoted by the root-associated microbiome as a water stress-dependent trait. Environ Microbiol 17:316–331

Rosier A, Bishnoi U, Lakshmanan V, Sherrier DJ, Bais HP (2016) A perspective on inter-kingdom signaling in plant-beneficial microbe interactions. Plant Mol Biol 90:537–548

Rousk J, Smith AR, Jones DL (2013) Investigating the long-term legacy of drought on the soil microbial community across five European shrubland ecosystems. Global Change Biol 19:3872–3884

Ruiz-Lozano JM, Aroca R, Zamarreño ÁM, Molina S, Andreo-Jiménez B, Porcel R, García-Mina JM, Ruyter-Spira C, López-Ráez JA (2016) Arbuscular mycorrhizal symbiosis induces strigolactone biosynthesis under drought and improves drought tolerance in lettuce and tomato. Plant Cell Environ 39:441–452

Ryu H, Cho H, Choi D, Hwang I (2012) Plant hormonal regulation of nitrogen-fixing nodule organogenesis. Mol Cells 34:117–126

Sadeghi A, Karimi E, Dahazi PA, Javid MG, Dalvand Y, Askari H (2012) Plant growth promoting activity of an auxin and siderophore producing isolate of *Streptomyces* under saline soil condition. World J Microbiol Biotechnol 28:1503–1509

Saeed W, Naseem S, Ali Z (2017) Strigolactones biosynthesis and their role in abiotic stress resilience in plants: a critical review. Front Plant Sci 8:1487. https://doi.org/10.3389/fpls.2017.01487

Sarkar A, Reinhold-Hurek B (2014) Transcriptional profiling of nitrogen fixation and the role of NifA in the diazotrophic endophyte *Azoarcus* sp. strain BH72. PLoS One 9:e86527

Sarwar N, Ishaq W, Farid G, Shaheen MR, Imran M, Geng M, Hussain S (2015) Zinc-cadmium interactions: impact on wheat physiology and mineral acquisition. Ecotoxicol Environ Saf 122:528–536

Sarwar N, Imran M, Shaheen MR, Ishaque W, Kamran MA, Matloob A, Rehim A, Hussain S (2017) Phytoremediation strategies for soils contaminated with heavy metals: modifications and future perspectives. Chemosphere 171:710–721

Sbrana C, Giovannetti M (2005) Chemotropism in the arbuscular mycorrhizal fungus *Glomus mosseae*. Mycorrhizae 15:539–545

Scaffidi A, Waters MT, Bond CS, Dixon KW, Smith SM, Ghisalberti EL, Flematti GR (2012) Exploring the molecular mechanism of karrikins and strigolactones. Bioorg Med Chem Lett 22:3743–3746

Scervino JM, Ponce MA, Erra-Bassells R, Bompadre MJ, Vierheilig H, Ocampo JA, Godeas A (2006) Glycosidation of apigenin results in a loss of activity on different growth parameters of arbuscular mycorrhizal fungi from the genus *Glomus* and *Gigaspora*. Soil Biol Biochem 38:2919–2922

Scharf KD, Berberich T, Ebersberger I, Nover L (2012) The plant heat stress transcription factor (Hsf) family: structure, function and evolution. Biochim Biophys Acta 1819:104–119

Schikora A, Schenk ST, Stein E, Molitor A, Zuccaro A, Kogel KH (2011) N-acyl-homoserine lactone confers resistance toward biotrophic and hemibiotrophic pathogens via altered activation of AtMPK6. Plant Physiol 157:1407–1418

Schimel J, Balser TC, Wallenstein M (2007) Microbial stress-response physiology and its implications for ecosystem function. Ecology 88:1386–1394

Schmidt R, Cordovez V, De Boer W, Raaijmakers J, Garbeva P (2015) Volatile affairs in microbial interactions. ISME J 9:2329. https://doi.org/10.1038/ismej.2015.42

Schulz S, Dickschat JS (2007) Bacterial volatiles: the smell of small organisms. Nat Prod Rep 24:814–842

Schulz S, Brankatschk R, Dümig A, Kögel-Knabner I, Schloter M, Zeyer J (2013) The role of microorganisms at different stages of ecosystem development for soil formation. Biogeosciences 10:3983–3996

Selvakumar G, Joshi P, Suyal P, Mishra PK, Joshi GK, Bisht JK, Bhatt JC, Gupta HS (2011) *Pseudomonas lurida* M2RH3 (MTCC 9245), a psychrotolerant bacterium from the Uttarakhand Himalayas, solubilizes phosphate and promotes wheat seedling growth. World J Microbiol Biotechnol 27:1129–1135

Seneviratne M, Seneviratne G, Madawala HM, Vithanage M (2017) Role of rhizospheric microbes in heavy metal uptake by plants. In: Singh JS, Seneviratne G (eds) Agro-environmental sustainability. Springer, Cham, pp 147–163

Seto Y, Yamaguchi S (2014) Strigolactone biosynthesis and perception. Curr Opin Plant Biol 21:1–6

Sexstone AJ, Revsbech NP, Parkin TB, Tiedje JM (1985) Direct measurement of oxygen profiles and denitrification rates in soil aggregates. Soil Sci Soc Am J 49:645–651

Sharma SS, Dietz KJ (2006) The significance of amino acids and amino acid-derived molecules in plant responses and adaptation to heavy metal stress. J Exp Bot 57:711–726

Sharma P, Sharma N, Deswal R (2005) The molecular biology of the low-temperature response in plants. BioEssays 27:1048–1059

Sharma A, Shankhdhar D, Shankhdhar SC (2013) Enhancing grain iron content of rice by the application of plant growth promoting rhizobacteria. Plant Soil Environ 59:89–94

Sheng M, Tang M, Zhang F, Huang Y (2011) Influence of arbuscular mycorrhiza on organic solutes in maize leaves under salt stress. Mycorrhiza 21:423–430

Sigel A, Sigel H (1998) Iron transport and storage in microorganisms, plants, and animals, vol 35. Marcel Dekker, New York. ISBN: 0824799844

Silvertown J, Dodd ME, Gowing DJG, Mountford JO (1999) Hydrologically defined niches reveal a basis for species richness in plant communities. Nature 400:61–63

Singh RP, Jha PN (2016) Mitigation of salt stress in wheat plant (*Triticum aestivum*) by ACC deaminase bacterium *Enterobacter* sp. SBP-6 isolated from *Sorghum bicolor*. Acta Physiol Plant 38:110. https://doi.org/10.1007/s11738-016-2123-9

Singh BK, Bardgett RD, Smith P, Reay DS (2010) Microorganisms and climate change: terrestrial feedbacks and mitigation options. Nat Rev Microbiol 8:779–790

Singh RP, Shelke GM, Kumar A, Jha PN (2015) Biochemistry and genetics of ACC deaminase: a weapon to "stress ethylene" produced in plants. Front Microbiol 6:937. https://doi.org/10.3389/fmicb.2015.00937

Skorupska A, Janczarek M, Marczak M, Mazur A, Król J (2006) Rhizobial exopolysaccharides: genetic control and symbiotic functions. Microb Cell Fact 5:7. https://doi.org/10.1186/1475-2859-5-7

Smith SM, Li J (2014) Signalling and responses to strigolactones and karrikins. Curr Opin Plant Biol 21:23–29

Smith SE, Christophersen HM, Pope S, Smith FA (2010) Arsenic uptake and toxicity in plants: integrating mycorrhizal influences. Plant Soil 327:1–21

Smith DL, Subramanian S, Lamont JR, Bywater-Ekegärd M (2015a) Signaling in the phytomicrobiome: breadth and potential. Front Plant Sci 6:709. https://doi.org/10.3389/fpls.2015.00709

Smith P, Cotrufo MF, Rumpel C, Paustian K, Kuikman PJ, Elliott JA, McDowell R, Griffiths RI, Asakawa S, Bustamante M, House JI (2015b) Biogeochemical cycles and biodiversity as key drivers of ecosystem services provided by soils. Soil 1:665–685

Soliman AS, Shanan NT, Massoud ON, Swelim DM (2012) Improving salinity tolerance of *Acacia saligna* (Labill.) plant by arbuscular mycorrhizal fungi and *Rhizobium* inoculation. Afr J Biotechnol 11:1259–1266

Soundappan I, Bennett T, Morffy N, Liang Y, Stanga JP, Abbas A, Leyser O, Nelson DC (2015) SMAX1-LIKE/D53 family members enable distinct MAX2-dependent responses to strigolactones and karrikins in *Arabidopsis*. Plant Cell 27:3143–3159

Steinkellner S, Lendzemo V, Langer I, Schweiger P, Khaosaad T, Toussaint JP, Vierheilig H (2007) Flavonoids and strigolactones in root exudates as signals in symbiotic and pathogenic plant-fungus interactions. Molecules 12:1290–1306

Stirnberg P, Furner IJ, Ottoline Leyser HM (2007) MAX2 participates in an SCF complex which acts locally at the node to suppress shoot branching. Plant J 50:80–94

Stoodley P, Sauer K, Davies DG, Costerton JW (2002) Biofilms as complex differentiated communities. Annu Rev Microbiol 56:187–209

Stotzky G (1997) Soil as an Environment for Microbial Life. In: van Elsas JD, Trevors JT, Wellington EMH (eds) Modern soil microbiology. Marcel Dekker, New York, pp 1–20

Sturz AV, Christie BR, Nowak J (2000) Bacterial endophytes: potential role in developing sustainable systems of crop production. Crit Rev Plant Sci 19:1–30

Subramanian KS, Balakrishnan N, Senthil N (2013) Mycorrhizal symbiosis to increase the grain micronutrient content in maize. Aust J Crop Sci 7:900

Sudhakar P, Chattopadhyay GN, Gangwar SK, Ghosh JK (2000) Effect of foliar application of *Azotobacter*, *Azospirillum* and *Beijerinckia* on leaf yield and quality of mulberry (*Morus alba*). J Agric Sci 134:227–234

Sun XG, Tang M (2013) Effect of arbuscular mycorrhizal fungi inoculation on root traits and root volatile organic compound emissions of *Sorghum bicolor*. S Afr J Bot 88:373–379

Sun H, Tao J, Liu S, Huang S, Chen S, Xie X, Xu G (2014) Strigolactones are involved in phosphate-and nitrate-deficiency-induced root development and auxin transport in rice. J Exp Bot 65:6735–6746

Suresh CK, Bagyaraj DJ (2002) Arbuscular mycorrhizae: interactions in plants, rhizosphere and soils. Oxford and IBH, New Delhi, pp 7–28

Suseela V, Tharayil N, Xing B, Dukes JS (2013) Labile compounds in plant litter reduce the sensitivity of decomposition to warming and altered precipitation. New Phytol 200:122–133

Talaat NB, Shawky BT (2011) Influence of arbuscular mycorrhizae on yield, nutrients, organic solutes, and antioxidant enzymes of two wheat cultivars under salt stress. J Plant Nutr Soil Sci 174:283–291

Talaat NB, Shawky BT (2013) Modulation of nutrient acquisition and polyamine pool in salt-stressed wheat (*Triticum aestivum* L.) plants inoculated with arbuscular mycorrhizal fungi. Acta Physiol Plant 35:2601–2610

Talaat NB, Shawky BT (2014) Protective effects of arbuscular mycorrhizal fungi on wheat (*Triticum aestivum* L.) plants exposed to salinity. Environ Exper Bot 98:20–31

Taniguchi H, Wendisch VF (2015) Exploring the role of sigma factor gene expression on production by *Corynebacterium glutamicum*: sigma factor H and FMN as example. Front Microbiol 6:740. https://doi.org/10.3389/fmicb.2015.00740

Thorn RM, Reynolds DM, Greenman J (2011) Multivariate analysis of bacterial volatile compound profiles for discrimination between selected species and strains in vitro. J Microbiol Methods 84:258–264

Timmusk S, El-Daim IAA, Copolovici L, Tanilas T, Kännaste A, Behers L, Niinemets Ü (2014) Drought-tolerance of wheat improved by rhizosphere bacteria from harsh environments: enhanced biomass production and reduced emissions of stress volatiles. PLoS One 9:1–13

Tiwari S, Thakur R, Shankar J (2015) Role of heat-shock proteins in cellular function and in the biology of fungi. Biotechnol Res Int. https://doi.org/10.1155/2015/132635

Turnau K, Kottke I (2005) Fungal activity as determined by microscale methods with special emphasis on interactions with heavy metals. In: Dighton J, White JF (eds) The fungal community. CRC, Boca Raton, pp 287–305

Ullah A, Heng S, Munis MFH, Fahad S, Yang X (2015) Phytoremediation of heavy metals assisted by plant growth promoting (PGP) bacteria: a review. Environ Exper Bot 117:28–40

Upreti KK, Bhatt RM, Panneerselvam P, Varalakshmi LR (2016) Morpho-physiological responses of grape rootstock 'Dogridge' to arbuscular mycorrhizal fungi inoculation under salinity stress. Int J Plant Sci 16:191–209

Vivas A, Barea JM, Azcón R (2005) Interactive effect of *Brevibacillus brevis* and *Glomus mosseae*, both isolated from Cd contaminated soil, on plant growth, physiological mycorrhizal fungal characteristics and soil enzymatic activities in Cd polluted soil. Environ Pollut 134:257–266

Vodnik D, Grčman H, Maček I, van Elteren JT, Kovačevič M (2008) The contribution of glomalin-related soil protein to Pb and Zn sequestration in polluted soil. Sci Total Environ 392:130–136

von Bodman SB, Bauer WD, Coplin DL (2003) Quorum sensing in plant-patho-genic bacteria. Annu Rev Phytopathol 41:455–482

Vurukonda SS, Vardharajula S, Shrivastava M, SkZ A (2016) Enhancement of drought stress tolerance in crops by plant growth promoting rhizobacteria. Microbiol Res 184:13–24

Wang FY, Liu RJ, Lin XG, Zhou JM (2004) Arbuscular mycorrhizal status of wild plants in saline-alkaline soils of the Yellow River Delta. Mycorrhiza 14:133–137

Wang X, Li W, Li M, Welti R (2006) Profiling lipid changes in plant response to low temperatures. Physiol Plant 126:90–96

Wani SH, Singh NB, Devi TR, Haribhushan A, Jeberson SM, Malik CP (2013) Engineering abiotic stress tolerance in plants: extricating regulatory gene complex. In: Malik CP, Sanghera GS, Wani SH (eds) Conventional and non-conventional interventions in crop improvement. CABI, New Delhi, pp 1–19

Warkentin BP (1995) The changing concept of soil quality. J Soil Water Conserv 50:226–228

White PJ (2003) Ion transport. In: Thomas B, Murphy DJ, Murray BG (eds) Encyclopaedia of applied plant sciences. Academic, London, pp 625–634

Wu QS, Zou YN (2017) Arbuscular mycorrhizal fungi and tolerance of drought stress in plants. In: Wu Q-S (ed) Arbuscular mycorrhizas and stress tolerance of plants. Springer, Singapore, pp 25–41

Wu QS, Xia RX, Zou YN (2008) Improved soil structure and citrus growth after inoculation with three arbuscular mycorrhizal fungi under drought stress. Eur J Soil Biol 44:122–128

Wu QS, He XH, Zou YN, Liu CY, Xiao J, Li Y (2012) Arbuscular mycorrhizas alter root system architecture of *Citrus tangerine* through regulating metabolism of endogenous polyamines. Plant Growth Regu 68:27–35

Xie X, Yoneyama K, Yoneyama K (2010) The strigolactone story. Annu Rev Phytopathol 48:93–117

Xie CM, Wei W, Sun Y (2013) Role of SKP1-CUL1-F-box-protein (SCF) E3 ubiquitin ligases in skin cancer. J Genet Genomics 40:97–106

Xie Y, Fan J, Zhu W, Amombo E, Lou Y, Chen L, Fu J (2016) Effect of heavy metals pollution on soil microbial diversity and Bermuda grass genetic variation. Front Plant Sci 7:755. https://doi.org/10.3389/fpls.2016.00755

Yang Y, Han X, Liang Y, Ghosh A, Chen J, Tang M (2015) The combined effects of arbuscular mycorrhizal fungi (AMF) and lead (Pb) stress on Pb accumulation, plant growth parameters, photosynthesis, and antioxidant enzymes in *Robinia pseudoacacia* L. PLoS One 10:e0145726. https://doi.org/10.1371/journal.pone.0145726

Yang A, Akhtar SS, Iqbal S, Amjad M, Naveed M, Zahir ZA, Jacobsen SE (2016) Enhancing salt tolerance in quinoa by halotolerant bacterial inoculation. Funct Plant Biol 43:632–642

Yao X, Liu Q (2007) Changes in photosynthesis and antioxidant defenses of *Picea asperata* seedlings to enhanced ultraviolet-B and to nitrogen supply. Physiol Plant 129:364–374

Ye S, Yanga Y, Xin G, Wang Y, Ruan L, Ye G (2015) Studies of the Italian ryegrass–rice rotation system in southern China: arbuscular mycorrhizal symbiosis affects soil microorganisms and enzyme activities in the *Lolium multiflorum* L. rhizosphere. Appl Soil Ecol 90:26–34

Yin N, Zhang Z, Wang L, Qian K (2016) Variations in organic carbon, aggregation, and enzyme activities of gangue-fly ash-reconstructed soils with sludge and arbuscular mycorrhizal fungi during 6-year reclamation. Environ Sci Pollut Res 23:17840–17849

Yoneyama K, Yoneyama K, Takeuchi Y, Sekimoto H (2007) Phosphorus deficiency in red clover promotes exudation of orobanchol, the signal for mycorrhizal symbionts and germination stimulant for root parasites. Planta 225:1031–1038

Yooyongwech S, Phaukinsang N, Cha-um S, Supaibulwatana K (2013) Arbuscular mycorrhiza improved growth performance in *Macadamia tetraphylla* L. grown under water deficit stress involves soluble sugar and proline accumulation. Plant Growth Regul 69:285–293

Young IM, Ritz K (2000) Tillage, habitat space and function of soil microbes. Soil Tillage Res 53:201–213

Zhang JL, Flowers TJ, Wang SM (2010) Mechanisms of sodium uptake by roots of higher plants. Plant Soil 326:45. https://doi.org/10.1007/s11104-009-0076-0

Zhao LH, Zhou XE, Yi W, Wu Z, Liu Y, Kang Y, Hou L, De Waal PW, Li S, Jiang Y, Scaffidi A (2015) Destabilization of strigolactone receptor DWARF14 by binding of ligand and E3-ligase signaling effector DWARF3. Cell Res 25:1219–1236

Zhou C, Ma Z, Zhu L, Xiao X, Xie Y, Zhu J, Wang J (2016) Rhizobacterial strain *Bacillus megaterium* BOFC15 induces cellular polyamine changes that improve plant growth and drought resistance. Int J Mol Sci 17:976. https://doi.org/10.3390/ijms17060976

Printed by Printforce, the Netherlands